ASP.NET MVC
企业级实战

邹琼俊 著

清華大學出版社
北京

内容简介

ASP.NET MVC 是微软官方提供的以 MVC 模式为基础的 ASP.NET Web 应用程序框架。MVC 将一个 Web 应用分解为 Model、View 和 Controller，有助于管理复杂的应用程序，并简化了分组开发，使得复杂项目更易维护。

本书共分为 12 章，以符合初学者思维的方式系统地介绍 ASP.NET MVC 的应用技巧，并结合实际项目详细地介绍如何基于 ASP.NET MVC 构建企业项目。通过本书的学习，读者可以全面掌握 ASP.NET MVC 的开发，并从代码中获取软件开发与架构设计的经验与灵感。

本书具有很大的参考价值，既适合 ASP.NET MVC 开发初学者阅读，也适合有一定基础的 ASP.NET MVC 开发人员进行技术强化和经验积累，同时还适合作为高等院校和培训机构相关专业的教学参考书。

图书在版编目（CIP）数据

ASP.NET MVC 企业级实战 / 邹琼俊著. —北京：清华大学出版社，2017 (2023.9重印)
ISBN 978-7-302-46504-1

I. ①A… II. ①邹… III. ①网页制作工具—程序设计 IV. ①TP393.092.2

中国版本图书馆 CIP 数据核字（2017）第 025431 号

责任编辑：夏毓彦
封面设计：王　翔
责任校对：闫秀华
责任印制：杨　艳

出版发行：清华大学出版社
网　　址：http://www.tup.com.cn，http://www.wqbook.com
地　　址：北京清华大学学研大厦 A 座　　**邮　　编：**100084
社 总 机：010-83470000　　**邮　　购：**010-62786544
投稿与读者服务：010-62776969，c-service@tup.tsinghua.edu.cn
质量反馈：010-62772015，zhiliang@tup.tsinghua.edu.cn

印 装 者：三河市铭诚印务有限公司
经　　销：全国新华书店
开　　本：190mm×260mm　　**印　　张：**30.5　　**字　　数：**781 千字
版　　次：2017 年 4 月第 1 版　　**印　　次：**2023 年 9 月第 9 次印刷
定　　价：89.00 元

产品编号：070811-01

推荐语

本书是一本非常接地气的 .NET Web 开发指导书籍。有别于市面上的同类书籍，它既不像微软官方资料和 Demo 那样照本宣科，也不像大学教材那样乏味说教。它凝聚了作者在 ASP.NET MVC 及其相关技术的实际经验，将实战与理论相结合，介绍如何使用 ASP.NET MVC 来进行企业应用开发，是初学者难得的引路书籍。

——微云数聚（北京）科技有限公司 CEO　张帜

本书所涉及的知识面非常广，内容涵盖了 ASP.NET MVC 开发中最常使用的技术和框架，由浅入深，易于读者理解和掌握，配合作者精心挑选的例子，实用性强，很具实战参考价值。相信本书讲解的内容能够使您的 Web 开发水平提升一个台阶。

——微软高级架构师　Terry Zhou

作者年轻、充满活力，给我最大的印象是公司年会上表演的双截棍节目，让我充分相信程序员是多么优秀的职业。作者能够在家人身体欠佳的状况下如期完成本书的编写，可见其毅力多么坚韧。

本书能够让初学 MVC 的读者有很清晰的技术路线，而且配套的 Demo 基本都是在实际项目中的运用，非常值得研究。

——深圳市跨境翼电子商务股份有限公司 CEO　李君

能够与作者在同一家公司，并一起完成书中的财务对账系统，是一件让人短暂痛苦但回想起来非常开心的事情。作者能坚持写作多年，并将平时学习到的知识在日常工作的项目中进行实战，这需要非常坚韧的毅力。希望本书能够让更多读者看到在企业项目实践中 MVC 的运用过程，同时也希望作者能够继续坚持拥抱变化。

——深圳市思创信息技术有限公司总经理　何成

我在公司实际项目中把《ASP.NET MVC 企业级实战》列为我们重要的中文参考资料。它覆盖了工作中所有的 ASP.NET MVC 技术要点，让我们的新员工顺利过渡到项目中，作者有深厚的编程和教学经验，也尽力把相关问题由浅入深分析清楚，这在 MVC 的同类书籍中是非常难得的。

——长沙道好信息科技有限公司技术总监　周尹

前　言

为什么要写这本书

ASP.NET MVC4 是目前大部分 IT 企业所使用的 Web 开发技术，许多互联网公司招聘都明确要求熟悉 ASP.NET MVC。

最近公司在招.NET 程序员，我发现好多来公司面试的.NET 程序员没有 ASP.NET MVC 项目经验，其中包括一些工作四五年了，甚至八九年的。有一些.NET 程序员对 ASP.NET MVC 的认知也只是停留在大学老师教过，自己学过，以前公司用过。然而，这样的话在公司项目开发过程中往往无法完成一些稍微复杂一点的开发任务或者只能以一种很 low 的方式实现一些功能。显然，这样的话是无法适应.NET Web 开发浪潮的。所以，我打算针对公司真实项目用到的 ASP.NET MVC 技术写一本书，供打算从事.NET Web 开发、无 MVC 项目经验、MVC 不熟或者打算提升.NET Web 开发水平的读者学习参考，以便快速适应工作。我是自学.NET 的，所以国内外的.NET 电子书翻阅过的不下百本，感觉书本上说的和企业里面用的完全是两码事，作为一线码农，我希望同大家分享工作当中真正有用的知识、技术及技巧。

本书特点

本书以符合初学者思维的方式，系统介绍了 ASP.NET MVC 的应用技巧，并结合作者实际参与过的项目，详细介绍了如何基于 ASP.NET MVC 构建企业项目，学以致用是本书最大的特点。通过本书的学习，读者可全面掌握 ASP.NET MVC 的开发，并可从本书代码中获取软件开发与架构设计的经验与灵感。

如何阅读本书

本书适合 ASP.NET MVC 的初学者，欲深入了解 ASP.NET MVC 开发的软件工程师，系统架构师，以及任何对 ASP.NET MVC 相关技术感兴趣的读者。书中所载技术均为一线城市中互联网企业所流行的.NET Web 开发技术，具有很强的参考价值。由于本书的结构是层进式的，部分章节之间有一定的关联，因此，对于初学者，建议按章节的顺序逐章阅读，在实践时，要理解核心的代码，自己开发相似的功能应用，并在其上进行扩展，从而真正掌握 ASP.NET MVC 开发技术。临渊羡鱼，不如退而结网，一定要多动手，多总结。

源码下载

本书配套源代码请扫描右边二维码下载。如果下载有问题，请电子邮件联系 booksaga@163.com，邮件主题为“ASP.NET MVC 企业级实战”。

修订与技术支持

写作本书耗时大半年，其中积累了笔者数年心得与技术感悟，希望本书能给读者带来思路上的启发与技术上的提升，使每位读者能够从中获益。同时，也非常希望借此机会能够与国内热衷于 ASP.NET MVC 的开发者们进行交流。由于时间和本人水平有限，书中难免存在一些纰漏和错误，希望大家批评、指正。如果大家发现了问题，可以直接和我联系，我会第一时间在本人的技术博客中发表加以改正，万分感谢。

由于有高校老师把本书作为教材，本书提供 PPT 教学课件，可在出版社网站下载或者联系本人索取。

本人博客和 QQ 技术交流群参见下载资源中的相关文件。

致 谢

本书能顺利地出版首先我要感谢的是夏毓彦老师，没有他耐心的指导，本书不可能出版；其次是清华出版社的其他编辑，正是他们在写作过程中的全程指导，才使得整个创作不断被完善，从而确保了本书顺利完稿。

写一本书所费的时间和精力都是巨大的，写书期间，我占用了太多本该陪家人的时间，在这里，要特别感谢我的爱人王丽丽，谢谢她帮我处理了许多生活上面的琐事；还要感谢我的父母，是他们含辛茹苦的把我培养成人；同时感谢我两位姐姐无微不至的关怀，正是家人们的理解与默默支持，我才能全身心投入写作，顺利完成本书的编写。

能有今天的成果，离不开我恩师周尹的悉心栽培，最后感谢曾经帮助过我的领导、同事、朋友、同学，感谢张帜、周纯星、李君、何成、饶成龙等，喝水不忘挖井人，祝你们身体健康，家庭美满。

编 者
2017 年 2 月于深圳

目　录

第 1 章 MVC开发前奏

俗话说："工欲善其事，必先利其器"，在进行 ASP.NET MVC 开发前，我们有必要安装好开发工具、配置好开发环境。

我一向认为程序员要学会"偷懒"，因为当你想要偷懒就不得不去思考其他高效的解决方案，那么怎样才能偷懒呢？好的开发工具、熟练的开发技巧（快捷键等）、巧妙的解决方案等将能有助于我们偷懒，学会偷懒才能高效地工作，当然也不能误解"偷懒"。有些人喜欢不写注释（如果能做到代码即注释例外）；做数据迁移的时候喜欢直接运行工具生成的一系列单条 insert 的 SQL 脚本，懒得去把脚本修改成批量插入，几十万条记录就这么一条一条插入，一运行就是好几个小时，而修改成批量可能就几分钟，修改后批量插入可能就几十秒；喜欢随意复制粘贴代码，又懒得去做重构……这样就走偏了。

1.1 开发环境搭建

1.1.1 操作系统和开发工具

本书中所使用的开发系统和软件版本如下所述。

1. Windows 10 64bit

如果你使用的是 Windows 7 系统，本书云盘提供系统激活软件 HEU_KMS_Activator_v7.8.6.exe，当然你也可以使用该软件激活 office 产品。目前许多互联网公司的开发系统已经是 Windows 8 和 Windows 10。

2. Microsoft Visual Studio Ultimate 2012

下载地址：https://www.microsoft.com/zh-cn/download/details.aspx?id=30678

在笔者电脑上面安装了 VS2012、VS2013、VS2015 这 3 个版本，考虑到目前公司使用的是 VS2012，所以本书中所有的示例都采用 VS2012。VS2012 应该是目前比较主流的.NET 开发工具之一，当然也有许多互联网公司已经在使用 VS2013 甚至 VS2015 进行开发了，不过使用 VS2015 作为开发工具的公司目前还是比较少的。即便如此，我们也应该有一个意识，那就是软件系统的升级是一种趋势，相比于一线城市，二三线城市开发工具和开发技术的升级相对较慢。从以前的 VS2003 到 VS2005，再到 VS2008，微软基本上是两年更新一次 VS 版本，再看现在的 VS2012、VS2013，发现微软的 VS 产品迭代更新的周期减少了，VS 的每一次升级都是在不断地优化和持续改进，并随之新增一些功能。

3. SQL Server 2012

下载地址：http://www.msdn.hk/7/177/

当前市面上的主流版本是 SQL Server 2008R2 和 SQL Server 2012，而使用 SQL Server 2014 和 SQL Server 2015 来开发的公司相对较少，因为 SQL Server 不便宜，而且版本越高价格越贵。（以前公司买过一张 SQL Server 2012 的正版光盘，50 多万。）

4. MySQL5.6

下载地址：http://www.mysql.com/products/

本书中使用的是 MySQL5.6，不是最新版本，云盘中提供了 MySQL 安装包 mysql5.6.msi 和 Mysql 数据库管理工具 Navicat for MySQL。

安装顺序：先安装 mysql5.6.msi，然后解压 Navicat for MySQL.zip，运行 navicat.exe。

当然，对于这些开发工具，大家都可以直接百度一下，然后选择自己想要的版本进行下载，尽量不要低于这些软件版本。不过为了减少差异性、方便大家学习，大家最好安装和本书中一样的软件版本。关于开发工具的具体安装细节，由于没有什么技术含量，而且篇幅有限，这里就不做过多介绍了，大家可以从网上查找到大把的软件安装图解教程。

5. Git

Git 是 Linux 的第二个伟大作品。2005 年 BitKeeper 软件公司对 Linux 社区停止了免费使用权，Linux 迫不得已自己开发了一个分布式版本控制工具——Git 诞生了。

目前 Git 越来越火，使用 Git 作为版本控制的公司也越来越多，所以我们有必要掌握 Git 的使用。但是 Git 的学习成本相对较高，尤其是那些抓狂的命令。如果你觉得这样的命令操作起来很烦琐，习惯了过去 TortoisSVN 乌龟壳式的可视化客户端工具，那么你也可以下载 Git 的可视化客户端工具。因为笔者的电脑是 64bit 的，所以这里只提供 64bit 的下载（TortoiseGit_1.8.9.0_x64.zip），同时提供 SourceTree 这款 Git 可视化操作工具的安装包（SourceTreeSetup_1.4.0.zip），你可以根据个人喜好安装使用。安装方法是先安装 Git-1.9.4-preview20140815.exe，再安装可视化操作工具。

6. Reflector

Reflector 是由微软员工 Lutz Roeder 编写的免费程序。Reflector 的出现使.NET 程序员眼前豁然开朗，因为这个免费工具可以将.NET 程序集中的中间语言反编译成 C#或者 Visual Basic 代码。除了能将 IL 转换为 C#或 Visual Basic 以外，Reflector 还能够提供程序集中类及其成员的概要信息、查看程序集中 IL 的能力以及对第三方插件的支持，名副其实的.NET 开发神器。安装后，可以直接在 VS 中查看程序集源码，使用非常方便。

本书源码中提供了 Reflector，安装 Reflector 之后，在 VS 中鼠标定位到.NET 类库中的类或者方法、属性上面后，按 F12 还可以直接在 VS 中查看其源码。

所有软件安装包都会放在云盘中的“各种开发工具”目录下面。

1.1.2　开发环境配置

软件安装好了，接下来就是对软件进行配置了。这里只讲一下 VS2012 的配置，因为 VS2012 在开发中用得最多且最频繁。

1. 显示行号

打开 VS，选择“工具→选项→文本编辑器→所有语言”，在右侧“显示栏”中勾选“行号”复选框，如图 1-1 所示。

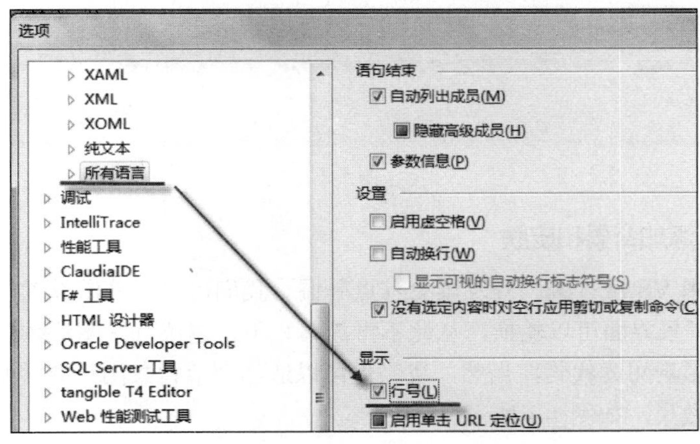

图 1-1

显示行号可以方便我们今后在开发过程中快速定位代码行。例如，使用火狐中的 Firebug 调用 CSS 样式的时候是可以看到样式所在行的，当我们调好样式后，可以把修改后的样式直接更新到 CSS 样式文件中；又或者当程序出现异常时，会看到异常显示某一行报错，我们可以使用 Ctrl+G 快捷键快速定位到某一行；还有就是解决代码冲突的时候，也能很直观地看到代码冲突的位置。

2. 设置屏幕保护色

像咱们软件开发人员，每天长时间对着电脑屏幕，不可避免地会遭受屏幕辐射所带来的视力问题，并引起眼睛的不适，我们可以通过设置保护色来缓解显示屏给眼睛带来的刺激。

打开 VS，选择“工具→选项→环境→字体和颜色→项背景→自定义”，设置“R：204，G：232，B：204”，如图 1-2 所示。

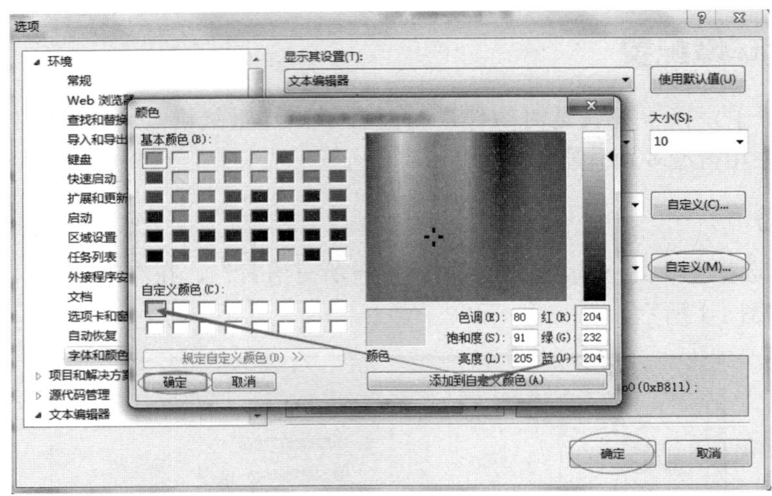

图 1-2

3. 为 VS2012 添加背景和皮肤

每天对着电脑用 VS 进行编码开发难免有点单调。我们可以试着像设置电脑桌面壁纸一样设置 VS 皮肤，而且背景界面可以轮换，从此不再孤单！不过建议背景图片颜色最好比较单一，而且是浅色系，以免影响阅读代码，当然，我们也可以适当调节背景图片的透明度。

我们先看一下效果，如图 1-3 所示。

图 1-3

操作步骤：

（1）打开 VS2012，选择“工具→扩展和更新”菜单，如图 1-4 所示。

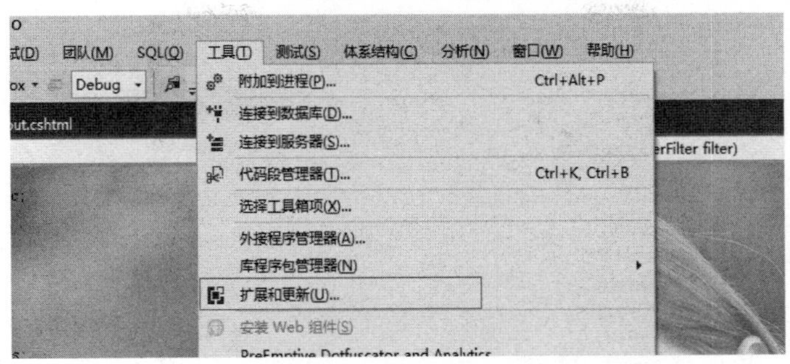

图 1-4

（2）选择联机，搜索并分别安装如下两个插件（见图 1-5）。

- Visual Studio 2012 Color Theme Editor：修改编辑器背景颜色。
- IDE Text Background：修改编辑器背景图片，支持轮播。

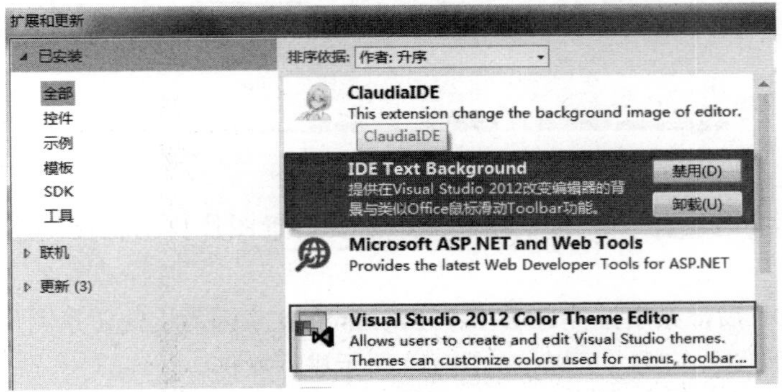

图 1-5

（3）安装完成后打开“我的文档”，找到文件 MaxZhang.VsixTheme.ini，右击，选择“编辑”命令，设置 ImageDirectory 属性，指定背景图片存放路径（ImageDirectory=C:\Users\Administrator\Pictures），如图 1-6 所示、图 1-7 所示，那么这一组图片将会自动成为 VS 的背景图片，并定时轮播。如果要改变背景图片的透明度，可以修改配置属性 ImageOpacity=0.60 的值，然后重启 VS。

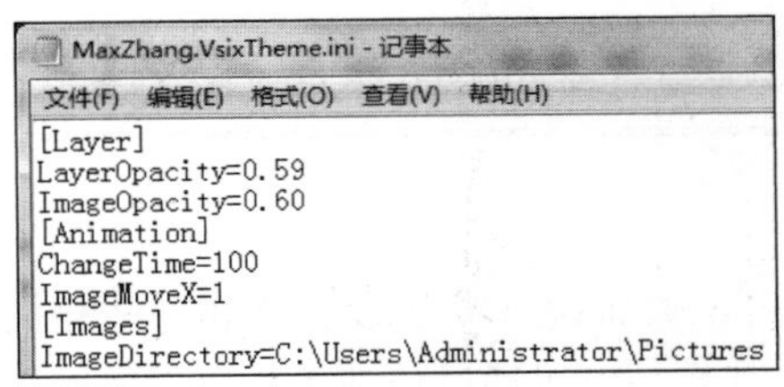

图 1-6

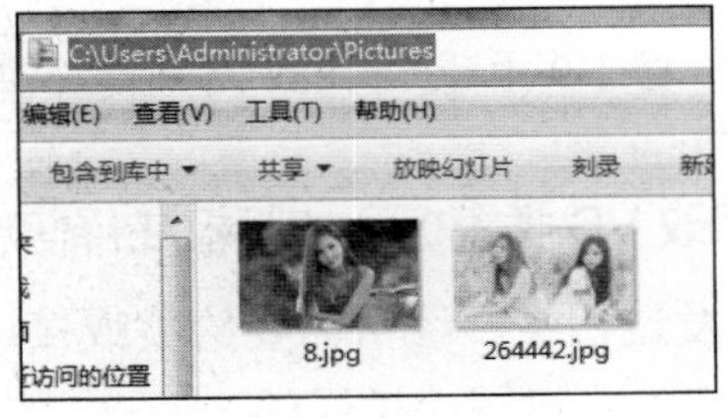

图 1-7

（4）重启 VS，奇迹就出现在你的面前。

安装完皮肤插件后，可以按照如下操作进行皮肤更换，如图 1-8 所示。

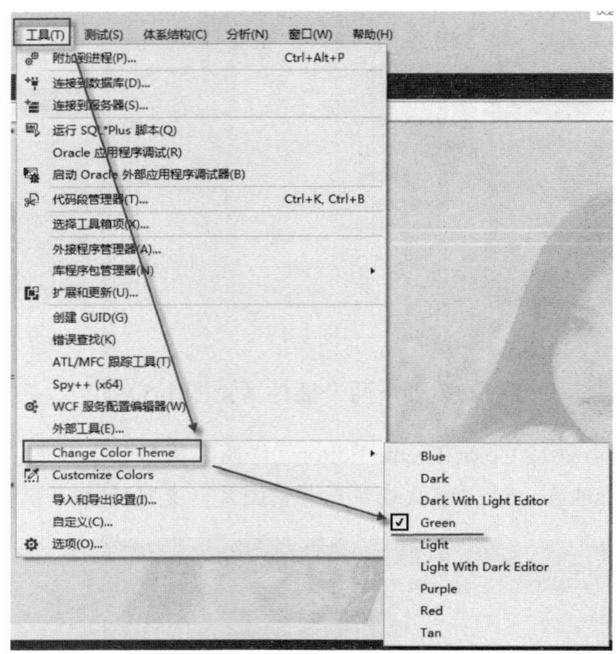

图 1-8

如果只是想单纯地设置 VS 背景图片，不需要背景图片自动轮播功能，也可以安装插件 ClaudiaIDE。

安装 ClaudiaIDE 后，禁用插件 IDE Text Background（因为这些背景图片插件同时使用会冲突），想要更换背景图片的话，可以按如图 1-9 所示进行操作。

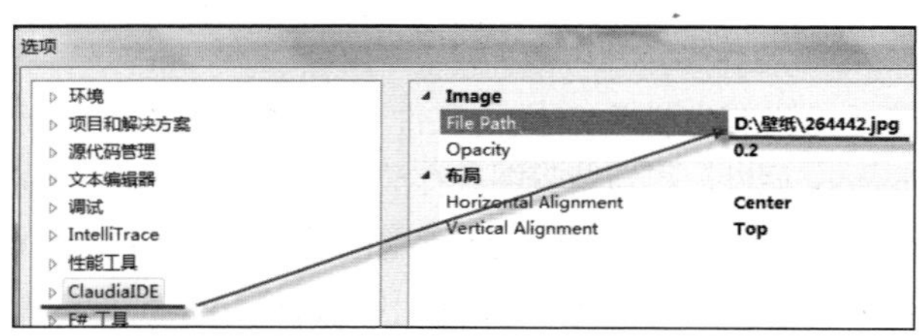

图 1-9

图 1-9 中的 Opacity 用来设置图片透明度，取值范围为 0~1，值越低越透明。

4. 修改 VS 类模板添加版权注释信息

在开发过程中，经常需要给类或接口添加 public 修饰符（默认没有）和一些相关的注释信息，这个工作是机械而枯燥的，而这个简单的需求其实是可以通过修改 VS 自带的类模板来实现的。下面给出详细的修改步骤。

（1）笔者电脑上面的 VS2012 是安装在 D 盘中的，所以找到目录 D:\Program Files (x86)\Microsoft Visual Studio 11.0\Common7\IDE\ItemTemplates\CSharp\Code\2052。如果你电脑上的 VS2012 默认安装在 C 盘，那么就要找到目录 C:\Program Files (x86)\Microsoft Visual Studio

11.0\Common7\IDE\ItemTemplates\CSharp\Code\2052。

（2）找到 Class 、Interface 、WebClass 这 3 个目录下面的 cs 文件后分别打开，并在文件的最前面加上如下代码：

```
/**********************************************************************************
 * Copyright (c) $year$$registeredorganization$ All Rights Reserved.
 * CLR 版本： $clrversion$
 *机器名称：$machinename$
 *公司名称：$registeredorganization$
 *命名空间：$rootnamespace$
 *文件名：  $safeitemname$
 *版本号：  V1.0.0.0
 *唯一标识：$guid10$
 *当前的用户域：$userdomain$
 *创建人：  $username$
 *电子邮箱：zouqiongjun@kjy.com
 *创建时间：$time$

 *描述：
 *
 *=====================================================================
 *修改标记
 *修改时间：$time$
 *修改人： $username$
 *版本号： V1.0.0.0
 *描述：
 *
 *********************************************************************************/
```

（3）在 class $safeitemrootname$前面添加 public 访问修饰符。这样我们每次在 VS 中新建类的时候，就不需要再手动去类前面添加 public 修饰符了。

```
namespace $rootnamespace$
{
  public class $safeitemrootname$
   {
   }
}
```

1.1.3　VS 常用快捷键

快捷键使用得熟练将极大地提高我们的开发效率，所以我们有必要记住开发中常用的快捷

键。VS 中一些常用的快捷键如下：

- F4：打开属性面板。
- F5：调试。
- Ctrl+F5：直接执行不调试。
- F9：设置、切换断点。
- F10：逐过程。
- F11：逐语句。
- F12：转到定义。
- Alt+F12：查看定义。
- Ctrl+F：查找。
- Ctrl+A：全选界面代码。
- Ctrl+K + Ctrl+K：设置书签。
- Ctrl+K + Ctrl+N：跳转到下一个书签。
- Ctrl+K + Ctrl+F：格式化选中代码。
- Ctrl+K + Ctrl+S：外侧代码，如添加#region 等。
- Ctrl+R + Ctrl+E：封装字段。
- Ctrl+R + Ctrl+M：提取方法。
- Ctrl+R + Ctrl+I：提取接口。
- Shift+Alt+F10，然后按回车键：添加命名空间引用。
- Shift+Home：选中当前行。
- Shift+方向键：向各个方向选中。

1.1.4 VS 技巧

1. 回到上一个光标位置

- 使用 Ctrl + - 组合键表示 Navigate BackWard。
- 使用 Ctrl + Shift + - 组合键表示 Forward。

2. 删除多余的 using 指令并排序

当我们新建一个类的时候，Visual Studio 会将常用的命名空间用 using 放在类的头部。当写完一个类的时候，有些 using 将是多余的，删除多余的 using，再排一下序，可以使代码看起来更清晰。VS2012 已经为我们做好了这一切。在代码编辑区右击，可以看到“组织 using”菜单，这就是我们需要的了。

3. 复制或删除一行代码时不用先选中

如果想复制一行代码，只需要简单地按 Ctrl+C 组合键，然后按 Ctrl+V 组合键粘贴就可以了，而不需要选择整行的代码。如果想删除一行代码，只需按 Ctrl+X 组合键就可以了。

4. 取代其他编辑器里 Ctrl+F 更方便的增量查找方法

Ctrl+F 的查找功能相信大家都用过，其实在 VS 里还有更方便的查找功能。操作方法如下：

（1）按 Ctrl+I 组合键。

（2）输入要搜索的文本。注意：这时你会看到光标跳至第一个匹配的地方，匹配的文本高亮显示。

（3）再次按下 Ctrl+I 组合键，光标将跳至下一个匹配的文本。

（4）按 Ctrl+Shift+I 组合键可向后搜索。

（5）要停止搜索，按 Esc 键。

5. 如何在编辑器中进行框式选择

你是否知道 VS 提供了流式和框式两种不同的选择模型？大家应该都熟悉流式选择模型了，只要使用 Shift+方向键即可（或者使用鼠标进行选择）。

框式选择允许你同时对行和列进行选择。只要同时按下 Shift+Alt+方向键，你就了解它的不同之处了。剪切、复制、粘贴这些功能都能使用，只是需要记住从哪里开始选择的。

也可以使用鼠标+Alt 键完成该操作。有时候我们复制网上的代码时会将行号一起复制过来，使用框式选择可以只选择行号部分并将其删除。

6. 如何使用快捷键在当前代码行的上面或下面插入一行

使用 Home 或 End，然后使用方向键，再使用回车键就能达到上面的效果。

7. 安装之后将 IDE 设置恢复到默认设置

如果 IDE 的设置在任何先前发布的版本中做了更改，那么它们都应该被恢复到默认设置。可以在 VS2012 中选择“工具（Tools）→导入导出设置向导（Import and ExportSettings...）→重置所有设置（Reset all settings）”，此外还有一些导入（Import）和导出（Export）的选项可用。

8. 通过按两次 Tab 键插入代码块

- 在编辑器中输入代码片段，比如“for”。
- 在这个状态下按两次 Tab 键将会插入代码块。

这样既快又不容易出现语法错误。

9. 使用 Ctrl+Tab 组合键打开 IDE 的导航，获得鸟瞰视图

同时在 VS 中导航到所有打开的文件和工具窗体，按 Ctrl+Tab 组合键，打开 IDE 导航窗口，按住 Ctrl 键，同时用方向键或鼠标选中一个文件或工具窗体来激活。

这时最好不要松开 Ctrl+Tab 组合键，按方向键看鸟瞰图，全部松开后就定位到需要的文件或工具窗体。说实在的，这个窗口挺酷的。

10. 查找匹配的标记

某些标识总是成对出现的。例如，“{”标识必须用对应的“}”标识关闭。虽然单击一个“{”，和它匹配的“}”就会高亮显示，但是代码过长的话就不好找了。同样，编译器指示符“#region”必须有对应的“#endregion”指示符。当导航代码时，有时需要查找对应的标识，可以通过按 Ctrl+]组合键完成。这个快捷键只有当光标在这些标识符的任何一个下面时才起作用，将会立即跳转到对应的标识符而不管它是开的还是闭的标识。

如果想显亮两个匹配的标识之间的所有代码，可以按 Ctrl+Shift+]组合键显亮整个块，并移动光标到开的标识处。这个快捷键只有当光标在任意标识的下面时才起作用（如光标在区域内将不会起作用）。

11. 添加命名空间引用

平时我们添加命名空间引用时，要么是直接手写，要么是用鼠标双击类，然后单击下拉框进行引用的，其实我们可以直接按 Shift+Alt+F10 组合键，然后按回车键就可以了。

12. 创建类的快捷方式

用鼠标选中需要创建快捷方式的类，然后同时按 Shift+Alt 组合键，并拖动到指定位置就可以了。

1.2 常用辅助开发工具介绍

1.2.1 Firebug

Firebug 是一个开源的 Web 开发工具，是网页浏览器 Mozilla Firefox 下的一款开发类插件，现属于 Firefox 的五星级强力推荐插件之一。Firebug 集 HTML 查看和编辑、Javascript 控制台、网络状况监视器于一体，是开发 JavaScript、CSS、HTML 和 Ajax 的得力助手。Firebug 如同一把精巧的瑞士军刀，从各个不同的角度剖析 Web 页面内部的细节层面，给 Web 开发者带来很大的便利。当然谷歌和 IE 也都自带 Web 开发工具，可以根据个人喜好选择一款熟练使用，不过笔者非常热衷于使用 Firebug。Firebug 并不是火狐浏览器独有的，可以在任意支持 Firebug 的浏览器上面安装。

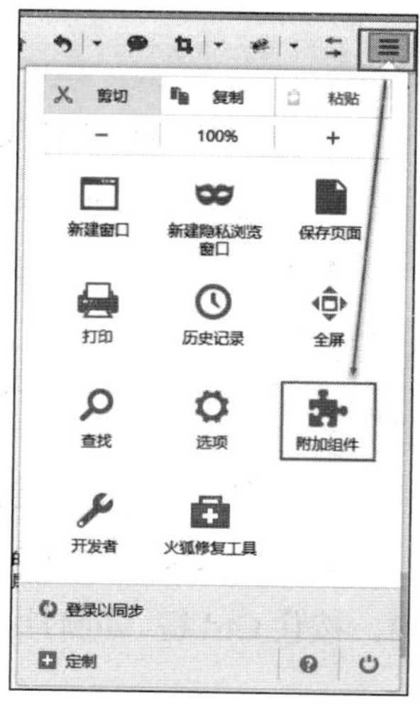

图 1-10

Firebug 也是一个除错工具。用户可以利用它除错、编辑甚至删改任何网站的 CSS、HTML、DOM 以及 JavaScript 代码，还可以在火狐浏览器中直接安装 Firebug 插件，如图 1-10、图 1-11 所示。

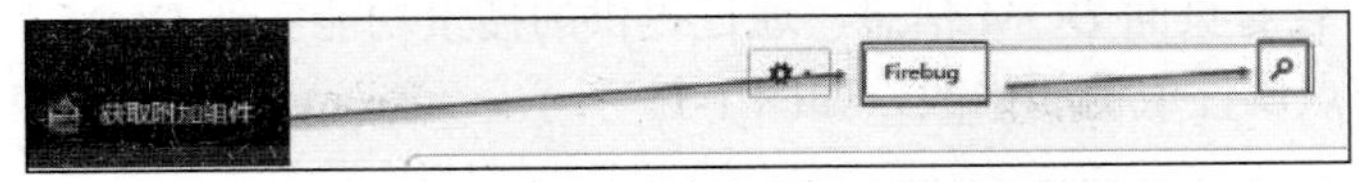

图 1-11

安装后重启 Firefox 浏览器，就可以在 Firefox 浏览器中看到 Firebug 的图标，如图 1-12 所示。单击 Firebug 图标（位于 Firefox 浏览器右上角）或者按 F12 键即可激活 Firebug 插件。

图 1-12

接下来我们介绍一下 Firebug 的功能。Firebug 有 5 个主要的 Tab 按钮，这里将主要介绍这几方面的功能。

（1）控制台

控制台如图 1-13 所示。

图 1-13

（2）HTML

利用 Inspect 检查功能，我们可以用鼠标在页面中直接选择一些区块，查看相应的 HTML 源代码和 CSS 样式表，真正做到所见即所得。如果你使用外部编辑器修改了当前网页，可以单击 Firebug 的 reload 图片重新载入网页，它会继续跟踪之前用 Inspect 选中的区块，方便调试，如图 1-14 所示。

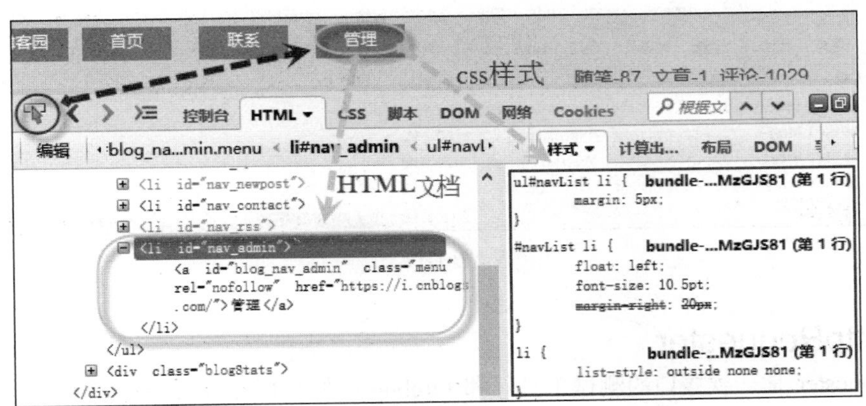

图 1-14

（3）DOM

该功能主要用于查看页面 DOM 信息，通过提供的搜索功能实现 DOM 的快速准确定位，并可双击实现 DOM 节点属性或值的修改，如图 1-15 所示。

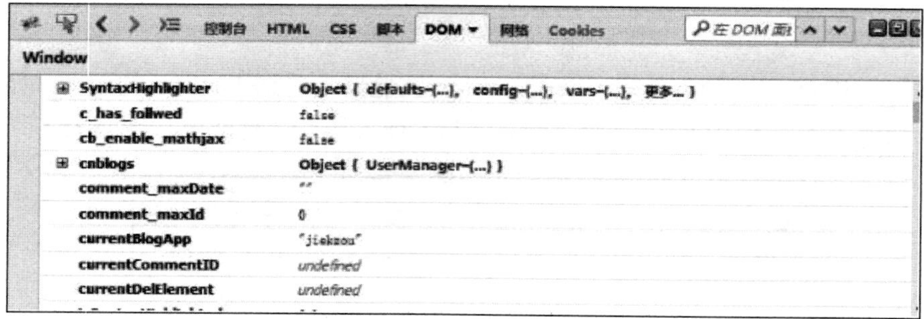

图 1-15

（4）脚本（Javascript）

脚本功能主要是一个脚本调试器，可以进行单步调试、断点设置、变量查看等功能，同时通过右边的监控功能来实现脚本运行时间的查看和统计，提高运行效率，如图 1-16 所示。

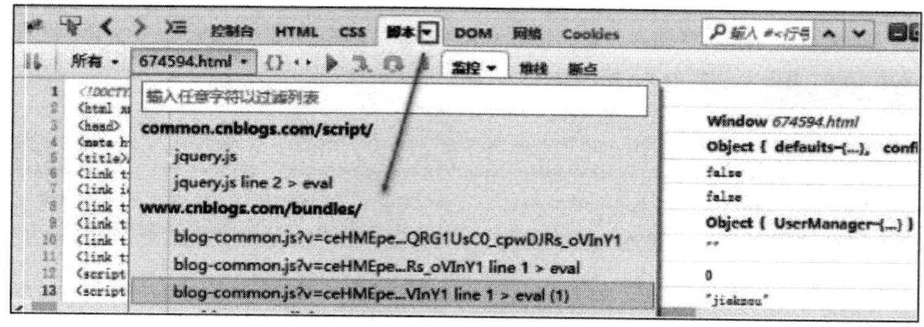

图 1-16

（5）网络（Net）

该标签功能主要用来监控网页各组成元素的运行时间信息，方便找出其中运行时间较慢的部分，进一步优化运行效率，如图 1-17 所示。

图 1-17

1.2.2 HttpRequester

HttpRequester 是一款接口的测试工具，和 Firebug 一样可以在火狐浏览器中通过附加组件的形式进行安装，如图 1-18~图 1-20 所示。

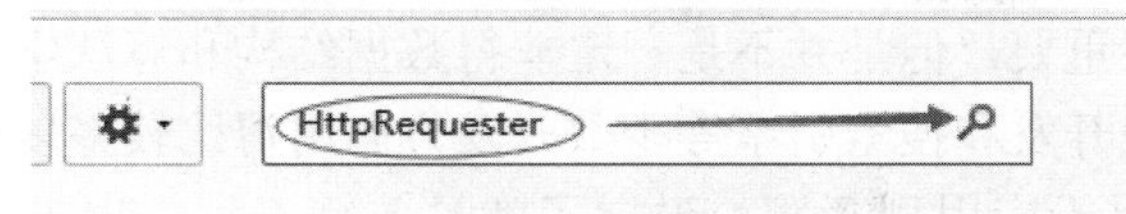

图 1-18

图 1-19

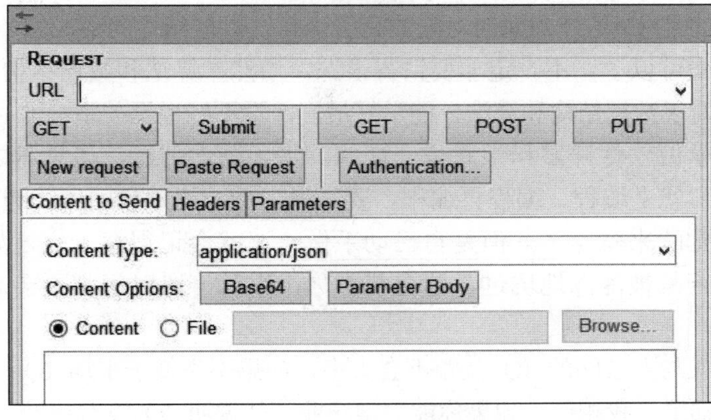

图 1-20

可以模拟各种请求方式，并且可以自定义添加需要提交的请求报文和请求头信息。另外，还有很多的功能，希望大家多去尝试，这里只是为了抛砖引玉。当我们开发 Wcf 和 WebAPI 程序的时候，这个插件非常有用。

1.3 知识储备

1.3.1　必备知识介绍

在学习 ASP.NET MVC 之前，要有 C#、ADO.Net、SQL、HTML、CSS、javascript、Jquery、ASP.NET WebForm 的基础，那些所谓零基础快速精通 ASP.NET MVC 的都是假的。没有扎实的基础，即便教会了精妙的剑招，耍出来也不会有什么杀伤力，反之，如果基础够扎实、内功够深厚，哪怕简简单单、平平无奇的一招也能有势不可挡的威力。

1.3.2　树立软件开发信心

在正式学习软件开发之前，你一定要树立信心，相信自己一定可以学会、可以学好，这样在遇到困难的时候才能坚持下去，毕竟求知之路较长，每当快坚持不下去的时候要学会安慰自己——有许多梦想总是遥不可及的，除非你坚持。

有人说，程序是由数据结构和算法组成的，所以数据结构和算法非常重要，大学里面学的 C 语言和高等数学课程必须要学好；还有人觉得软件开发适合理工专业的人，学文科的不适合做软件开发……

数据结构和算法是很重要，但是并不是一定非得数据结构和算法学得很熟才能胜任软件开发工作，算法除了做游戏开发和搜索引擎等少数开发方向用得比较多以外，做一般的软件开发很少用到，更多的是侧重于应用型开发，说白了就是熟练工种，也难怪许多程序员自嘲为码农、搬砖工。算法的重要性显得并不是那么重要，更重要的反倒是面向对象的思想，因为大家使用的是 C#这样的面向对象的高级语言。数据结构的话，做一般的应用开发知道基本原理和怎么用就可以了，不一定非要钻得很深。当然，算法和数据结构学得好，对做开发还是大有裨益的。对于数据结构和算法，能学好最好，学不好也没关系。

学做软件开发和学武是一样的，除了勤学苦练、多思考之外，没有捷径，如果硬要说捷径，那就是编程、编程再编程。要知道，看一遍跟自己做一遍的效果是完全不同的。看一遍往往一觉醒来就忘了，但是自己动手做一遍就会记得深一点，如果再思考总结过，印象就更深了，若是被深深地坑过，那么恭喜你，很长一段时间都不会忘记了，以后也很难犯同样的错误。

人与人之间的 IQ 差距不会太多，如果不幸低于平均 IQ、悟性差，那就只能像郭靖一样以勤补拙了。我相信，如果你能大学毕业，就说明 IQ 没有问题，既然 IQ 没有问题，那学软件开发也就不存在任何问题了，前提是你肯学。

我一直觉得学软件开发无外乎三板斧：一抄，二仿，三思考、创新、钻研+总结。最终目标是青出于蓝而胜于蓝。

最后，做软件开发的并不只有做技术这一条出路。在职业生涯中，我们会遇到各种机会和挑战，不要把自己框死在 just do coding 上，先学做人，再学做事，要学会与人打交道，而不是只和机器打交道。

1.4 C#语法新特性

在学习 ASP.NET MVC 之前，有必要先了解一下 C#2.0~C# 4.0 版本所带来的新的语法特性。这一点尤为重要，因为在 MVC 项目中我们利用 C#的新特性将会大大提高开发效率；同时，在 MVC 项目中随处可见 C#新特性的身影。其实，其本质都是“语法糖”，由编译器在编译时转成原始语法。

1.4.1 C# 2.0 新特性

1. 泛型（Generics）

微软官方定义：2.0 版 C# 语言和公共语言运行时（CLR）中增加了泛型。泛型将类型参数的概念引入 .NET Framework，类型参数使得设计如下类和方法成为可能：这些类和方法将一个或多个类型的指定推迟到客户端代码声明并实例化该类或方法的时候。

下面给出一个简单的泛型例子：

```
public class List<T>{ }
```

其中，T 就是 System.Collections.Generic.List<T>实例所存储类型的占位符。当定义泛型类的实例时，必须指定这个实例所存储的实际类型：

```
List<string> lst = new List<string>();
```

泛型允许程序员将一个实际的数据类型规约延迟至泛型的实例被创建时才确定。

泛型主要有两个优点：

- 编译时可以保证类型安全。
- 不用做类型转换，获得一定的性能提升。

2. 泛型方法、泛型委托、泛型接口

除了泛型类之外，还有泛型方法、泛型委托、泛型接口：

```
//泛型委托
public delegate void Del<T>(T item);
public static void Notify(int i) { }
Del<int> m1 = new Del<int>(Notify);
//泛型接口
public class MyClass<T1, T2, T3> : MyInteface<T1, T2, T3> { public T1 Method1(T2
param1, T3 param2) { throw new NotImplementedException(); } }
interface MyInteface<T1, T2, T3> { T1 Method1(T2 param1, T3 param2); }
//泛型方法
static void Swap<T>(ref T t1, ref T t2) { T temp = t1; t1 = t2; t2 = temp; }
public void Interactive ()
{
    string str1 = "a"; string str2 = "b";
    Swap<string>(ref str1, ref str2);
    Console.WriteLine(str1 + "," + str2);
}
```

3. 泛型约束（constraints）

可以给泛型的类型参数上加约束，要求这些类型参数满足一定的条件，如表 1-1 所示。

表 1-1　泛型的类型参数

约束	说明
where T: struct	类型参数需是值类型
where T : class	类型参数需是引用类型
where T : new()	类型参数要有一个 public 的无参构造函数
where T : <base class name>	类型参数要派生自某个基类
where T : <interface name>	类型参数要实现某个接口
where T : U	这里 T 和 U 都是类型参数，T 必须是或者派生自 U

4. 部分类（partial）

在申明一个类、结构或者接口的时候用 partial 关键字可以让源代码分布在不同的文件中。过去部分类是为了在 ASPX 页面和 ASPX.cx 页面进行 Code-Behind 的。在 EF 中使用 T4 模板自动生成代码的时候部分类的作用非常重要。

部分类仅是编译器提供的功能，在编译的时候会把 partial 关键字定义的类合在一起去编译。

5. 匿名方法

匿名方法的本质其实就是委托，函数式编程的最大特点之一就是把方法作为参数和返回值。ConsoleWrite→MulticastDelegate(intPtr[])→Delegate(object,intPtr)匿名方法：编译后会生成委托对象，生成方法，然后把方法装入委托对象，最后赋值给声明的委托变量。匿名方法可以省略参数：编译的时候会自动为这个方法按照委托签名的参数添加参数。

```
public delegate void ConsoleWrite(string strMsg);
//匿名方法测试
ConsoleWrite delCW1 = new ConsoleWrite(WriteMsg);
delCW1("天下第一");
ConsoleWrite delCW2 = delegate(string strMsg)
{
   Console.WriteLine(strMsg);
};
delCW2("天下第二");
```

1.4.2　C# 3.0/C# 3.5 新特性

1. 自动属性

这个概念很简单，简化了我们在做 C#开发的时候手写一堆私有成员+属性的编程方式，我们只需要使用如下方式声明一个属性，编译器就会自动生成所需的成员变量。

回顾一下传统属性概念,属性的目的一是封装字段，二是控制读写权限及字段的访问规则（如年龄、生日范围），平时主要是用来封装读写权限。

我们来看一下基本用法：

```
public class User
{
   public int Id { get; set; } //自动属性
   public string Name { get; set; }
   public int Age { get; set; }
   public Address Address { get; set; }
}
```

在 C# 3.0 之前，我们是这样来实现属性的：

```
private int id;
public int Id
{
    get
    {
        return id;
    }
    set
    {
        id = value;
    }
}
```

读者可以思考一下：使用自动属性的话程序员写的代码少了，机器做的事情就多了，那我们到底要不要使用它？

如果是针对读写权限的封装，就推荐使用，因为它是在编译的时候产生了负担，并不是在运行的时候，所以不会影响客户运行程序时的效率！但是编译时生成的代码也有一个显而易见的缺点，语法太完整，编译后的程序集会比较大，不过这对于现今的硬件配置而言也算不上什么了。

2. 隐式推断类型 Var

你可能对这个名称比较陌生，但是 var 这个关键字应该很熟悉，在 C#中使用 var 声明一个对象时编译器会自动根据赋值语句推断这个局部变量的类型。赋值以后，这个变量的类型也就已经确定并且不可以进行更改。另外，var 关键字也可用于匿名类的声明。

应用场合：var 主要用于表示一个 LINQ 查询的结果。这个结果既可能是 ObjectQuery<>或 IQueryable<>类型的对象，也可能是一个简单的实体类型的对象或者是一个基本类型对象，这时使用 var 声明这个对象可以节省很多代码书写上的时间。

```
var customer = new User();
var i = 1;
```

var 隐式类型的限制：

- 被声明的变量必须是一个局部变量，而不是静态或实例字段。
- 变量必须在声明的同时被初始化，因为编译器要根据初始化值推断类型。
- 初始化的对象不能是一个匿名函数。
- 初始化表达式不能是 null。
- 语句中只声明一次变量，声明后不能更改类型。
- 赋值的数据类型必须是可以在编译时确定的类型。

3. 对象集合初始化器

在.NET 2.0 中构造一个对象的方法一是提供一个重载的构造函数，二是用默认的构造函数生成一个对象，然后对其属性进行赋值。在.NET 3.5/C# 3.0 中，我们有一种更好的方式来进行对象的初始化，那就是使用对象初始化器。这个特性也是匿名类的一个基础，所以放在匿名类之前介绍。

对象初始化：

```
User user = new User { Id = 1, Name = "Zouqj", Age = 27 };
```

集合初始化：

```
List<Dog> dogs = new List<Dog>() { new Dog() { Name = "Tom", Age = 1 }, new Dog()
{ Name = "Lucy", Age = 3 } };
```

创建并初始化数组：

```
string[] array = { "西施", "貂蝉" };
```

4. 匿名类

有了前面初始化器的介绍，匿名类就简单了。匿名类型提供了一种方便的方法，可用来将一组只读属性封装到单个对象中，而无须首先显式定义一个类型。类型名由编译器生成，并且不能在源代码级使用，每个属性的类型由编译器推断。我们可以使用 new { object initializer }或 new[]{ object, …}来初始化一个匿名类或不确定类型的数组。匿名类的对象需要使用 var 关键字声明。示例代码：

```
var p = new { Id = 1, Name = " Zouqj ", Age = 27 };//属性名字和顺序不同会生成不同类
```

在编译后会生成一个“泛型类”，我们可以使用反编译工具 Reflector 来查看这个生成的泛型类中都有些什么，如图 1-21 所示。

```
[DebuggerDisplay(@"\{ Id = {Id}, Name = {Name}, Age = {Age} }", Type="<Anonymous Type>"), CompilerGenerated]
internal sealed class <>f_AnonymousType0<<Id>j_TPar, <Name>j_TPar, <Age>j_TPar>
{
    // Fields
    [DebuggerBrowsable(DebuggerBrowsableState.Never)]
    private readonly <Age>j_TPar <Age>i_Field;
    [DebuggerBrowsable(DebuggerBrowsableState.Never)]
    private readonly <Id>j_TPar <Id>i_Field;
    [DebuggerBrowsable(DebuggerBrowsableState.Never)]
    private readonly <Name>j_TPar <Name>i_Field;

    // Methods
    [DebuggerHidden]
    public <>f_AnonymousType0(<Id>j_TPar Id, <Name>j_TPar Name, <Age>j_TPar Age);
    [DebuggerHidden]
    public override bool Equals(object value);
    [DebuggerHidden]
    public override int GetHashCode();
    [DebuggerHidden]
    public override string ToString();

    // Properties
    public <Age>j_TPar Age { get; }
    public <Id>j_TPar Id { get; }
    public <Name>j_TPar Name { get; }
}
```

图 1-21

可以看到包含了如下信息：

- 获取所有初始值的构造函数，顺序与属性顺序一样。
- 属性的私有只读字段。
- 重写 Object 类中的 Equals、GetHashCode、ToString()方法。
- 包含公共只读属性，属性不能为 null、匿名函数、指针类型。

用处：

- 避免过度的数据累积。
- 为一种情况特别进行的数据封装。
- 避免进行单调重复的编码。

应用场合：

直接使用 select new { object initializer }这样的语法就是将一个 LINQ 查询的结果返回到一个匿名类中。

注意：

- 当出现“相同”的匿名类时，编译器只会创建一个匿名类。
- 编译器如何区分匿名类是否相同。
- 属性名、属性值（因为这些属性是根据值来确定类型的）、属性个数、属性的顺序。
- 匿名类的属性是只读的，可放心传递，并且可用在线程间共享数据。

5. 扩展方法

扩展方法是一种特殊的静态方法，可以像扩展类型上的实例方法一样进行调用，能向现有类型“添加”方法，而无须创建新的派生类型、重新编译或以其他方式修改原始类型。

例如，在编译时直接将 str.WriteSelf(2016) 替换成：

```
StringUtil.WriteSelf(str,2016);
```

想为一个类型添加一些成员时，可以使用扩展方法：

```
public static class StringUtil
{
       public static void WriteSelf(this string strSelf, int year)
       {
          Console.WriteLine(string.Format("我是{0}人，今年是{1}年。", strSelf, year));
       }
}
```

测试：

```
//扩展方法
string str = "冷水江";
str.WriteSelf(2016);
```

编译器认为一个表达式要使用一个实例方法，但是没有找到，需要检查导入的命名空间和当前命名空间里所有的扩展方法，并匹配到适合的方法。

注意：

（1）实例方法优先于扩展方法（允许存在同名实例方法和扩展方法）。

（2）可以在空引用上调用扩展方法。

（3）可以被继承。

（4）并不是任何方法都能作为扩展方法使用，必须有以下特征：

- 它必须放在一个非嵌套、非泛型的静态类中。
- 它至少有一个参数。
- 第一个参数必须附加 this 关键字。
- 第一个参数不能有任何其他修饰符（out/ref）。
- 第一个参数不能是指针类型，其类型决定是在何种类型上进行扩展。

6. 系统内置委托

Func / Action 委托使用可变性：

```
Action<object> test=delegate(object o){Console.WriteLine(o);};
Action<string> test2=test;
Func<string> fest=delegate(){return Console.ReadLine();};
fest2=fest;

public delegate void Action();
public delegate bool Predicate<in T>(T obj);
public delegate int Comparison<in T>(T x, T y);
```

协变指的是委托方法的返回值类型直接或间接继承自委托签名的返回值类型，逆变则是参数类型继承自委托方法的参数类型 System.Func，代表有返回类型的委托。

```
public delegate TResult  Func<out TResult>();
public delegate TResult  Func<in T, out TResult>(T arg);
.....
```

注：输入泛型参数-in 最多可以有 16 个，输出泛型参数-out 只有一个。System.Action 代表无返回类型的委托。

```
public delegate void Action<in T>(T obj);    //list.Foreach
public delegate void Action<in T1, in T2>(T1 arg1, T2 arg2);
......
```

注：最多有 16 个参数，System.Predicate<T> 代表返回 bool 类型的委托，用作执行表达式。

```
public delegate bool Predicate<in T>(T obj);  //list.Find
```

System.Comparison<T> 代表返回 int 类型的委托，用于比较两个参数的大小。

```
public delegate int Comparison<in T>(T x, T y); //list.Sort
```

为什么要定义这么多简单的委托？方便!

7. Lambda 表达式

Lambda 表达式的本质就是匿名函数，Lambda 表达式基于数学中的 λ 演算得名，直接对应于其中的 lambda 抽象(lambda abstraction)，是一个匿名函数，即没有函数名的函数。“Lambda 表达式”是一个匿名函数，可以包含表达式和语句，并且可用于创建委托或表达式树类型。

Lambda 表达式的运算符为=>，读作“goes to”。=>运算符具有与赋值运算符 (=) 相同的优先级。

Lambda 的基本形式是：

```
(input parameters) => expression
```

只有在 Lambda 有一个输入参数时，括号才是可选的，否则括号是必需的。两个或更多输入参数由括在括号中的逗号分隔：(x, y) => x == y。

有时，编译器难于或无法推断输入类型。 如果出现这种情况，您可以按以下示例中所示的方式显式指定类型：

```
(int x, string s) => s.Length > x
```

使用空括号指定零个输入参数：

```
() => SomeMethod()
```

最常用的场景是 IEnumerable 和 IQueryable 接口的 Where(c=>c.Id>3)。

下列规则适用于 Lambda 表达式中的变量范围：

- 捕获的变量将不会被作为垃圾回收，直至引用变量的委托超出范围为止。
- 在外部方法中看不到 Lambda 表达式内引入的变量。
- Lambda 表达式无法从封闭方法中直接捕获 ref 或 out 参数。
- Lambda 表达式中的返回语句不会导致封闭方法返回。
- Lambda 表达式不能包含其目标位于所包含匿名函数主体外部或内部的 goto 语句、break 语句或 continue 语句。

8. Lambda 表达式缩写推演

Lambda 表达式缩写推演如图 1-22 所示。

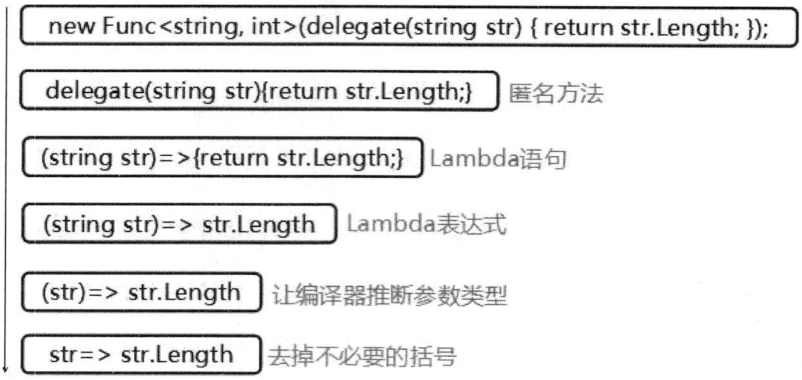

new Func<string, int>(delegate(string str) { return str.Length; });

delegate(string str){return str.Length;} 匿名方法

(string str)=>{return str.Length;} Lambda语句

(string str)=> str.Length Lambda表达式

(str)=> str.Length 让编译器推断参数类型

str=> str.Length 去掉不必要的括号

图 1-22

例如：

```
delegate int AddDel(int a, int b); //定义一个委托
#region lambda

AddDel fun = delegate(int a, int b) { return a + b; }; //匿名函数
//Console.WriteLine(fun(1, 3));
//lambda  参数类型可以进行隐式推断，可以省略类型 lambda 本质就是匿名函数
AddDel funLambda = (a, b) => a + b;
List<string> strs = new List<string>() {     "1","2","3"
                    };

var temp = strs.FindAll(s => int.Parse(s) > 1);
foreach (var item in temp)
{
    Console.WriteLine(item);
}
//Console.WriteLine(funLambda(1,3));

#endregion
static void Main(string[] args)
{
   List<int> nums = new List<int>() { 1, 2, 3, 4, 6, 9, 12 };
   //使用委托的方式
   List<int> evenNums = nums.FindAll(GetEvenNum);
   foreach (var item in evenNums)
  {
      Console.WriteLine(item);
  }
```

```
    Console.WriteLine("使用 lambda 的方式");

    List<int> evenNumLamdas = nums.FindAll(n => n % 2 == 0);
    foreach (var item in evenNumLamdas)
    {
        Console.WriteLine(item);
    }
    Console.ReadKey();
static bool GetEvenNum(int num)
{
  if (num % 2 == 0)
  {
      return true;
  }
  return false;
}
```

9. 标准查询运算符(SQO)

标准查询运算符是定义在 System.Linq.Enumerable 类中的 50 多个为 IEnumerable<T>准备的扩展方法，换句话说 IEnumerable<T>上的每个方法都是一个标准查询运算符，这些方法用来对操作的集合进行查询筛选。

标准查询运算符提供了包括筛选、投影、聚合、排序等功能在内的查询功能。

先准备一下测试数据和测试方法，代码如下。

```
private List<User> InitLstData()
{
    return new List<User>(){
        new User { Id = 1, Name = "Zouqj1", Age = 21 },
        new User { Id = 2, Name = "Zouqj2", Age = 22 },
        new User { Id = 3, Name = "Zouqj3", Age = 23 },
        new User { Id = 4, Name = "Zouqj4", Age = 24 },
        new User { Id = 5, Name = "Zouqj5", Age = 25 },
        new User { Id = 6, Name = "Zouqj6", Age = 26 },
        new User { Id = 7, Name = "Zouqj7", Age = 27 },
        new User { Id = 8, Name = "Zouqj8", Age = 28 },
        new User { Id = 9, Name = "Zouqj9", Age = 29 },
        new User { Id = 10, Name = "Zouqj10", Age = 30 },
        new User { Id = 11, Name = "Zouqj11", Age = 31 }
    };
}
```

（1）筛选集合 Where

Where 方法提供了我们对于一个集合的筛选功能，但需要提供一个带 bool 返回值的“筛选器”（匿名方法、委托、Lambda 表达式均可），从而表明集合中某个元素是否应该被返回。

这里筛选出所有年龄大于等于 30 岁的数据，代码如下：

```
var lst = InitLstData();
var result = lst.Where(x => x.Age >= 30).ToList();
result.ForEach(r => Console.WriteLine(string.Format("{0},{1},{2}", r.Id, r.Name, r.
Age)));
Console.ReadLine();
```

运行结果如下：

```
10,Zouqj10,30
11,Zouqj11,31
```

（2）查询投射 Select

返回新对象集合 IEnumerable<TSource>Select()。返回年龄大于等于 30 岁的人的名字，代码如下：

```
var result = lst.Where(x => x.Age >= 30).Select(s => s.Name).ToList();
result.ForEach(x => Console.WriteLine(x));
```

运行结果如下：

```
Zouqj10
Zouqj11
```

（3）统计数量 int Count()

```
lst.Where(x => x.Age >= 30).Count();
```

（4）多条件排序 OrderBy().ThenBy().ThenBy()

```
lst.OrderBy(x=>x.Age).OrderBy(x=>x.Id)
```

（5）集合连接 Join()

新建一个 Student 类，并初始化数据。

```
public class Student
{
      public int ID { get; set; }
      public int UserId { get; set; }
      public string ClassName { get; set; }
}
List<Student> lstStu = new List<Student>() {
      new Student{ID=1,UserId=1,ClassName="本科8班"},
```

```
        new Student{ID=1,UserId=3,ClassName="本科2班"},
        new Student{ID=1,UserId=2,ClassName="电信1班"}};
var result = lst.Join(lstStu, u => u.Id, p => p.UserId, (u, p) => new { UserId = u.Id, Name = u.Name, ClassName =p.ClassName});
```

（6）延迟加载 Where

在标准查询运算符中，Where 方法就是一个典型的延迟加载案例。在实际的开发中，我们往往会使用一些 ORM 框架例如 EF 去操作数据库，Where 方法的使用则是每次调用都只是在后续生成 SQL 语句时增加一个查询条件，EF 无法确定本次查询是否已经添加结束，所以没有办法在每个 Where 方法执行的时候确定最终的 SQL 语句，只能返回一个 DbQuery 对象，当使用到这个 DbQuery 对象的时候，才会根据所有条件生成最终的 SQL 语句去查询数据库。

```
IEnumerable<User> usr = lst.Where(x => x.Age >= 30);
```

（7）即时加载 FindAll

在开发中如果使用 FindAll 方法，EF 会根据方法中的条件自动生成 SQL 语句，然后立即与数据库进行交互获取查询结果，并加载到内存中去。

```
List<User> lstUsr = lst.FindAll(x => x.Age >= 30);
```

SQO 缺点：语句太庞大复杂。

10. LINQ

C# 3.0 新语法，查询表达式，和 SQL 风格接近的代码。

```
IEnumerable<Dog> listDogs = from dog in dogs
where  dog.Age>5
//let d=new{Name=dog.Name}
orderby dog.Age descending
select dog;
//select new{Name=dog.Name}
```

以 from 开始，以 select 或 group by 子句结尾。输出是一个 IEnumerable<T> 或 IQueryable<T> 集合。

注：T 的类型 由 select 或 group by 推断出来。

LINQ 分组：

```
IEnumerable<IGrouping<int, Dog>> listGroup = from dog in listDogs where dog.Age > 5 group dog by dog.Age;
```

遍历分组：

```
foreach (IGrouping<int, Dog> group in listGroup)
{
```

```
    Console.WriteLine(group.Key+"岁数: ");
    foreach (Dog d in group)
    {
        Console.WriteLine(d.Name + ",age=" + d.Age);
    }
}
```

注意：LINQ 查询语句编译后会转成标准查询运算符。

这里提一下 LINQPad 工具，LINQPad 支持 object 、xml、sql、to linq。通过这个工具，我们可以分析 LINQ 语句最终转换为 SQL 或者 lambda 表达式树时会是什么样子的。

1.4.3 C# 4.0 新特性

1. 可选参数和命名参数

可选参数：可选参数是 C# 4.0 提出来的，当我们调用方法，不给这个参数赋值时，它会使用我们定义的默认值。

需要注意的是：

（1）可选参数不能为参数列表的第 1 个参数，必须位于所有的必选参数之后（除非没有必选参数）；

（2）可选参数必须指定一个默认值，且默认值必须是一个常量表达式，不能为变量；

（3）所有可选参数以后的参数都必须是可选参数。

命名参数：通过命名参数调用，实参顺序可以和形参不同。

对于简单的重载，可以使用可选参数和命名参数混合的形式来定义方法，提高代码的运行效率。

```
public class Dog
{
    public string Name { get; set; }
    public int Age { get; set; }

    /// <summary>
    /// 参数默认值 和 命名参数
    /// </summary>
    /// <param name="name"></param>
    /// <param name="age"></param>
    public void Say(string name = "汪汪汪", int age = 1)
    {
        Console.WriteLine(name + "," + age);
    }
```

```
}
```

运行结果如图 1-23 所示。

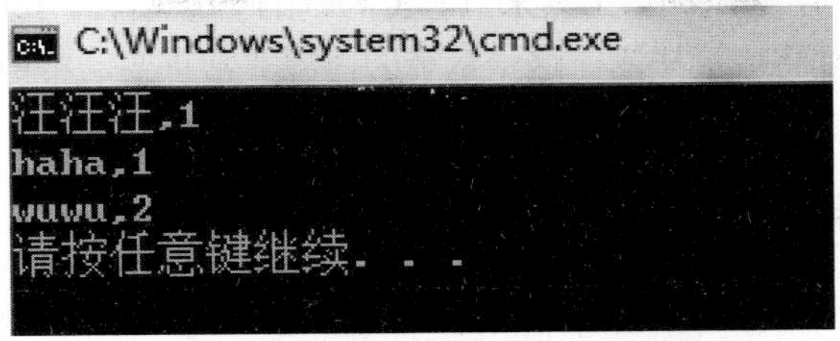

图 1-23

如果要让 name 使用默认值，age 怎么给值？

```
_dog.Say(age: 3); //输入结果：汪汪汪,3
```

2. Dynamic 特性

C# 4.0 的 Dynamic 特性需引用 System.Dynamic 命名空间：

```
using System.Dynamic;
//Dynamic
dynamic Customer = new ExpandoObject();
Customer.Name = "Zouqj";
Customer.Male = true;
Customer.Age = 27;
Console.WriteLine(Customer.Name + Customer.Age + Customer.Male);
Console.ReadKey();
```

另外，需要提一下 params。params 并非新的语法特性，但是这里也要简单介绍一下，因为使用 params 关键字作为方法参数可以指定采用数目可变的参数，可以发送参数声明中所指定类型用逗号分隔的参数列表或指定类型的参数数组，还可以不发送参数。若未发送任何参数，则 params 列表的长度为零。

在方法声明中的 params 关键字之后不允许有任何其他参数，并且在方法声明中只允许一个 params 关键字。

```
public void ParamsMethod(params int[] list)
{
    for (int i = 0; i < list.Length; i++)
    {
        Console.WriteLine(list[i]);
    }
    Console.ReadLine();
}
```

```
ParamsMethod( 25, 24, 21,15);
ParamsMethod(25, 24, 21, 15);
```

此外，建议大家多使用 reflector 工具来查看 C#源码和 IL 语言，reflector 就像一面照妖镜，任何 C#语法糖在它面前都将原形毕露。这一章节的内容非常重要，希望大家熟练掌握。

1.4.4 C#5.0 新特性

C#5.0 的新特性，最重要的就是异步和等待（async 和 await），其使用方式特别简单，就是在方法的返回值前面添加关键字 async，同时在方法体中需要异步调用的方法前面再添加关键字 await。需要注意的是这个异步方法必须以 Task 或者 Task<TResult>作为返回值。

更多内容请参考下面的网页：

http://www.cnblogs.com/zhili/archive/2013/05/15/Csharp5asyncandawait.html

第 2 章

◄ Entity Framework ►

在企业应用中做 ASP.NET MVC 开发，通常都搭配着 ORM 框架使用，而 Entity Framework 正是微软开发的一种 ORM 框架，因为都是微软的产品，所以通常它和 MVC 是黄金搭档。使用 EF 的目的就是为了高效地开发，如果项目中对性能要求非常高，业务又十分复杂的话，就不建议使用 EF。你可以考虑使用一些轻量级 ORM 框架或者直接使用原生的 SQL 或存储过程。

2.1 Entity Framework 简介

Entity Framework 的全称为 ADO.NET Entity Framework，简称为 EF，我们一般很少用全称，而是直接用 EF。

2.1.1 与 ADO.NET 的关系

Entity Framework（实体框架）是微软以 ADO.NET 为基础所发展出来的对象关系对应 (O/R Mapping) 解决方案，早期被称为 ObjectSpace，目前最新版本是 EF7（CodeOnly 功能得到了更好的支持）。

Entity Framework 是 ADO.NET 中的一组支持开发面向数据的软件应用程序的技术，是微软的一个 ORM 框架。其他的一些基于.NET 开发的 ORM 框架有 Nibernate、PetaPoco 等。

Entity Framework 的特点：

- 支持多种数据库（MSSQL、Oracle、Mysql 和 DB2）。
- 强劲的映射引擎，能很好地支持存储过程。
- 提供 Visual Studio 集成工具，进行可视化操作。
- 能够与 ASP.NET、WPF、WCF、WCF Data Services 进行很好的集成。

2.1.2 什么是 O/R Mapping

广义上，ORM 指的是面向对象的对象模型和关系型数据库的数据结构之间的相互转换。

狭义上，ORM 可以被认为是基于关系型数据库的数据存储，实现一个虚拟的面向对象的数据访问接口。理想情况下，基于这样一个面向对象的接口，持久化一个 OO 对象应该不需要了解任何关系型数据库存储数据的实现细节。EDM 设计器如图 2-1 所示。

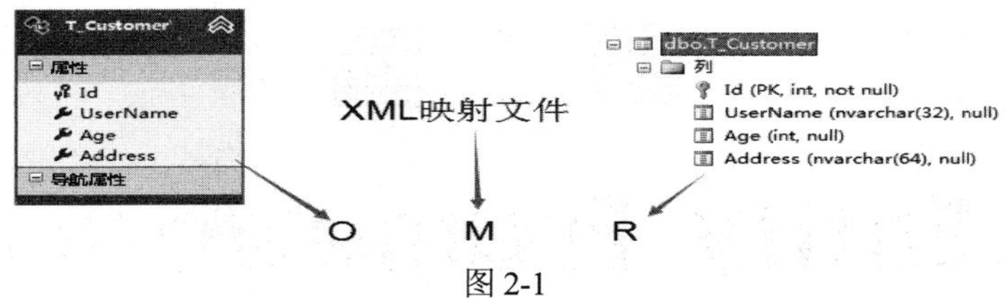

图 2-1

在面向对象的世界里，我们使用单向关联，然而在关系数据库的世界里，我们使用外键作为双向关联。

面向对象有继承的概念。例如，车辆类有很多继承类，小汽车是一种车辆，大卡车也是一种车辆，这种都是继承关系。

在关系数据库世界里，没有继承的关系。

ORM 是对象世界和关系世界的一座桥梁，通过映射关系，简化了大量操作数据库的代码。

2.1.3 ORM in EF

Entity Framework 利用了抽象化数据结构的方式，将每个数据库对象都转换成应用程序对象（entity），而数据字段都转换为属性（property），关系则转换为结合属性（association），让数据库的 E/R 模型完全转成对象模型，如此让程序设计师能用最熟悉的编程语言来调用访问。在抽象化的结构之下，则是高度集成与对应结构的概念层、对应层和储存层，以及支持 Entity Framework 的数据提供者（provider），让数据访问的工作得以顺利、完整地进行。如图 2-2 所示。

- 概念层：负责向上的对象与属性显露与访问。
- 对应层：将上方的概念层和底下的储存层的数据结构对应在一起。
- 储存层：依不同数据库与数据结构而显露出实体的数据结构体，和 Provider 一起，负责实际对数据库的访问和 SQL 的产生。

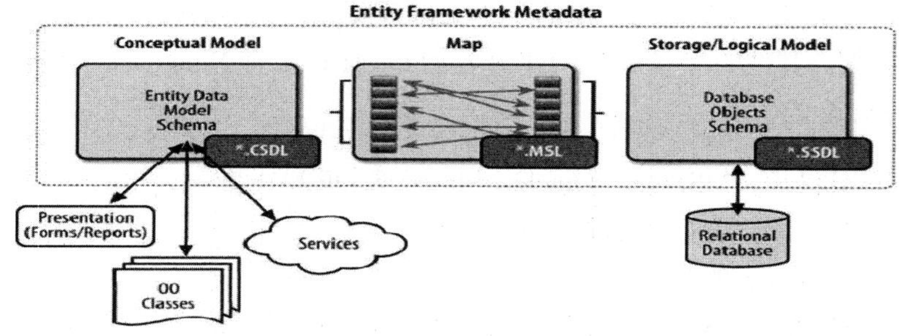

图 2-2

2.1.4 EF 的优缺点

1. EF 的优点

- 极大地提高开发效率。EF 是微软自己的产品，跟 VS 开发工具集成度比较好，开发中代码都是强类型的，写代码效率非常高，自动化程序非常高，采用命令式的编程。

- EF 提供的模型设计器非常强大，不仅仅带来了设计数据库的革命，其附带来的自动化生成模型代码的功能也极大地提高了开发和架构设计的效率。
- EF 跨数据库支持是 ORM 框架的主要功能点之一，有着仅仅通过改变配置就可以做到跨数据库的能力，更换数据库非常方便。

2. EF 的缺点

- EF 性能不好，性能有损耗。（生成 SQL 脚本阶段）在复杂查询的时候生成的 SQL 脚本效率不是很高。
- 数据库端性能损耗是一样的，但是在将对象状态转换为 SQL 语句时会损失性能。

2.2 Database First 开发方式

DatabaseFirst 又叫数据库优先的开发方式，是一种比较旧的开发方式，现在越来越多的企业已经不再使用此种开发方式。当然，对于一些旧项目进行升级，在已经有了数据库的情况下，使用此种方式还是十分方便的。

2.2.1 创建 Dtabase First Demo

1. 创建控制台项目

选择“新建→项目→Windows→控制台应用程序”，如图 2-3 所示。

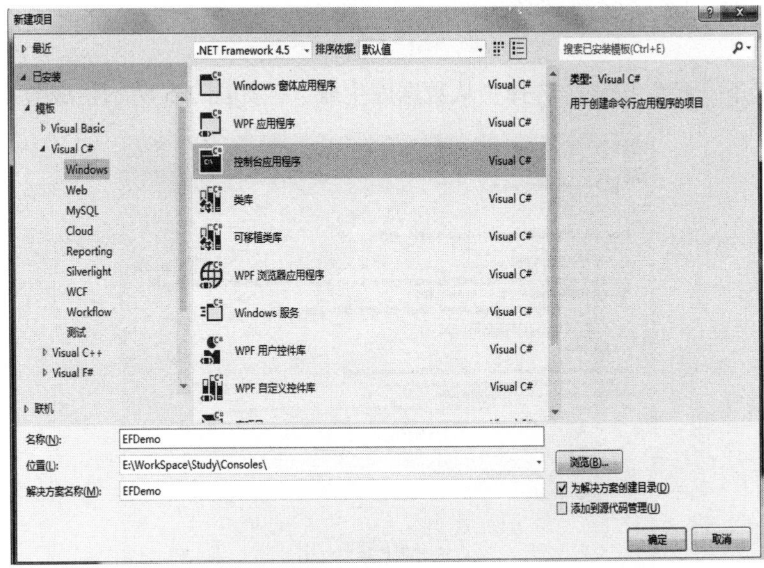

图 2-3

2. 创建数据库（添加表）

```
CREATE TABLE [dbo].[T_Customer](
    [Id] [int] IDENTITY(1,1) NOT NULL,
    [UserName] [nvarchar](32) NULL,
    [Age] [int] NULL,
    [Address] [nvarchar](64) NULL,
 CONSTRAINT [PK_T_Customer] PRIMARY KEY CLUSTERED
(
    [Id] ASC
)WITH (PAD_INDEX  = OFF, STATISTICS_NORECOMPUTE  = OFF, IGNORE_DUP_KEY = OFF,
ALLOW_ROW_LOCKS  = ON, ALLOW_PAGE_LOCKS  = ON) ON [PRIMARY]
) ON [PRIMARY]
```

3. 在项目中添加“数据实体模型”

（1）右击项目，选择“添加新项”，如图 2-4 所示添加新项。

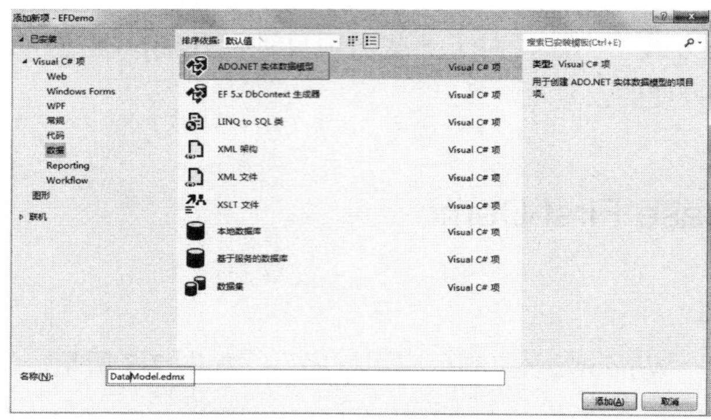

图 2-4

（2）单击“下一步”按钮，选择“从数据库生成”，如图 2-5 所示生成数据。

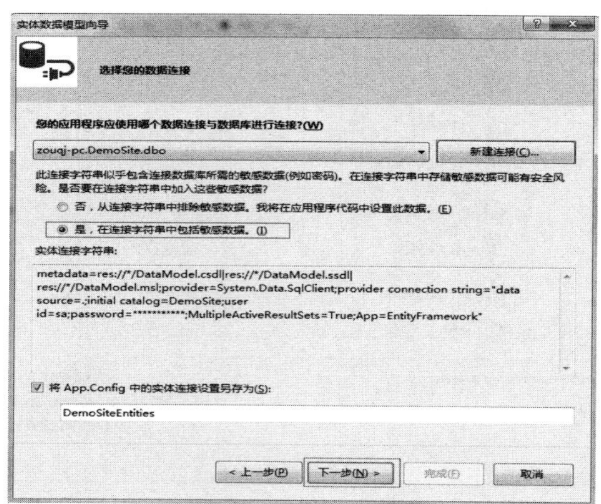

图 2-5

（3）新建数据库连接，如图 2-6 所示。

（4）在 EF 中是可以直接调用存储过程、视图、函数的，这里先只选择一张表，如图 2-7 所示。单击“完成”按钮，最后的展示效果如图 2-8 所示。

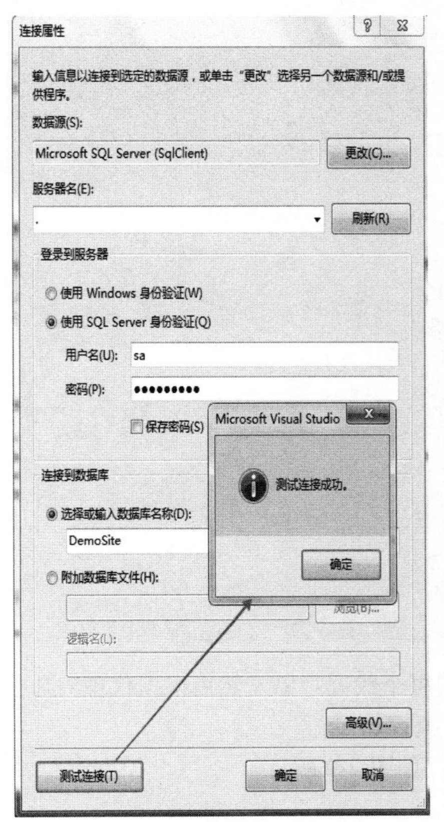

图 2-6

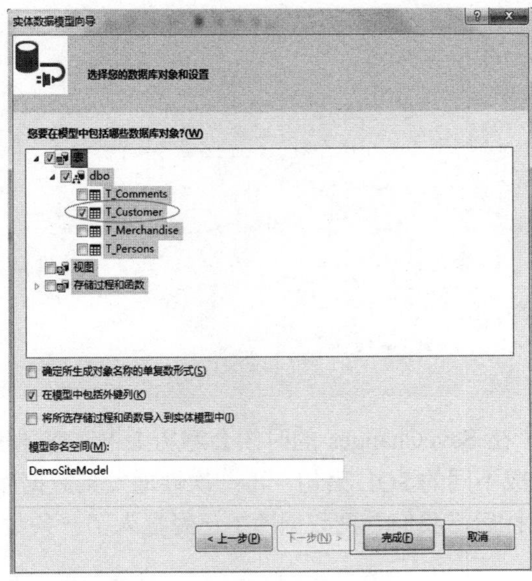

图 2-7

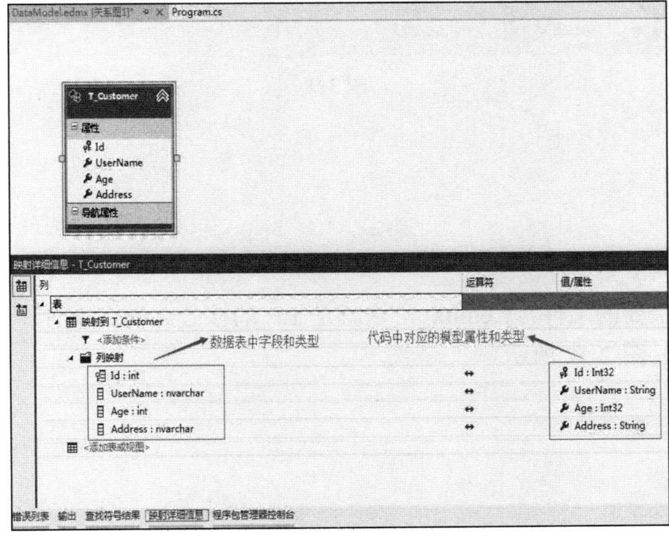

图 2-8

（5）在代码中添加访问上下文保存到数据库的代码：

```
using System;
using System.Collections.Generic;
using System.Linq;
using System.Text;
using System.Threading.Tasks;
namespace EFDemo
{
    class Program
    {
        static void Main(string[] args)
        {
            DemoSiteEntities entity = new DemoSiteEntities();
            T_Customer customer = new T_Customer { Address=" 东 海 五 彩 金 轮
",Age=27,UserName="楚留香"};
            entity.T_Customer.Add(customer); //这里只相当于构造 SQL 语句
            entity.SaveChanges(); //这里才进行数据库操作，相当于按 F5执行
        }
    }
}
```

EF 在 SaveChanges 的时候会遍历上下文容器里的每个代理对象，然后根据代理对象的 State 属性生成不同的 SQL 语句，再一次性地发到数据库中执行。

执行后，我们看到数据库中已经插入了一条数据，如图 2-9 所示。

ZOUQJ-PC.DemoSite - dbo.T_Customer

	Id	UserName	Age	Address
▶	1	楚留香	27	东海五彩金轮
*	NULL	NULL	NULL	NULL

图 2-9

2.2.2 EF 原理

1. 实体数据模型（EDM）

EF 中存在一个主要的文件：*.edmx ，这就是 EF 的核心。我们来看一下 edmx 文件里面到底有什么，如图 2-10~图 2-12 所示。

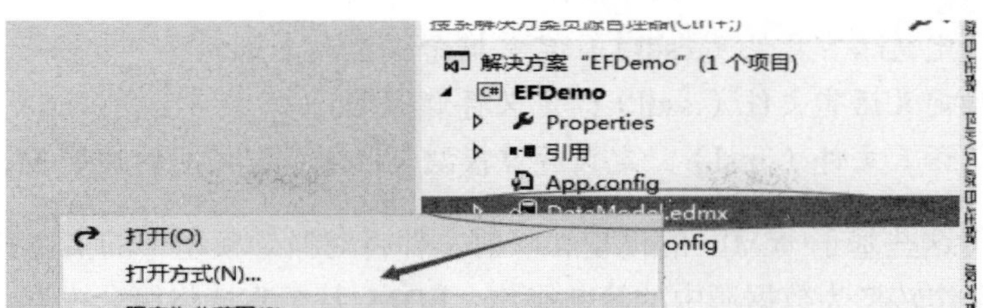

图 2-10

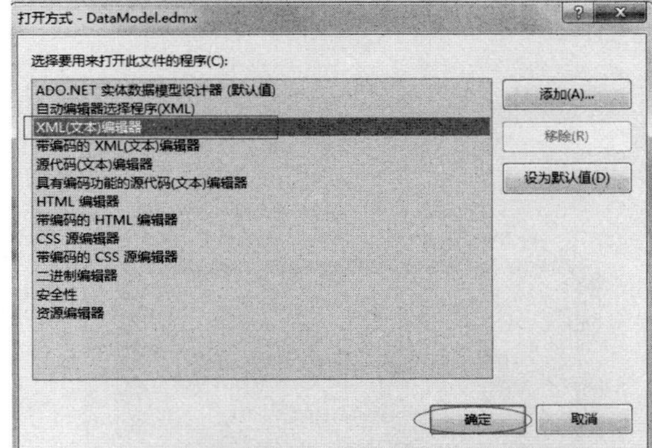

图 2-11

DataModel.edmx　DataModel.tt　DataModel.Designer.cs　Program.cs

```
<?xml version="1.0" encoding="utf-8"?>
<edmx:Edmx Version="3.0" xmlns:edmx="http://schemas.microsoft.com/ado/2009/11/edmx">
  <!-- EF Runtime content -->
  <edmx:Runtime>
    <!-- SSDL content -->
    <edmx:StorageModels>...</edmx:StorageModels>
    <!-- CSDL content -->
    <edmx:ConceptualModels>...</edmx:ConceptualModels>
    <!-- C-S mapping content -->
    <edmx:Mappings>...</edmx:Mappings>
  </edmx:Runtime>
  <!-- EF Designer content (DO NOT EDIT MANUALLY BELOW HERE) -->
  <Designer xmlns="http://schemas.microsoft.com/ado/2009/11/edmx">
    <Connection>
      <DesignerInfoPropertySet>
        <DesignerProperty Name="MetadataArtifactProcessing" Value="EmbedInOutputAssembly" />
      </DesignerInfoPropertySet>
    </Connection>
    <Options>
      <DesignerInfoPropertySet>
        <DesignerProperty Name="ValidateOnBuild" Value="true" />
        <DesignerProperty Name="EnablePluralization" Value="False" />
        <DesignerProperty Name="IncludeForeignKeysInModel" Value="True" />
        <DesignerProperty Name="CodeGenerationStrategy" Value="无" />
      </DesignerInfoPropertySet>
    </Options>
    <!-- Diagram content (shape and connector positions) -->
    <Diagrams></Diagrams>
  </Designer>
</edmx:Edmx>
```

图 2-12

在 EF 中的实体数据模型（EDM）由以下 3 种模型和具有相应文件扩展名的映射文件进行定义。

- 概念架构定义语言文件（.csdl）：定义概念模型。
- 存储架构定义语言文件（.ssdl）：定义存储模型。
- 映射规范语言文件（.msl）：定义存储模型与概念模型之间的映射 M。

实体框架使用这些基于 XML 的模型和映射文件将对概念模型中的实体和关系的创建、读取、更新和删除操作转换为数据源中的等效操作。EDM 甚至支持将概念模型中的实体映射到数据源中的存储过程。EF 以 EDM（Entity Data Model）为主，其上还有 Entity Client、Object Service 以及 LINQ 可以使用，如图 2-13、2-14 所示。

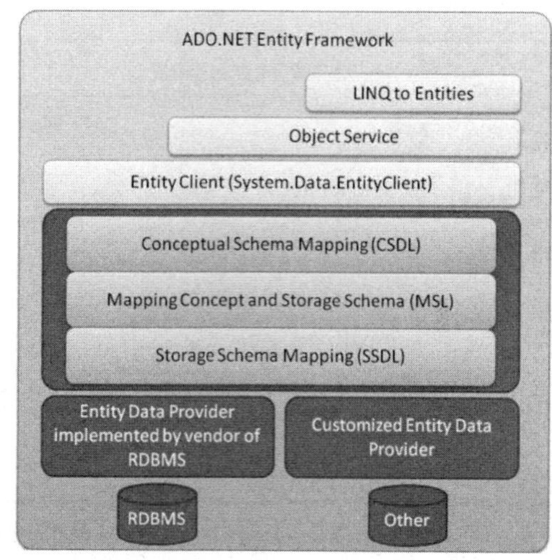

图 2-13

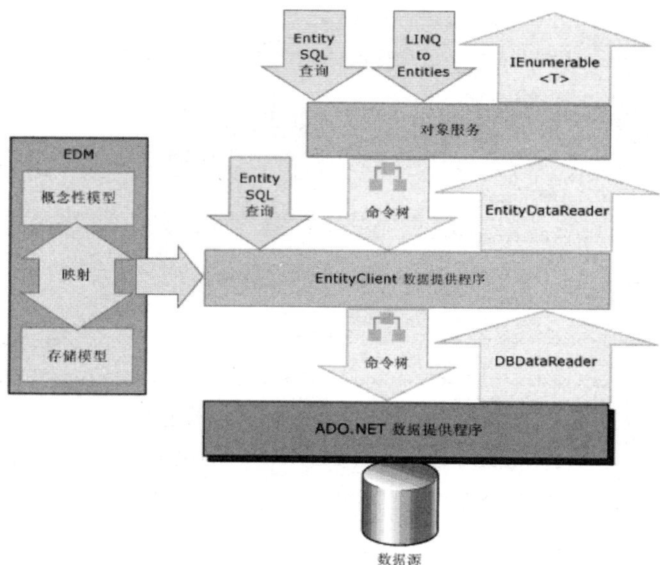

图 2-14

2. EF 中操作数据库的网关：ObjectContext

ObjectContext 封装 .NET Framework 和数据库之间的连接。此类用作创建、读取、更新和删除操作的网关。如图 2-15 所示。

ObjectContext 类为主类，用于与作为对象（这些对象为 EDM 中定义的实体类型的实例）的数据进行交互。

ObjectContext 类的实例封装以下内容：

- 到数据库的连接，以 EntityConnection 对象的形式封装。
- 描述该模型的元数据，以 MetadataWorkspace 对象的形式封装。
- 用于管理缓存中持久保存的对象的 ObjectStateManager 对象。

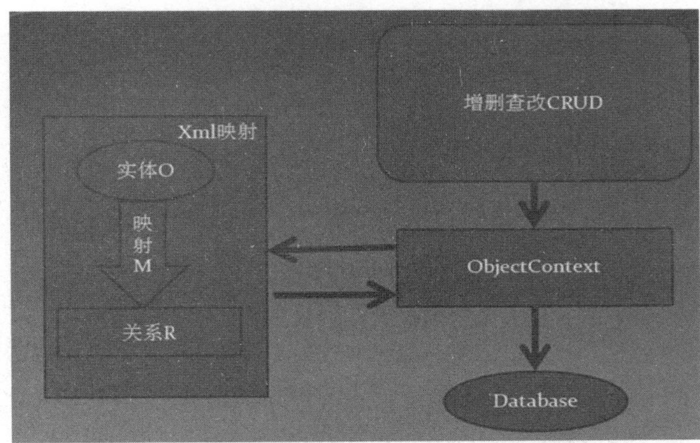

图 2-15

3. EF 执行原理

在.NET 中有两项重要的技术：反射和特性。

EDM 中利用特性来标识实体映射到具体数据库中的 TableName，属性对应的具体表的 ColumnName，还有主键、外键、默认值等，都用特性来标识。然后通过反射技术，从 EF 中 edmx 元数据获取数据库表的结构的描述，再根据增删改查操作方法，就可以产生对应的 SQL 语句，然后发送给 ADO.NET，最终由 ADO.NET 负责从数据库中读取数据，返回给我们的 EF。

2.3 Entity Framework 增删改查

2.3.1　附加数据库

从云盘中找到 Northwind 数据库文件，附加到 SQL Server 2012，或者从网上下载 Northwind 数据库，下载地址为 http://files.cnblogs.com/files/jiekzou/northwnd.zip。

附加方式：打开 SQL Server 数据，右击“数据库”，选择“附加”命令，再选择数据库文件。

如果使用的是低于 SQL Server 2012 版本的数据库，就会附加失败。你可以在网上下载低版本的 Northwind 数据库。

关于 Northwind 表字段的说明如下：

```
-------------------------Categories:种类表
--相应字段:CategoryID:类型 ID; CategoryName:类型名;Description:类型说明;Picture:产品样本

-------------------------CustomerCustomerDemo:客户类型表1
--相应字段：CustomerID：客户 ID；CustomerTypeID：客户类型 ID

-------------------------CustomerDemographics:客户类型表2
--相应字段：CustomerTypeID：客户类型 ID；CustomerDesc：客户描述

-------------------------Customers:客户表
--相应字段：CustomerID:客户 ID；CompanyName:所在公司名称
--ContactName:客户姓名;ContactTitle:客户头衔；Address：联系地址
--City：所在城市;Region:所在地区;PostalCode:邮编;Country:国家
--Phone:电话；Fax：传真

-------------------------Employees：员工表
--相应字段：EmployeeID:员工代号;LastName + FirstName:员工姓名
--Title:头衔;TitleOfCourtesy:尊称;BirthDate:出生日期;HireDate:雇用日期
--Address：家庭地址；City:所在城市;Region:所在地区;PostalCode:邮编
--Country:国家用；HomePhone：宅电;Extension:分机;Photo:手机
--notes:照片；ReportsTo：上级;PhotoPath:照片

-------------------------EmployeeTerritories：员工部门表
--相应字段：EmployeeID:员工编号；TerritoryID：部门代号

-------------------------Order Details：订单明细表
--相应字段：OrderID:订单编号；ProductID：产品编号；UnitPrice：单价
--Quantity:订购数量；Discount:折扣

-------------------------Orders:订单表
--相应字段：OrderID:订单编号；CustomerID:客户编号;EmployeeID:员工编号
--OrderDate：订购日期；RequiredDate：预计到达日期;ShippedDate:发货日期
--ShipVia:运货商;Freight:运费;ShipName:货主姓名；ShipAddress:货主地址
--ShipCity:货主所在城市；ShipRegion：货主所在地区；ShipPostalCode：货主邮编
--ShipCountry:货主所在国家

-------------------------Products:产品表
```

```
--相应字段：ProductID:产品 ID；ProductName：产品名称；SupplierID：供应商 ID
--CategoryID：类型 ID；QuantityPerUnit：数量；UnitPrice：单价
--UnitsInStock：库存数量；UnitsOnOrder：订购量；ReorderLevel：再次订购量
--Discontinued：中止
---------------------------Region:地区表
--RegionID：地区 ID；RegionDescription：地区描述

---------------------------Shippers:运货商
--相应字段：ShipperID:运货商 ID；CompanyName：公司名称;Phone:联系电话

---------------------------Suppliers：供应商表
--相应字段：ShipperID：供应商 ID； CompanyName：供应商姓名； Phone；联系电话

---------------------------Territories：地域表
--相应字段： TerritoryID：地域 ID；TerritoryDescription：地域描述；RegionID：地区 ID
```

2.3.2　新建项目

右击 EFDemo 解决方案，选择“添加新项目→Windows→控制台应用程序”，修改项目名称为 EFCRUD，然后按照 2.2 节的方式添加一个 ADO.NET 实体数据模型 Northwind.edmx，再选择所有的表。

2.3.3　新增

在 Program 类中添加命名空间引用：

```
using System.Data.Entity.Infrastructure;
#region 新增
        static int Add()
        {
            using (NorthwindEntities db = new NorthwindEntities())
            {
                Customers _Customers = new Customers
                {
                    CustomerID = "zouqj",
                    Address = "南山区新能源创新产业园",
                    City = "深圳",
                    Phone = "15243641131",
                    CompanyName = "深圳跨境翼电商商务有限公司",
                    ContactName = "邹琼俊"
                };
                //方法一
```

```
            //db.Customers.Add(_Customers);

            //方法二
            DbEntityEntry<Customers> entry = db.Entry<Customers>(_Customers);
            entry.State = System.Data.EntityState.Added;

            return db.SaveChanges();
        }
    }
#endregion
```

2.3.4 简单查询和延时加载

为了查看延时加载，要使用 SqlProfiler 跟踪 SQL 语句，加断点、加 SqlProfiler 跟踪事件。

EF 生成 SQL 语句发送给数据库执行，我们可以使用 SQL Server 的跟踪监测工具查看。（可监测 SQL 语句、CPU 占用、执行时间等。）

“延时加载”有两种形式：

（1）EF 本身查询方法返回的都是 IQueryable 接口，此时并未查询数据库；只有当调用接口方法获取数据时，才会查询数据库。

```
static void QueryDelay1()
{
    using (NorthwindEntities db = new NorthwindEntities())
    {
        DbQuery<Customers> dbQuery = db.Customers.Where(u => u.ContactName
== "邹琼俊").OrderBy(u => u.ContactName).Take(1) as DbQuery<Customers>;
        //获得延迟查询对象后，调用对象的获取方法，此时，就会根据之前的条件生成 SQL 语句，
        //查询数据库了！
        Customers _Customers = dbQuery.FirstOrDefault();
        // 或者 SingleOrDefault ()
        Console.WriteLine(_Customers.ContactName);
    }
}
```

在代码 Customers _Customers 处添加断点，然后打开 SQL Server 的跟踪监测工具 SQL Server Profiler，新建跟踪，然后运行，如图 2-16 所示。

图 2-16

由于我们这里只需要跟踪 SQL 查询语句，因此我们可以在事件里面进行过滤，只选中 TSQL 事件类，在跟踪属性窗体中，切换到“事件选择”选项卡，进行如下操作，如图 2-17 所示。

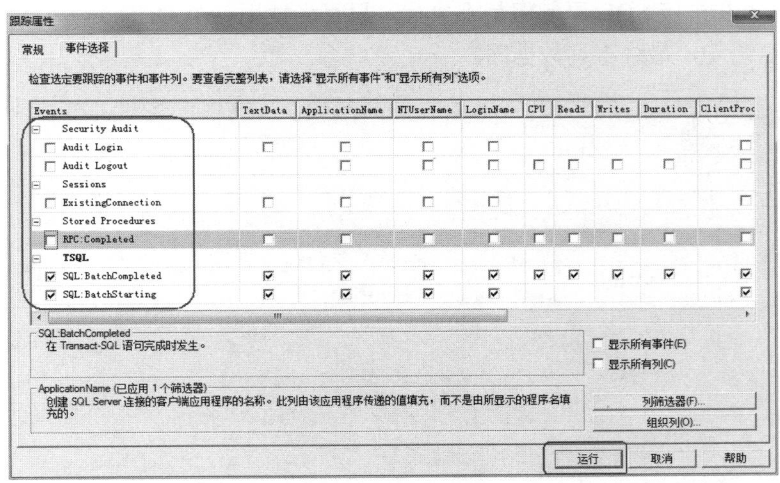

图 2-17

用到的事件说明如表 2-1 所示。

表 2-1　事件选择器说明

事件类	事件	说明
T-SQL	SQL:BatchCompleted	T-SQL 批处理完成事件
	SQL:StmtCompleted	一条 T-SQL 语句完成事件

当我们运行到断点的时候，监听窗体中没有监听到 SQL 执行语句，当执行下一步到 dbQuery.FirstOrDefault()的时候，可以看到执行了如下 SQL 语句，如图 2-18 所示。

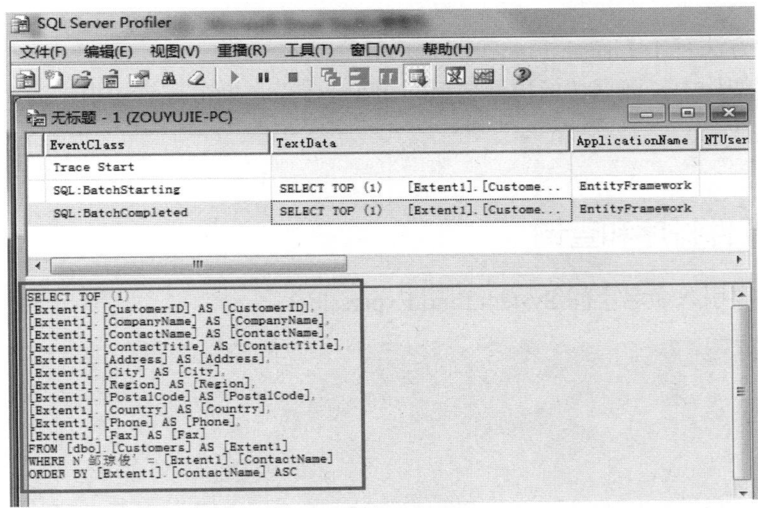

图 2-18

（2）当前可能通过多个标准查询运算符（SQO）方法来组合查询条件，那么每个方法都只是添加一个查询条件而已，无法确定本次查询条件是否已经添加结束，所以没有办法在执行每个 SQO 方法的时候确定 SQL 语句是什么，只能返回一个包含了所有添加条件的 DBQuery 对象，就相当于一直在拼接 SQL 语句但是不执行，只有当使用这个 DBQuery 对象的时候才根据所有条件生成 SQL 语句，最终查询数据库。

```
static void QueryDelay2()
{
      using (NorthwindEntities db = new NorthwindEntities())
      {
              IQueryable<Orders> _Orders = db.Orders.Where(a => a.CustomerID ==
"zouqj");//真实返回的 DbQuery 对象，以接口方式返回
              //此时只查询了订单表
              Orders order = _Orders.FirstOrDefault();
              //当访问订单对象里的外键实体时，EF 会查询订单对应的用户表，查到之后，再将数据装入
              //这个外键实体
              Console.WriteLine(order.Customers.ContactName);
              IQueryable<Orders> orderList = db.Orders;
              foreach (Orders o in orderList)
              {
                  Console.WriteLine(o.OrderID + ":ContactName=" +
o.Customers.ContactName);
              }
      }
}
```

延迟加载的缺点是每次调用外键实体时都会去查询数据库，不过 EF 有小优化，即相同的外键实体只查一次。

禁用延迟的方法有 ToList()、FirstOrDefault()、Include()等。

通常在多层架构中，数据访问层都会返回 IQueryable，业务逻辑层根据需要转为 List，这样可以有更多的选择。

2.3.5 根据条件排序和查询

首先我们需要引入命名空间 System.Linq.Expressions。

```
#region 根据条件排序和查询
      /// <summary>
      /// 根据条件排序和查询
      /// </summary>
      /// <typeparam name="TKey">排序字段类型</typeparam>
      /// <param name="whereLambda">查询条件 lambda 表达式</param>
```

```
        /// <param name="orderLambda">排序条件 lambda 表达式</param>
        /// <returns></returns>
        public List<Customers> GetListBy<TKey>(Expression<Func<Customers, bool>>
        whereLambda, Expression<Func<Customers, TKey>> orderLambda)
        {
            using (NorthwindEntities db = new NorthwindEntities())
            {
                return db.Customers.Where(whereLambda).OrderBy(orderLambda).ToList();
            }
        }
#endregion
```

2.3.6　分页查询

```
#region 分页查询
        /// <summary>
        /// 分页查询
        /// </summary>
        /// <param name="pageIndex">页码</param>
        /// <param name="pageSize">页容量</param>
        /// <param name="whereLambda">条件 lambda 表达式</param>
        /// <param name="orderBy">排序 lambda 表达式</param>
        /// <returns></returns>
        public List<Customers> GetPagedList<TKey>(int pageIndex, int pageSize,
        Expression<Func<Customers, bool>> whereLambda,
        Expression<Func<Customers, TKey>> orderBy)
        {
            using (NorthwindEntities db = new NorthwindEntities())
            {
                // 分页时一定注意： Skip 之前一定要 OrderBy
                return db.Customers.Where(whereLambda).OrderBy(orderBy).Skip
                ((pageIndex - 1) * pageSize).Take(pageSize).ToList();
            }
        }
#endregion
```

2.3.7　修改

关于数据修改，微软官方推荐的修改方式是先查询再修改。

```
#region 官方推荐的修改方式（先查询，再修改）
```

```
        /// <summary>
        /// 官方推荐的修改方式（先查询，再修改）
        /// </summary>
        static void Edit()
        {
            using (NorthwindEntities db = new NorthwindEntities())
            {
                //1.查询出一个要修改的对象 -- 注意：此时返回的是一个 Customers 类的代理类对象
                //（包装类对象）
                Customers _Customers = db.Customers.Where(u => u.CustomerID ==
                "zouqj").FirstOrDefault();
                Console.WriteLine("修改前：" + _Customers.ContactName);
                //2.修改内容 -- 注意：此时其实操作的是代理类对象的属性，这些属性会将值设置给内部
                //的 Customers 对象对应的属性，同时标记此属性为已修改状态
                _Customers.ContactName = "邹玉杰";
                //3.重新保存到数据库 -- 注意：此时 EF 上下文会检查容器内部所有的对象，先找到标记
                //为修改的对象，然后找到标记为修改的对象属性，生成对应的 update 语句执行
                db.SaveChanges();
                Console.WriteLine("修改成功：");
                Console.WriteLine(_Customers.ContactName);
            }
        }
#endregion

#region 自己优化的修改方式（创建对象，直接修改）
        /// <summary>
        /// 自己优化的修改方式（创建对象，直接修改）
        /// </summary>
        static void Edit2()
        {
            //1.查询出一个要修改的对象
            Customers _Customers = new Customers()
            {
                CustomerID = "zouqj",
                Address = "南山区新能源创新产业园",
                City = "深圳",
                Phone = "15243641131",
                CompanyName = "深圳跨境翼电商商务有限公司",
                ContactName = "邹玉杰"
            };
            using (NorthwindEntities db = new NorthwindEntities())
            {
```

```
            //2.将对象加入 EF 容器，并获取当前实体对象的状态管理对象
            DbEntityEntry<Customers> entry = db.Entry<Customers>(_Customers);
            //3.设置该对象为被修改过
            entry.State = System.Data.EntityState.Unchanged;
            //4.设置该对象的 ContactName 属性为修改状态，同时 entry.State 被修改为
            //Modified 状态
            entry.Property("ContactName").IsModified = true;

            //var u = db.Customers.Attach(_Customers);
            //u.ContactName = "郭富城";
            //5.重新保存到数据库 -- EF 上下文会根据实体对象的状态 entry.State =Modified
            //值生成对应的 update sql 语句
            db.SaveChanges();
            Console.WriteLine("修改成功：");
            Console.WriteLine(_Customers.ContactName);
        }
    }
#endregion
```

2.3.8　删除

```
#region 删除 -void Delete()
    /// <summary>
    ///  删除
    /// </summary>
    static void Delete()
    {
        using (NorthwindEntities db = new NorthwindEntities())
        {
            //1.创建要删除的对象
            Customers u = new Customers() { CustomerID = "zouqj" };
            //2.附加到 EF 中
            db.Customers.Attach(u);
            //3.标记为删除--注意：此方法就是标记当前对象为删除状态
            db.Customers.Remove(u);

            /*
                也可以使用 Entry 来附加和修改
                DbEntityEntry<Customers> entry = db.Entry<Customers>(u);
                entry.State = System.Data.EntityState.Deleted;
            */
```

```
            //4.执行删除 SQL
            db.SaveChanges();
            Console.WriteLine("删除成功！");
        }
    }
#endregion
```

2.3.9 批处理

在批处理中，我们能深深体会到上下文对象中 SaveChanges 方法的好处。

```
#region 批处理 -- 上下文 SaveChanges 方法的好处!
    /// <summary>
    /// 批处理 -- 上下文 SaveChanges 方法的好处!
    /// </summary>
    static void SaveBatched()
    {
        //1.新增数据
        Customers _Customers = new Customers
        {
            CustomerID = "zouyujie",
            Address = "洛阳西街",
            City = "洛阳",
            Phone = "1314520",
            CompanyName = "微软",
            ContactName = "邹玉杰"
        };
        using (NorthwindEntities db = new NorthwindEntities())
        {
            db.Customers.Add(_Customers);

            //2.新增第二个数据
            Customers _Customers2 = new Customers
            {
                CustomerID = "zhaokuanying",
                Address = "洛阳西街",
                City = "洛阳",
                Phone = "1314520",
                CompanyName = "微软",
                ContactName = "赵匡胤"
            };
```

```
            db.Customers.Add(_Customers2);

            //3.修改数据
            Customers usr = new Customers() { CustomerID = "zhaomu", ContactName
            = "赵牧" };
            DbEntityEntry<Customers> entry = db.Entry<Customers>(usr);
            entry.State = System.Data.EntityState.Unchanged;
            entry.Property("ContactName").IsModified = true;

            //4.删除数据
            Customers u = new Customers() { CustomerID = "zouyujie" };
            //5.附加到 EF 中
            db.Customers.Attach(u);
            //6.标记为删除--注意：此方法就是标记当前对象为删除状态
            db.Customers.Remove(u);

            db.SaveChanges();
            Console.WriteLine("批处理 完成~~~~~~~~~~~~! ");
        }
    }
#endregion

#region  批处理 -- 一次新增 50条数据 -void BatcheAdd()
    /// <summary>
    ///  批处理 -- 一次新增 50条数据
    /// </summary>
    static void BatcheAdd()
    {
        using (NorthwindEntities db = new NorthwindEntities())
        {
            for (int i = 0; i < 50; i++)
            {
                Customers _Customers = new Customers
                {
                    CustomerID = "zou" + i,
                    Address = "洛阳西街",
                    City = "洛阳",
                    Phone = "1314520",
                    CompanyName = "微软",
                    ContactName = "邹玉杰" + i
                };
                db.Customers.Add(_Customers);
```

```
            }
            db.SaveChanges();
        }
    }
#endregion
```

2.4 EF 查询相关

2.4.1 IQueryable 与 IEnumberable 接口的区别

IQueryable 接口与 IEnumberable 接口的区别是：IEnumerable<T> 泛型类在调用自己的 SKip 和 Take 等扩展方法之前数据就已经加载在本地内存里了，而 IQueryable<T> 是将 Skip、take 这些方法表达式翻译成 T-SQL 语句之后再向 SQL 服务器发送命令，也就是延迟在要真正显示数据的时候才执行。IQueryable 其实继承了 IEnumberable。

在 Linq To EF 中使用 IEnumberable 与 IQueryable 的区别要用到 SQL Server Profiler 工具，那么我们什么时候使用 IQueryable、什么时候使用 IEnumberable 呢？通常，在数据访问层都是使用 IQueryable，因为可以把对数据的加载延时到业务逻辑层来处理。很多时候，业务层调用数据访问层的方法时并不要求马上从数据库中加载数据再存到内存中。这时在业务逻辑层依旧可以使用延时加载，当真正需要加载数据的时候，再在业务逻辑层把 IQueryable 转换成 IEnumberable 把数据加载进来就可以了。

2.4.2 LINQ To EF

这里使用 northwnd.mdf 数据库。

1. 简单查询

```
var result = from c in Entities.Customer select c;
```

2. 条件查询

普通 LINQ 写法：

```
var result = from c in Entities.Customer where c.Gender =='w' select c;
```

Lambda 表达式写法：

```
var result = Entities.Customer.Where<Customer>(c =>c.Gender=='w');
```

3. 排序分页

```
IQueryable<Customers> cust10 = (from c in customers
                    orderby c.CustomerID
```

```
                        select c).Skip(0).Take(10);
```

4. 聚合

可使用的聚合运算符有 Average、Count、Max、Min 和 Sum。

```
using (var edm = new NorthwindEntities())
{
      var maxuprice = edm.Products.Max(p => p.UnitPrice);
      Console.WriteLine(maxuprice.Value);
}
```

5. 连接

可以使用的连接有 Join 和 GroupJoin 方法。GroupJoin 组连接等效于左外部连接，返回第一个（左侧）数据源的每个元素（即使其他数据源中没有关联元素）。

```
using (var edm = new NorthwindEntities())
{
      var query = from d in edm.Order_Details
                  join order in edm.Orders
                  on d.OrderID equals order.OrderID
                  select new
                   {
                       OrderId = order.OrderID,
                       ProductId = d.ProductID,
                       UnitPrice = d.UnitPrice
                     };

       foreach (var q in query)
        Console.WriteLine("{0},{1},{2}",q.OrderId,q.ProductId,q.UnitPrice);
}
```

EF 不支持复杂类型（如实体）的直接检索，只能用简单类型，比如常用的标量类型 string、int 和 guid。在执行这类查询时，会引发 NotSupportedException 异常，并显示消息“无法创建‘System.Object’类型的常量值”。

如下代码将会引发异常：

```
using (var edm = new NorthwindEntities())
{
      Customers customer = edm.Customers.FirstOrDefault();
      IQueryable<string> cc = from c in edm.Customers
                              where c == customer
                              select c.ContactName;
```

```
        foreach (string name in cc)
            Console.WriteLine(name);
}
```

在上面的代码中，由于 customer 是引用类型而不是 Int32、String、Guid 的标量类型，因此在执行到 where c==customer 这个地方时会报错。

6. 分组查询

```
var query = from c in db.Categories
        join p in db.Products
        on c.CategoryID equals p.CategoryID
        group new {c, p} by new {c.CategoryName} into g
        select new
        {
            g.Key.CategoryName,
            SumPrice = (decimal?)g.Sum(pt=>pt.p.UnitPrice),
            Count = g.Select(x=>x.c.CategoryID).Distinct().Count()
        };
```

2.4.3 关于 EF 对象的创建问题

在开发过程中，项目往往被划分为多层，而一个请求过来往往是从表示层开始一层一层向下调用，那么如果我们在不同的层中都使用到了 EF 上下文对象，而有好几层都这么创建一个 EF 对象然后对其进行操作，那么最终哪一层的 EF 对象是我们需要的最新的数据就很难确定了，这时就很容易产生脏读。在这种情况下，我们首先会想到使用单例模式，这样在整个应用程序的生命周期内只允许被创建一次。但是这样又会出现一个问题，所有的用户都访问同一个 EF 对象，随着访问的用户越来越多，这个 EF 对象的资源无法及时释放，导致占用的内存也会越来越大，虽然垃圾回收机制会自动回收掉大部分内存，但是有一些内存对象是需要我们手动回收的，可是在这里我们又不能使用 using 直接回收这个单例对象，因为如果把单例对象释放了，大家就都访问不了了，所以无法使用单例模式来解决这个 EF 对象的创建问题。

这时，创建线程对象就可以了，我们只需要保证 EF 对象在一个线程内唯一，一个请求就是一个线程，这样就无须关心多层直接操作 EF 对象的问题了。HttpContext 对象就是微软封装的一个线程对象，完全可以把创建的 EF 对象存到 HttpContext 对象中，实现 EF 对象在线程中唯一。使用示例如下：

```
public MvcFirstCodeContext DB2
{
    get {
            MvcFirstCodeContext db = null;
            if (HttpContext.Items["db1"] == null)
             {
                db = new MvcFirstCodeContext();
                HttpContext.Items["db1"] = db;
```

```
            }
            else
            {
                db = HttpContext.Items["db1"] as MvcFirstCodeContext;
            }
            return db;
        }
}
```

2.4.4 关于上下文的使用注意事项

- 不同的上下文实例直接控制对应的实体。
- 实体只能由一个上下文跟踪管理。
- EF 上下文的 ObjectStateMagner 管理实体。
- 批量操作时提交数据库的选择。
- 延迟加载机制的选择。
- 查询 Distinct 的使用数据量大小，适时地选择是在内存中操作还是在数据库中操作。

2.4.5 EF 跨数据库支持

目前已有数个数据库厂商或元件开发商宣布要支持 ADO.NET Entity Framework:

- Core Lab，支持 Oracle、MySQL、PostgreSQL 与 SQLite 数据库。
- IBM，实现 DB2 使用的 LINQ Provider。
- MySQL，发展 MySQL Server 所用的 Provider。
- Npqsql，发展 PostgreSQL 所用的 Provider。
- OpenLink Software，发展支持多种数据库所用的 Provider。
- Phoenix Software International，发展支持 SQLite 数据库的 Provider。
- Sybase，支持 Anywhere 数据库。
- VistaDB Software，支持 VistaDB 数据库。
- DataDirect Technologies，发展支持多种数据库所用的 Provider。
- Firebird，支持 Firebird 数据库。

2.5 Model First 开发方式

在项目一开始，没有数据库时，可以借助 EF 设计模型，然后根据模型同步完成数据库中表的创建，这就是 Model First 开发方式。总结一点就是先有模型再有表。

2.5.1 创建 Model First Demo

创建 Model First 的操作步骤如下：

（1）创建控制台项目

右击解决方案“EFDemo”，选择“添加项目→控制台应用程序”，并将项目命名为 ModelFirst。

（2）添加 ADO.NET 实体模型

右击 ModelFirst 项目，再选择“添加新建项→ADO.NET 实体数据模型”并命名为 ModelFirstModel.edmx，创建空模型后单击“完成”。

（3）添加实体 Customer

① 添加实体和几个必要的测试字段。切记，一定要添加主键。主键既可以是自增长的数字类型，也可以是 Guid 类型，如图 2-19、图 2-20 所示。

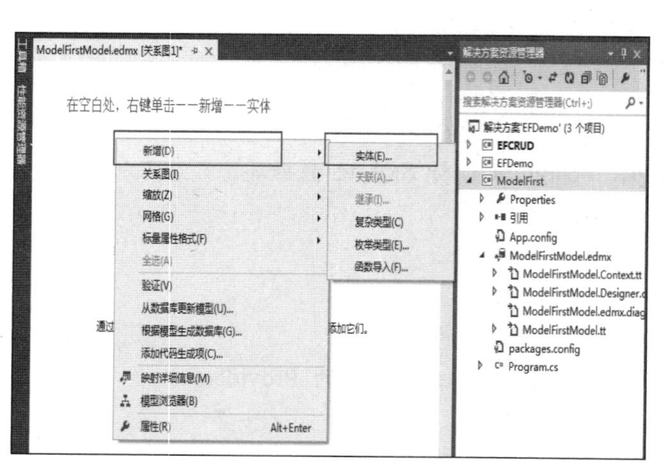

图 2-19

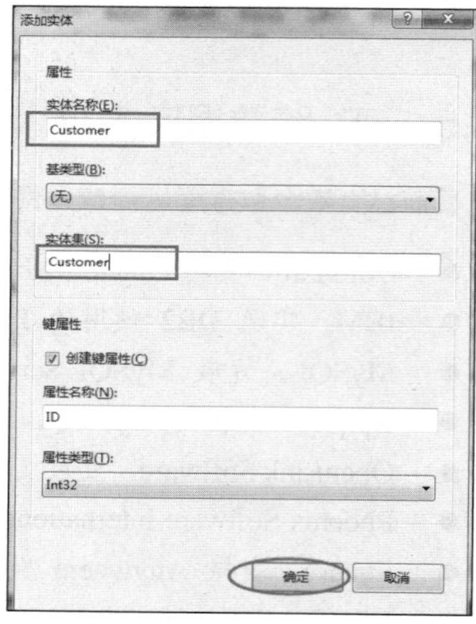

图 2-20

② 添加标量属性。标量属性可以看成数据库中的普通字段（主键和外键之外的），我们在设计字段属性的时候，一定要记得设置其最大范围，否则最终会生成一个比较大的默认范围，严重影响性能，并占用不必要的磁盘空间，如图 2-21、图 2-22 所示。

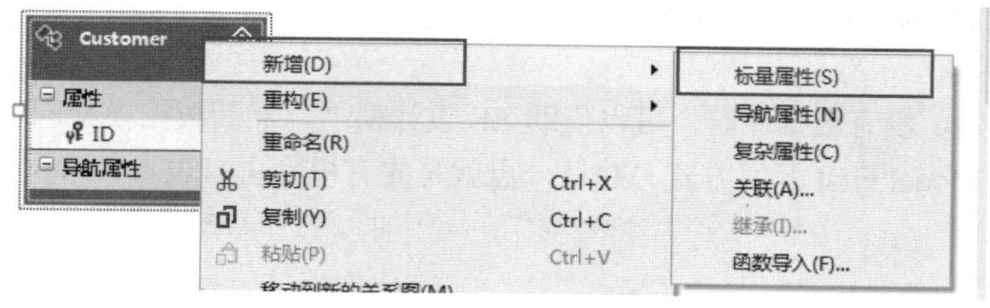

图 2-21

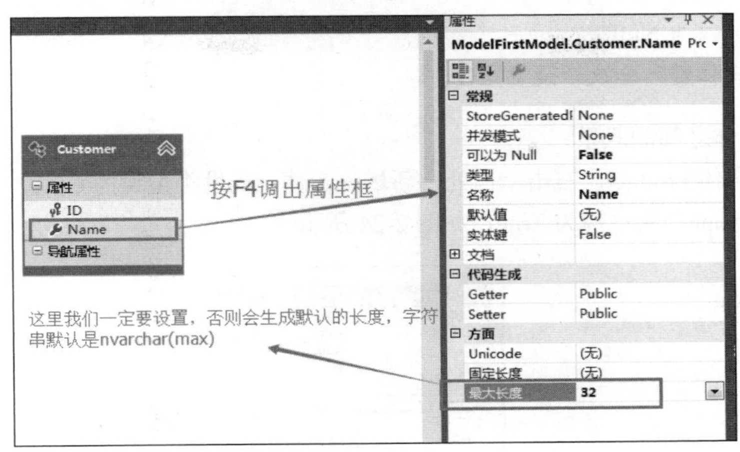

图 2-22

再依次添加标量属性 Telphone、CompanyName、Age，如表 2-2 所示。

表 2-2　Customer 实体

名称	类型	最大长度
Telphone	String	11
CompanyName	String	50
Age	Int16	

接着看一下 ID 属性，ID 是实体键，存储方式是自增长，如图 2-23 所示。

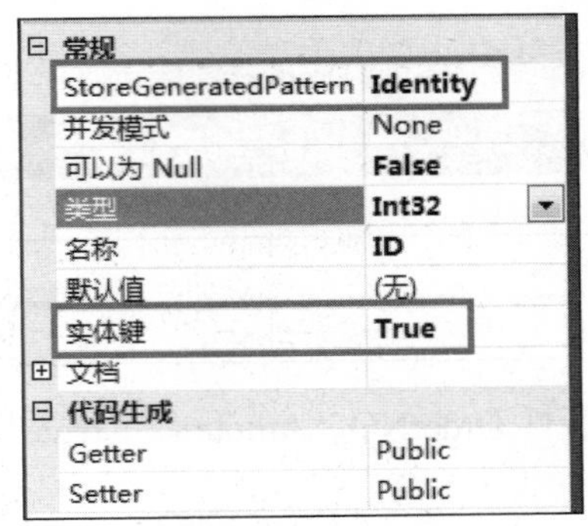

图 2-23

这里简单说明一下模型属性的类型。这里的类型都是 CTS 中的类型，即 IL 中使用的类型。这些类型可以是如下选项：

- Int32。
- String，可以选择是否采用 Unicode 编码，如果采用就对应 SQL Server 中的 nvarchar 类型。
- Decimal，表示指定小数位数及数据精度的类型，范围表示小数个数，精度显示总的数据位数。

- 属性“可以为 Null”。
- 属性“实体键”，表示设置主键。

（4）添加实体之间的联系

再添加一个实体 Product，右击空白处→新增→实体，属性类型设为 Guid，这时字段 ID 属性的存储方式为 None，实体键为 True，如图 2-24 所示。

图 2-24

这里涉及自增长主键和 Guid 主键。

Guid 比自增长要快，因为自增长会先查询表中最大的 ID，然后锁表，再在这个最大的 ID 基础上加 1，然后插入数据表。

自增长也有自己的长处，比如占用空间小，因为它是 int 类型的，而 Guid 一般是 32 或 64 个字符长度，还有就是比较清晰，不像 Guid 那么大。如果涉及数据迁移，自增长主键就会变得非常不方便，而 Guid 的优势则非常明显。

再依次添加标量属性 Name、Price、Weight，如表 2-3 所示。

表 2-3 Product 实体

名称	类型	精度	最大长度
Name	String		11
Price	Decimal	2	
Weight	Decimal	2	

再添加实体 Order 以及标量属性 OrderNO、Amount、CreateTime，如表 2-4 所示。

表 2-4 Order 实体

名称	类型	精度	最大长度
OrderNO	String		30
Amount	Decimal	2	
CreateTime	DateTime		

接着，我们添加两个实体之间的关联，右击工作面板空白处→新增→关联，在“添加关联”对话框（见图 2-25）中进行设置。

图 2-25

Customer 和 Order 是一个一对多的关系，Customer 实体和 Order 实体上面的导航属性不要去掉，因为后面用它来查询将会变得非常方便。导航属性，顾名思义，就是根据这个属性可以找到一个和它关联的对象实体。我们再添加 Customer 和 Product 多对多的关联，最终结果展示如图 2-26 所示。

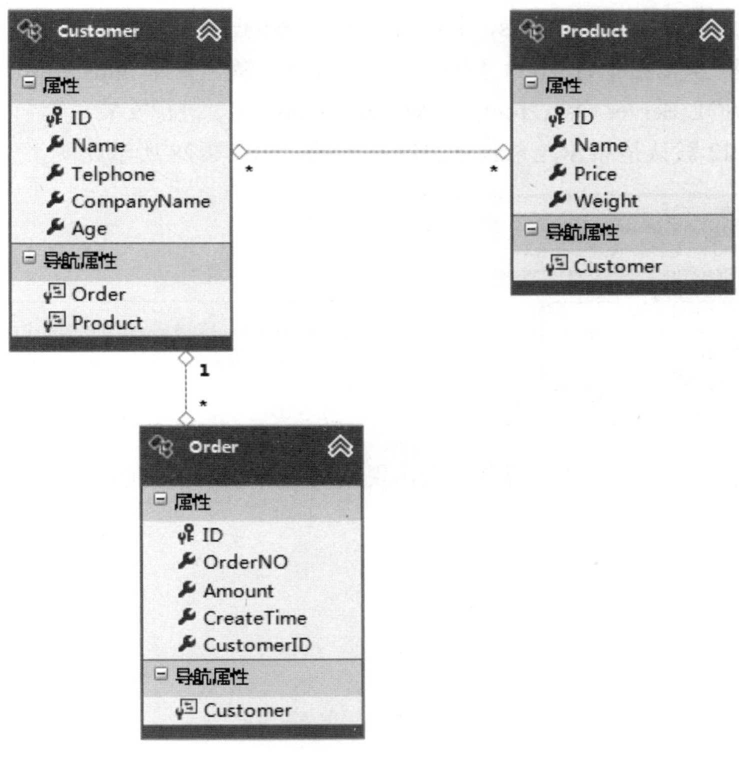

图 2-26

说明：这里只是为了演示，在实际应用中商品是和订单明细关联的。

对于“关联”的说明如下：

- 1:1，性能低（不会延迟加载，添加时必须同时创建两个对象），尽量不要使用，可以自己实现逻辑代码完成这种操作。查看一下表结构，可以发现本质还是 1: *m* 的结构。
- 1:*m*。
- *M*:*n*: 可以手动创建中间表采用 1:*m* 关系，也可以直接使用此种关系，EF 会自动创建中间表。
- 在创建关联时，可以选择是否要创建导航属性、外键。

对于“导航属性”的说明如下：

- 根据关系的不同，查看生成的导航属性的类型。
- 示例：在 1 对多关系中，对于多端表数据的插入。

添加完成之后，按 Ctrl+S 组合键，会在 ModelFirstModel.edmx 中 ModelFirstModel.tt 下面生成 Customer、Order 和 Product 三个类，如图 2-27 所示。.tt 后缀的文件为 T4 模板，后面会单独讲解。

（5）根据模型创建数据库

为了方便，这里直接使用 VS2012 中自带的 localDB，右击工作区空白处，根据模型生成数据库，将数据库命名为 ModelFirstDB，并执行 SQL 脚本创建数据库。

如果这里是接 SQL Server 2008 上面创建的数据库，就需要从 https://msdn.microsoft.com/en-us/jj650015 下载 SQL Server Data Tools in Visual Studio 2012 进行安装，否则后面无法直接执行 SQL，因为 VS2012 默认是和 SQL Server 2012 搭配的，如图 2-28 所示。

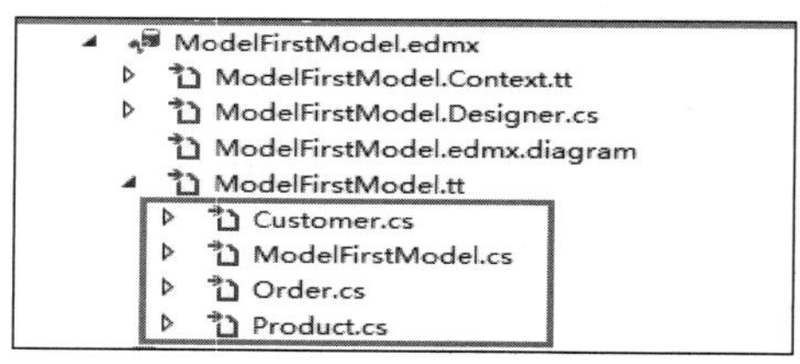

图 2-27

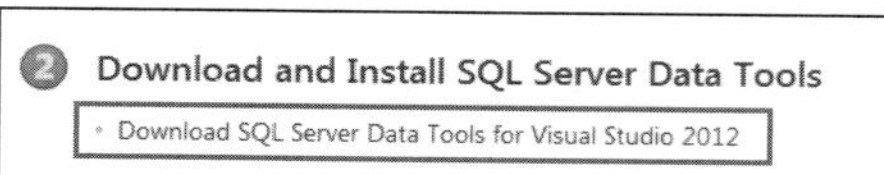

图 2-28

单击“根据模型生成数据库”操作后会出现如图 2-29 所示的界面。

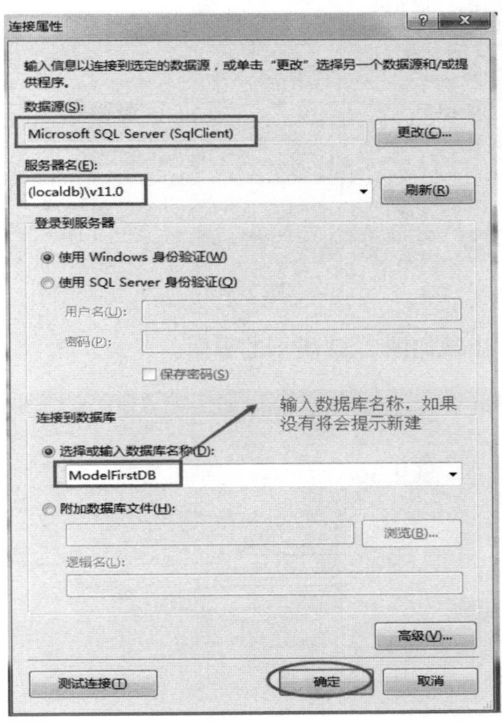

图 2-29

单击“确定”按钮，出现如图 2-30 所示的界面。

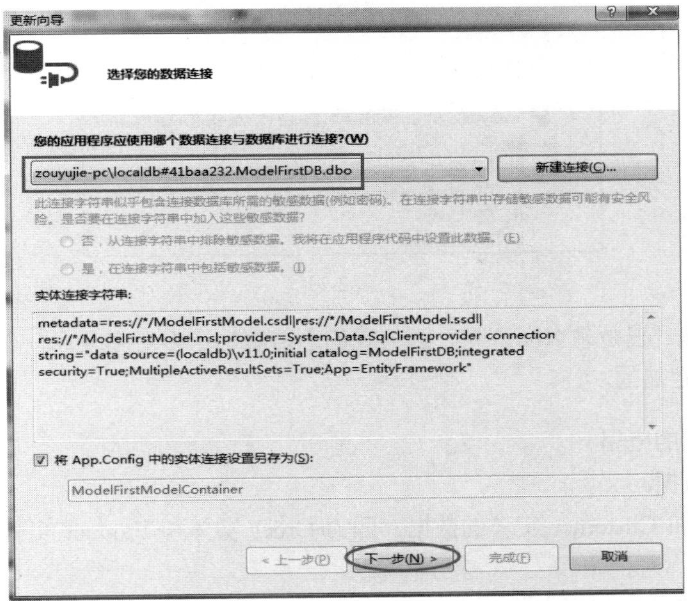

图 2-30

单击“下一步→完成”，生成 DLL 脚本，然后执行生成的 DLL 脚本，如图 2-31 所示。

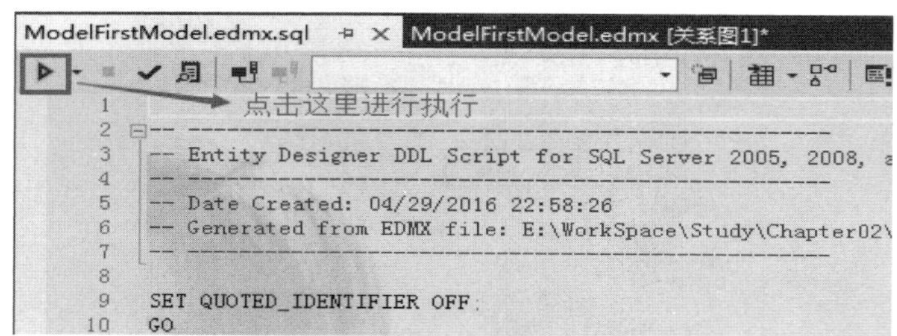

图 2-31

单击“执行”按钮后，出现如图 2-32 所示的界面。

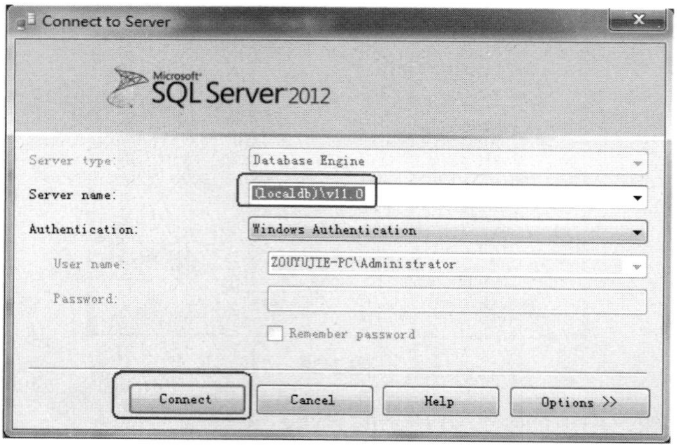

图 2-32

执行 DLL 脚本完成之后，我们会看到数据库中出现了 4 张表，如图 2-33 所示。

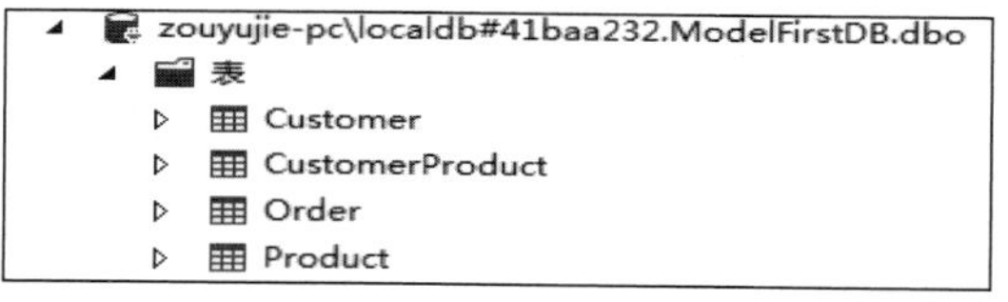

图 2-33

CustomerProduct 表是怎么回事呢？因为我们之前添加了多对多关联，而多对多关联就是通过一张新表来实现存储的。

（6）导航属性的应用

① 创建测试数据。

记住，要先添加 Customer 实体的数据，因为 Order 实体和 Product 实体都引用了 Customer 实体。

在 Program 类中执行如下方法：

```
static void AddTestData()
{
```

```
    using (ModelFirstModelContainer db = new ModelFirstModelContainer())
    {
        Customer _Customer = new Customer { Name="楚留香
        ",Age=27,CompanyName="大旗门",Telphone="15243641131"};
        Order _Order1 = new Order {Amount=15,CreateTime=DateTime.Now,
        OrderNO="2016043001",CustomerID=_Customer.ID};
        Order _Order2 = new Order { Amount = 16, CreateTime = DateTime.Now,
        OrderNO = "2016043002", Customer = _Customer };
        Product _Product = new Product {ID=Guid.NewGuid(), Name="牛栏1段
        ",Price=12,Weight=80,Customer=new List<Customer>(){_Customer}};

        db.Customer.Add(_Customer);
        db.Product.Add(_Product);
        db.Order.Add(_Order1);
        db.Order.Add(_Order2);

        if (db.SaveChanges() > 0)
        {
            Console.WriteLine("添加成功！");
        }
        else
        {
            Console.WriteLine("添加失败！");
        }
    }
}
```

注意：db.SaveChanges()默认是已经开启了事务的，而且在这之前都只进行了一次数据库的连接，这种类似于批处理的操作大大地提升了性能。

② 查询客户楚留香的所有订单信息：

```
static void SearchCusOrder()
{
    using (ModelFirstModelContainer db = new ModelFirstModelContainer())
    {
        var _orderList = from o in db.Order where o.Customer.Name == "楚留香"
        select o;
        Console.WriteLine("客户楚留香的所有订单如下：");
        _orderList.ToList().ForEach(o=>Console.WriteLine(string.Format("订单
        号：{0}，订单金额：{1},订单创建时间：{2}",o.OrderNO,o.Amount,o.CreateTime)));
        Console.ReadKey();
    }
```

```
}
```

③ 运行结果如图 2-34 所示。

图 2-34

这里先查 Order 表信息，然后直接通过导航属性 Customer 来过滤信息。

当然，我们也可以通过使用 Join 来查询，达到同样的查询效果。

```
var _orderList = from c in db.Customer join o in db.Order on c.ID equals
o.CustomerID where c.Name == "楚留香" select o;
```

那么什么情况下使用 join 查询、什么情况下使用导航属性查询呢？

导航属性查询就相当于 SQL 中的子查询。Join 查询和 SQL 中的 Inner Join 查询一样，所以当两张表的数据量都大的时候就使用导航属性查询，在数据量不大的情况下使用 join 查询。

2.5.2 经验分享

如果模型有更新，直接选择从模型更新数据库，那么生成的 SQL 语句将会先判断是否存在相应的表，存在的话将先删除再创建。这样我们的操作就要很谨慎了，不然后果很严重。

在生产环境下操作更是要谨慎再谨慎，通常生产环境是做了比较严格的权限控制的，万一需要去生产环境中执行 SQL 脚本，就必须保证在测试环境或者开发环境中执行没问题了，然后备份数据库（开启增量备份最佳），最后执行。如果 SQL 执行脚本有问题，那也没办法，只能恢复数据库了。像在 MySQL 中，可以根据日志中的时间点和位置来进行恢复。

一般而言，我们可以把改动地方的 SQL 语句单独抠出来执行。如果涉及只修改模型中某个字段的微小操作，就可以直接在数据库中改过来。除此之外，没有什么好的办法。

2.6 Code First 开发方式

本节介绍通过 Code First 开发建立新数据库。借助 Code First 可以选择使用类的特性和属性执行配置，或者使用 XML 配置文件来配置，当然也可以使用 Fluent API 执行配置。

Code First 使用场景：对于已经存在了模型类型的项目，怎么使用 EF 呢？Code first，也叫 POCO+Code Only。

Code only，顾名思义，只需要代码，不需要 Edmx 模型。EF 提供了通过类型的结构推断生成 SQL 并创建数据库中的表，而且能够通过类型的成员推断出实体间的关系，开发人员只需要编写实体类就可以进行 EF 数据库的开发。

优势：

- 使开发更进一步简洁化。
- 开发效率又一次提高。
- 自动化程度进一步提高。
- 可以适用于原有的老项目。

劣势：

- 性能不怎么好。
- 了解的人比较少。
- 学习成本相对较高，对开发人员的要求相对较高。

Code First 有两种配置数据库映射的方式，一种是使用数据属性 DataAnnotation，另外一种是使用 Fluent API。DataAnnotation 的配置方式需要给定义实体和值对象的类和类中的属性加上与数据库映射相关的配置标签。而 Code First Fluent API 是在 DbContext 中定义数据库配置的一种方式。要使用 Fluent API 就必须在自定义的继承自 DbContext 类中重载 OnModelCreating 方法。这个方法的签名如下：

```
protected override void OnModelCreating(DbModelBuilder modelBuilder)
```

通过 modelBuilder 这个对象的 Entity<>泛型方法来配置 DbContext 中每个类的数据库映射。

我们可以通过 Fluent API 配置数据表的名字：

```
protected override void OnModelCreating(DbModelBuilder modelBuilder)
{
    modelBuilder.Entity<Customer>().ToTable("CustomerInfo")
}
```

这里只简单讲解一下 DataAnnotation，关于 Code First Fluent API 的使用，大家可以自己查找相关资料学习。

2.6.1 创建 Code First Demo

创建 Code First 的操作步骤如下：

（1）创建应用程序

打开 Visual Studio，选择“文件→新建→项目→Visual C#→Web→ASP.NET MVC 4 Web 应用程序”，如图 2-35 所示。

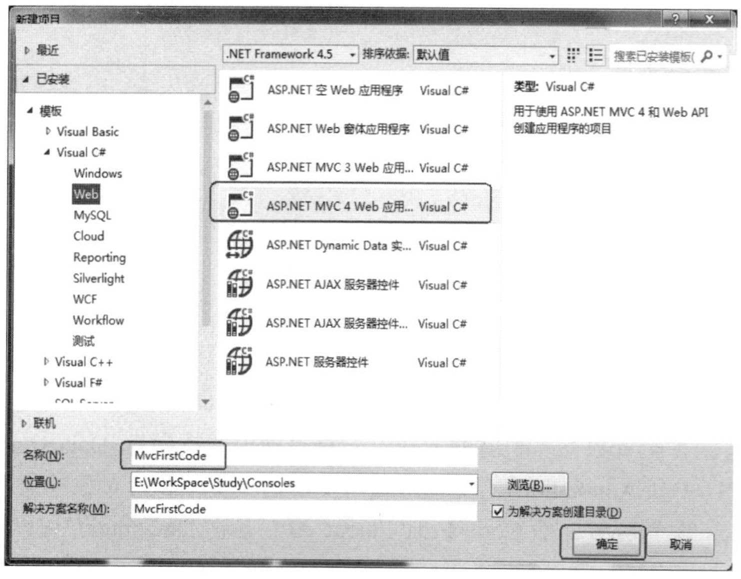

图 2-35

单击“确认”按钮后，项目模板选择“空”，然后再单击“确认”按钮。

（2）引入程序集 EntityFramework 和 System.Data.Entity

我们可以使用 NuGet 来进行安装，右击选择“引用→管理 NuGet 程序包”，再选择“联机”选项卡，选择“EntityFramework”程序包，单击“安装”按钮。

当然，如果觉得以上步骤比较麻烦，也可以使用一种简单的方式来引用：选中项目 MvcFirstCode，右击，选择“添加→新建项”，如图 2-36 所示。

图 2-36

单击“添加”按钮，弹出如图 2-37 所示的界面。

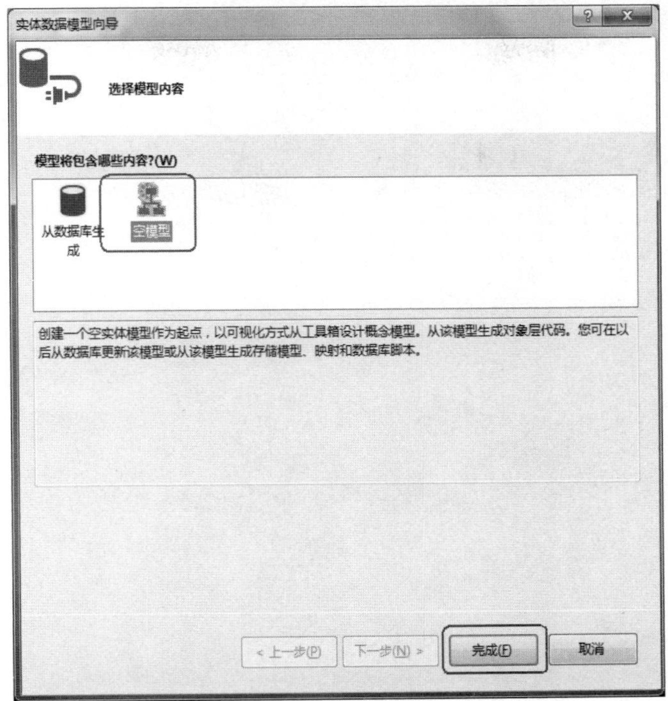

图 2-37

然后删除 Model1.edmx，可以在项目中看到这两个引用已经自动添加上，如图 2-38 所示。

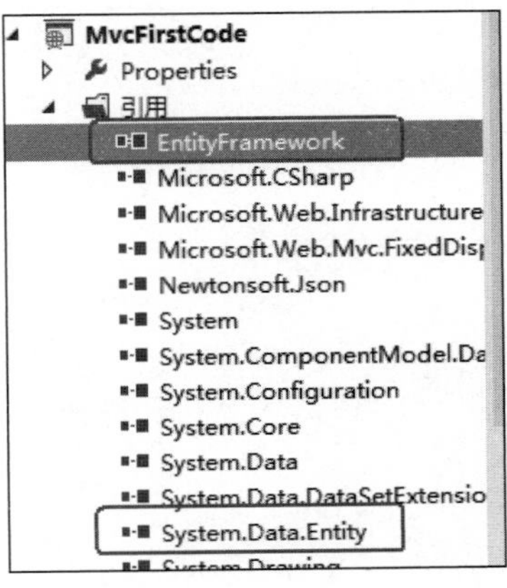

图 2-38

（3）创建模型

在 Models 文件夹下面新建 Order.cs 和 OrderDetail.cs 模型类文件，Order 代码如下：

```
using System;
using System.Collections.Generic;
using System.ComponentModel.DataAnnotations;
```

```
namespace MvcFirstCode.Models
{
    public class Order
    {
        /// <summary>
        /// 如果属性名后面包含 Id，则默认会当成主键，可以不用添加[Key]属性
        /// </summary>
        [Key]
        public int OrderId { get; set; }
        /// <summary>
        /// 订单号
        /// </summary>
        [StringLength(50)]
        public string OrderCode { get; set; }
        /// <summary>
        /// 订单金额
        /// </summary>
        public decimal OrderAmount { get; set; }
        /// <summary>
        /// 导航属性设置成 virtual，可以实现延迟加载
        /// </summary>
        public virtual List<OrderDetail> OrderDetail { get; set; }
    }
}
```

OrderDetail 代码如下：

```
using System;
using System.Collections.Generic;
using System.ComponentModel.DataAnnotations;
using System.ComponentModel.DataAnnotations.Schema;

namespace MvcFirstCode.Models
{
    public class OrderDetail
    {
        [Key]
        public int OrderDetailId { get; set; }
        /// <summary>
        /// 订单明细单价
        /// </summary>
        public decimal Price { get; set; }
```

```
        /// <summary>
        /// 订单明细数量
        /// </summary>
        public int Count { get; set; }
        /// <summary>
        /// 外键，如果属性名和 Order 主键名称一样，默认会当成外键，可以不加 ForeignKey 特性。注
        /// 意 ForeignKey 里面的值要和导航属性的名称一致
        /// </summary>
        [ForeignKey("Order")]
        public int OrderId { get; set; }

        /// <summary>
        /// 导航属性
        /// </summary>
        public virtual Order Order { get; set; }
    }
}
```

EF 支持的完整注释列表如下：

- KeyAttribute
- StringLengthAttribute
- MaxLengthAttribute
- ConcurrencyCheckAttribute
- RequiredAttribute
- TimestampAttribute
- ComplexTypeAttribute
- ColumnAttribute
- TableAttribute
- InversePropertyAttribute
- ForeignKeyAttribute
- DatabaseGeneratedAttribute
- NotMappedAttribute

（4）在配置文件中写连接字符串

在 Web.config 中添加如下配置节点，注意 providerName 属性必填，否则会报错。

```
  <connectionStrings>
    <add name="MvcFirstCodeContext" connectionString="server=.\MSSQLSERVER2012;
database=MvcFirstCode;uid=sa;pwd=yujie1127"
providerName="System.Data.SqlClient"/>
  </connectionStrings>
```

（5）创建上下文类 MvcFirstCodeContext.cs

继承自 DbContext，需要引入命名空间“using System.Data.Entity；”（引入命名空间的快捷方法是：将鼠标移动到 DbContext 上，然后按 Ctrl+Alt+F10 组合键，再按回车键即可）。调用父类构造方法，传递连接字符串"name=xx"，这个 xx 就是刚才配置文件里面配置的连接字符串 name 的名称。代码如下：

```
using System;
using System.Collections.Generic;
using System.Data.Entity;

namespace MvcFirstCode.Models
{
    public class MvcFirstCodeContext:DbContext
    {
        /// <summary>
        /// 注意这里的 name 要和配置文件里面配置的上下文连接字符串名称一致
        /// </summary>
        public MvcFirstCodeContext() : base("name=MvcFirstCodeContext") { }
        public DbSet<Order> Order { get; set; }
        public DbSet<OrderDetail> OrderDetail { get; set; }
    }
}
```

（6）根据类型创建数据库表

先生成项目 MvcFirstCode，然后在 Controllers 目录下面添加控制器 Home，如图 2-39 所示。

图 2-39

在 Index Action 中添加如下代码：

```
//当操作的表存在时不进行创建，如果不存在就创建
db.Database.CreateIfNotExists();
```

注意，在这里我们使用 context.Database.CreateIfNotExists()完成数据库中表的创建，调用 context.SaveChanges()方法完成数据保存。

再次生成项目 MvcFirstCode，按 Ctrl+F5 组合键运行项目，如图 2-40 所示。

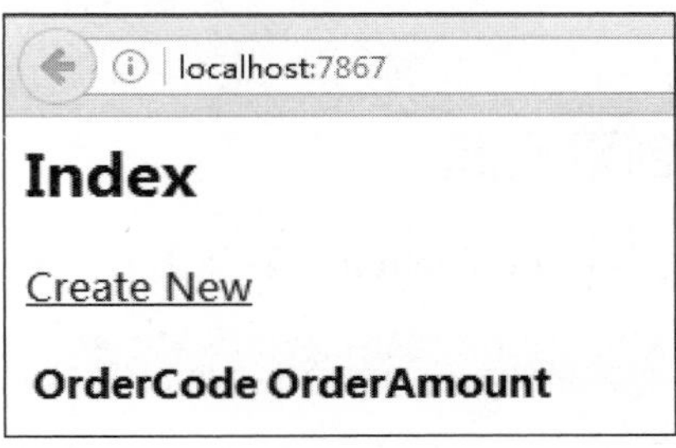

图 2-40

出现图 2-40 显示的内容说明代码已经执行成功了。

打开视图→服务器资源管理器，如图 2-41 所示。

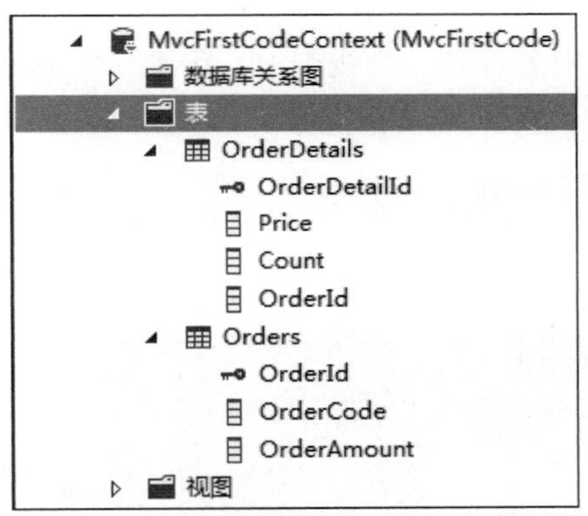

图 2-41

从图 2-41 中可以看到数据库和表也已经创建好了。下面新建一条订单数据，由于我们之前新建 MVC 项目的时候选择的是空项目，因此刚才新建 Home 控制器生成的 Create.cshtm 和 Edit.cshtml 视图要注释掉代码：

```
@*   @Scripts.Render("~/bundles/jqueryval")*@
```

因为是空项目，所以默认并没有启用文件合并和压缩功能，运行结果如图 2-42、图 2-43 所示。

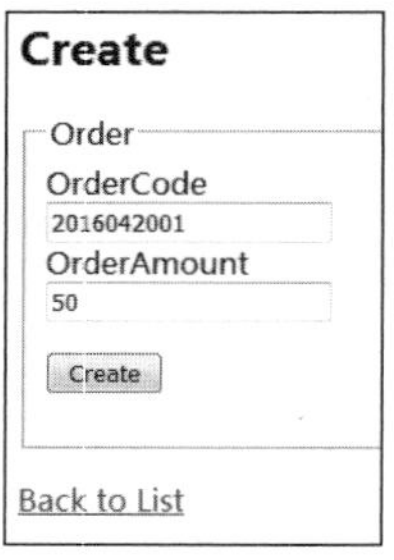

图 2-42

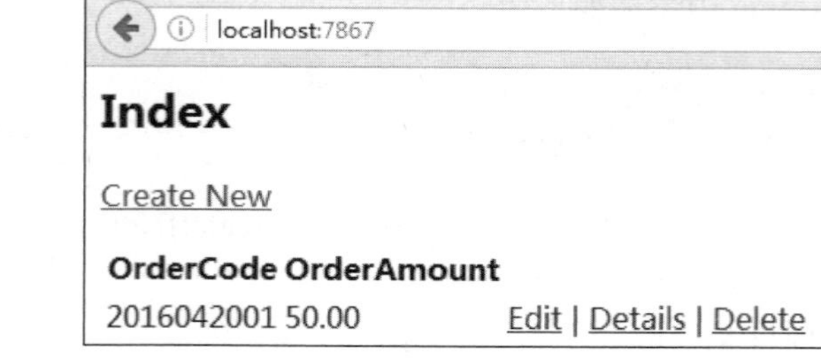

图 2-43

2.6.2 关于 EF 实例的创建问题

在 Home 控制器中，直接在类 HomeControll 中新建了一个 EF 数据库对象，代码如下：

```
private MvcFirstCodeContext db = new MvcFirstCodeContext();
```

什么时候释放这个 db 对象呢？既不能每次调用都直接放到 using()方法中新建一个 db 对象，也不能对 db 对象使用单例模式。因为每次调用都在 using()方法中新建一个 db 对象会频繁地连接关闭数据库，而且更要命的是在并发操作的情况下会产生脏数据。对 db 使用单例模式的话，有多个用户操作数据库的时候操作的就是同一个 db 对象了，这样也是不行的。

我们可以考虑把这个 EF 对象存到一个线程中，新建一个 Base 控制器，作为其他控制器的基类，然后我让 Home 控制器继承自 Base 控制器，Base 控制器中，添加命名空间引用：

```
using System.Runtime.Remoting.Messaging;
```

然后添加如下属性:

```
public MvcFirstCodeContext db
{
        get
        { //从当前线程中获取 MvcFirstCodeContext 对象
            MvcFirstCodeContext db = CallContext.GetData("DB") as MvcFirstCodeContext;
             if (db == null)
            {
                db = new MvcFirstCodeContext();
                CallContext.SetData("DB", db);
            }
            return db;
        }
}
```

这时就可以在 Home 控制器中注释掉如下代码：

```
//private MvcFirstCodeContext db = new MvcFirstCodeContext();
```

第 3 章

初识MVC

前面两章介绍了 ASP.NET MVC 的开发环境和一些必备知识以及 EF，主要是做一些准备工作，为接下来的章节做铺垫，从本章起，我们将正式走进 ASP.NET MVC 的世界。微软 ASP.NET MVC 框架的发展历史如表 3-1 所示。

表 3-1　ASP.NET MVC 框架发展历史

版本	时间
ASP.NET MVC CTP	2007/12/10
ASP.NET MVC 1.0	2009/3/31
ASP.NET MVC 2.0	2010/4/21
ASP.NET MVC 3.0 Razor 视图引擎	2011/1/13
ASP.NET MVC 4.0 ① Web API ② SingalR HTML5 WebSocket ③ 异步请求支持 ④ Mobile Web	2012/9/12
ASP.NET MVC 5.0 ① Bootstrap3.0 ② OWIN ③ ASP.NET Identity	2013/10/17
ASP.NET MVC 6.0	2015/7/21

ASP.NET MVC 4 的新特性：

- 改进 Razor 语法
- 捆绑
- 压缩
- 数据迁移
- ASP.NET Web API
- 移动平台 Web
- 实时双工 SignalR
- 异步支持 Task

3.1 MVC 简介与三层架构

3.1.1 MVC 简介

MVC 最早于 1978 年提出，是软件工程中的一种软件架构模式，这时距离微软在 1985 年推出 Windows 1.0 还有 7 年之久，当时的 MVC 即所有的输入、输出、逻辑控制，这些都要由软件开发者完全实现。

MVC 模式可以有两种理解：一种是表现模式，另外一种是架构模式。这里先将其理解为表现模式。

MVC 是模型（Model），视图（View）和控制（Controller）的缩写，其目的是实现 Web 系统的职能分工。其中，Model 层实现系统中的业务逻辑；View 层用于与用户的交互，通常用 Razor 和 APSX 来实现；Controller 层是 Model 与 View 之间沟通的桥梁，可以分派用户的请求并选择恰当的视图以用于显示，同时还可以解释用户的输入并将它们映射为模型层可执行的操作，如图 3-1 所示。

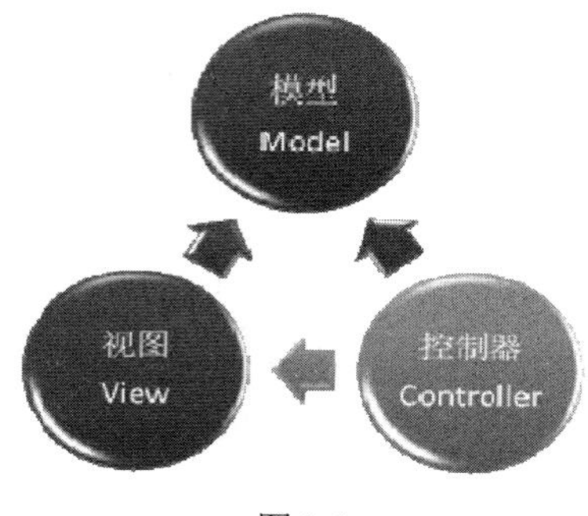

图 3-1

（1）控制器（Controller）

接收用户输入，并完成模型、视图的调用。Controller 处理用户交互，从 Model 中获取数据并将数据传给指定的 View 。

（2）视图（View）

View 是用户接口层组件，主要是将 Model 中的数据展示给用户。cshtml、ASPX 和 ASCX 文件提供处理视图的职责。

（3）模型（Model）

Model 主要是存储或者是处理数据的组件，实现业务逻辑层对实体类相应数据库的操作，如 CRUD（C:Create/R:Read/U:Update/D:Delete），包括数据、验证规则、数据访问和业务逻辑等应用程序信息。Model 具有两方面的含义：DomainModel 和 ViewModel。

- 领域模型 DomainModel：不仅仅是一个实体类，而是整个业务处理流程的一个规则，是实现业务逻辑层对实体类的相应操作，包括逻辑操作与数据库操作，如验证规则、数据访问和业务逻辑等应用程序信息。

- 视图模型 ViewModel：是与显示页面强关联的模型对象，用于实现页面强类型，如做了连接查询得到的结果要显示到前台，就没有相应的对象存在，这时就可以新建一个类来包含结果集中的行。

此外需要注意的是，Model 是独立的组件，并不知道 View 的存在，也不知道 Controller 的存在。

过去 MVC 模式并不适合小型甚至中等规模的应用程序，因为这样会带来额外的工作量，增加应用的复杂性。现在多数软件设计框架（如 ASP.NET MVC）都能直接快速提供 MVC 骨架，供中小型应用程序开发，此问题不再存在。对于存在大量用户界面并且逻辑复杂的大型应用程序，MVC 将会使软件在健壮性、代码重用和结构方面上一个新的台阶。尽管在最初构建 MVC 模式框架时会花费一定的工作量，但从长远的角度来看它会大大提高后期软件开发的效率。

很多公司都在努力让自己的产品遵循 MVC 思想，其中微软的 ASP.NET MVC 就是其中一款，这是一款对输入、输出进行分离的 UI 层框架。

ASP.NET MVC 是微软 2009 年对外公布的第一个开源的表示层框架，是微软的第一个开源项目。它将 Web 应用程序分成 3 个主要组件，即视图（View）、控制器（Controller）、模型（Model）。

Razor 是 MVC 3.0 以后的视图引擎。MVC 提供了 Razor、ASPX 两种方式。

3.1.2 三层架构

当把 MVC 当成一种架构模式来理解，就是所谓的三层架构，如图 3-2 所示。

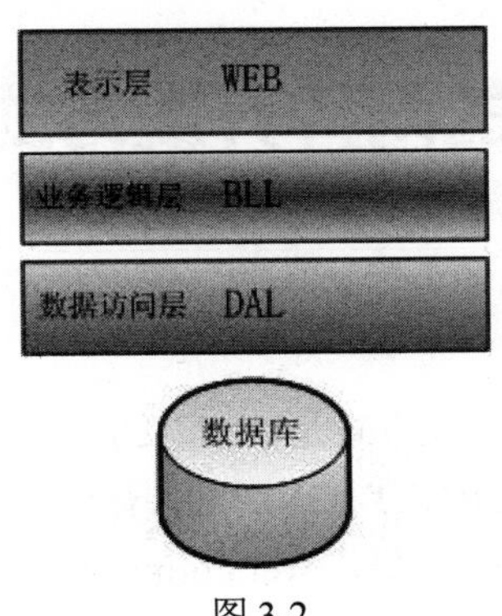

图 3-2

三层模式是软件工程中的程序设计模式，是 MVC 设计思想的一种实现。随着技术的发展，现在基本上已经将 MVC 模式等同于三层模式，包括数据访问层、业务逻辑层、表示层。再细一点可以分为 UI 层、业务逻辑层、数据访问层、模型层。如果要严格区分，那么 UI 层是指 View 与 Controller，业务逻辑层、数据访问层、模型层都被包括在 Model 中。三层之间有着非常强的依赖关系：表示层 ← 业务逻辑层 ← 数据访问层。而且它们之间的数据传递是双向的，并且通常借助实体类传递数据。

1. 各层的作用

（1）数据访问层：主要是对原始数据（数据库或者文本文件等存放数据的形式）的操作层，而不是指原始数据，也就是说，是对数据的操作，而不是数据库，具体为业务逻辑层或表示层提供数据服务。

（2）业务逻辑层：主要是针对具体问题的操作，也可以理解成对数据层的操作，对数据业务逻辑处理，如果说数据层是积木，那么逻辑层就是对这些积木的搭建。

（3）表示层：主要表示 Web 方式，也可以表示成 WinForms 方式。Web 方式也可以表现成 aspx，如果逻辑层相当强大和完善，那么无论表现层如何定义和更改，逻辑层都能完善地提供服务。

2. 具体的区分方法

（1）数据访问层：主要看数据层里面有没有包含逻辑处理，实际上它的各个函数主要完成对数据文件的操作，而不必管其他操作。

（2）业务逻辑层：主要负责对数据层的操作，也就是说把一些数据层的操作进行组合。

（3）表示层：主要接收用户的请求以及数据的返回，为客户端提供应用程序的访问。

使用三层架构的好处是：项目结构更清楚，分工更明确，有利于后期的维护和升级。虽然它会带来一定的性能损失（因为当子程序模块未执行结束时主程序模块只能处于等待状态，说明将应用程序划分层次会带来执行速度上的一些损失），但是从团队开发效率角度上来讲却可以感受到大不相同的效果。

在本书中，接下来的内容都是把 MVC 当成三层架构中的表示层来理解。

3.2 ASP.NET 的两种开发方式

3.2.1 ASP.NET 开发现状

目前，ASP.NET 中两种主流的开发方式是 ASP.NET WebForms 和 ASP.NET MVC。从图 3-3 可以看到 ASP.NET WebForms 和 ASP.NET MVC 是并行的，也就是说 MVC 不会取代 WebForms（至少目前是这样），而是多了一个选择，WebForms 在短期之内不会消亡，尽管存在许多缺点，但过去许多老项目依旧是使用 WebForms 进行开发的，虽然许多公司已经在积极地将 WebForms 项目迁移或升级到 MVC 项目，但是 WebFoms 项目依然会存在很长一段时间。

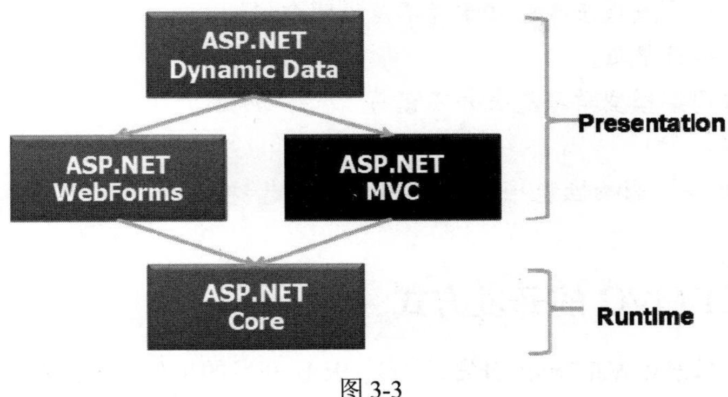

图 3-3

3.2.2　WebForms 的开发方式

WebForms 有以下 3 种开发方式：

- 服务器端控件。
- 一般处理程序+HTML 静态页+Ajax。
- 一般处理程序+HTML 模板。

WebForms 的请求模型如图 3-4 所示。WebForms 请求的是具体的某一个文件、具体的一个类。

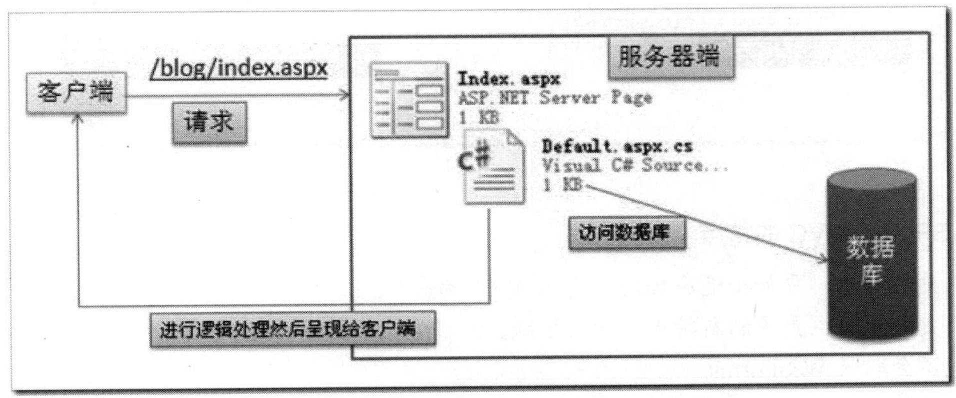

图 3-4

WebForms 的优点：

- 支持事件模型开发，得益于丰富的服务器端组件。WebFroms 开发可以迅速地搭建 Web 应用。
- 使用方便，入门容易。
- 控件丰富。

WebFroms 的缺点：

- 封装太强，很多底层东西让初学者不是很明白。
- 入门容易，提升很难。
- 复杂的生命周期模型学习起来并不容易。
- 控制不灵活。
- ViewState 处理，影响性能。

3.2.3 ASP.NET MVC 的开发方式

MVC 的目的不是取代 WebFrom 开发，而是 Web 开发的另外一种选择，但是目前来看，新项目基本上都是使用 ASP.NET MVC，很少有使用 WebForm 的了。这是一种趋势，很好理解，因为有更好的东西，当然要用更好的。

MVC 的请求模型如图 3-5 所示。ASP.NET MVC 请求的是 Controller 中具体的 Action 方法。

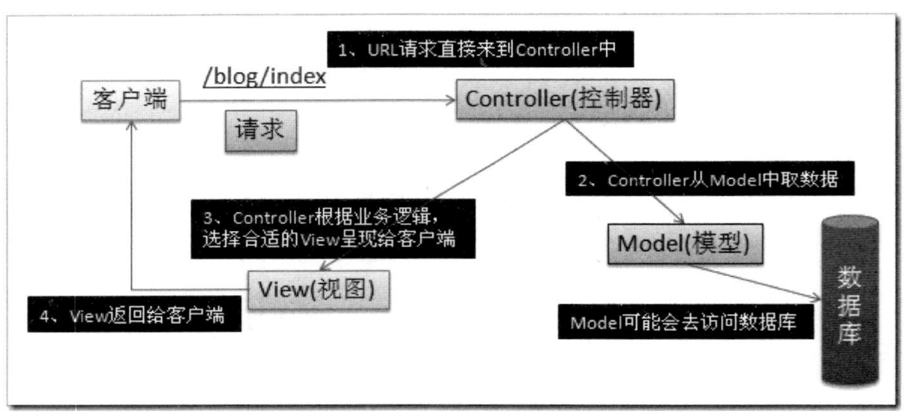

图 3-5

1. ASP.NET MVC 的特点

- 更加简洁，更加接近原始的“请求-处理-响应”。
- 更加开放、更多的新特点、社区活跃。
- 不会取代 WebForms。
- 底层跟 WebForms 是一样的，只是管道上处理不同而已。
- MVC 只是表示层的一种新方式。

2. MVC 的优点

- 很容易将复杂的应用分成 M、V、C 三个组件模型。通过 Model、View 和 Controller 有效地简化复杂的架构，体现了很好的隔离原则。
- 因为没有使用 server-based forms，所以程序员控制得更加灵活，页面更加干净。
- 可以控制生成自定义的 URL，对于 seo 友好的 URL 更是不在话下。WebForm 要额外做路由重写来实现伪静态的形式。

- 强类型 View 实现，更安全、更可靠、更高效。
- 让 Web 开发可以专注于某一层，有利于开发中的分工，更利于分工配合，适用于大型架构开发。
- 很多企业已经使用 MVC 作为项目开发框架，招聘明确要求熟悉 MVC 开发模式，有些项目架构就是 MVC+EF+WebAPI+…。
- 松耦合、易于扩展和维护。
- 有利于组件的重用。
- ASP.NET MVC 更好地支持单元测试（Unit Test）。
- 在团队开发模式下表现更出众。
- MVC 将代码和页面彻底分离，分离到了两个文件中，即视图和控制器。WebForm 的 Codebehind 技术没有完全对代码和前台页面进行分离，耦合度太高。ASPX 和 ASPX.cs 是典型的继承关系。

3. 学习 ASP.NET MVC 的原因

MVC 架构模式诞生 30 年后，因为其提供了良好的松耦合、易于扩展、高可维护性等优点，重新在开发社区火起来。作为微软全新的 Web 网站开发框架，ASP.NET MVC 提供了全新的开发模式，完美支持经典的 MVC 架构模式，为.NET 工程师提供了完全不同的开发体验。

此外，越来越多的公司和开发者开始加入到 MVC 开发模式中来，使它进入了一个高速发展的状态，而现在 MVC 已经变成 ASP.NET 下一种常见的开发模式，它能让你学习一种完全不同的架构，所以这是我们有理由也有必要掌握的一项开发技能。

接触了 MVC 后，就再也不想拖控件了，对 MVC 会有一种相见恨晚的感觉。

3.3 第一个 ASP.NET MVC 程序

本节我们通过新建一个 ASP.NET MVC 程序来了解 ASP.NET MVC 项目框架的一些基本概念和执行过程，让大家先对 ASP.NET MVC 有一个基本的认识。

3.3.1 创建项目

开发流程：

- 新建一个 MVC 项目。
- 新建 Controller。
- 创建 Action。
- 根据 Action 创建 View。
- 在 Action 获取数据并生产 ActionResult 传递给 View。

- View 是显示数据的模板。
- URL 请求→Controller.Action 处理→View 响应。

首先打开 VS，选择“文件→新建项目 FirstMVCApp”，如图 3-6 所示。

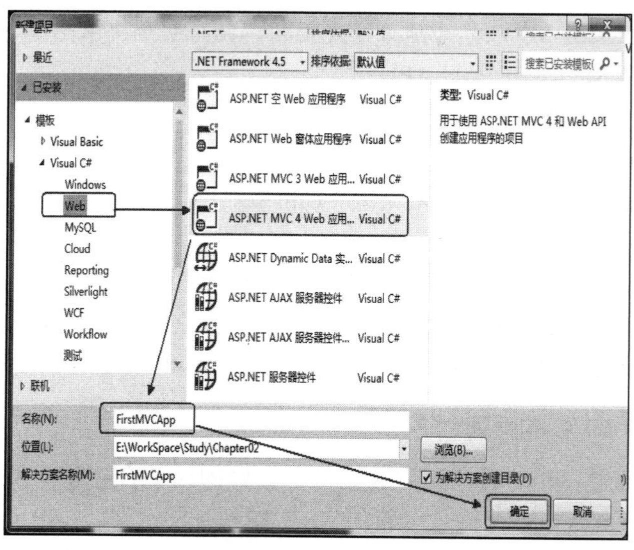

图 3-6

单击“确定”后，会出现如图 3-7 所示的对话框，进行模板的选择。在这里还可以选择是否创建单元测试，为了养成一个良好的习惯，这里勾选“创建单元测试项目”复选框。这里选择“基本”模板，在项目开发中一般选择“空”或“基本”，因为这样的项目环境相对比较干净。

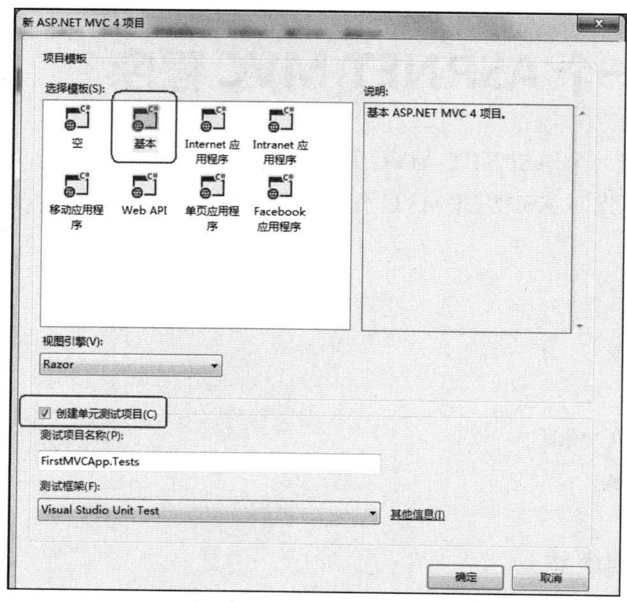

图 3-7

有关项目模板的说明：

- Internet 应用程序：外网访问的 Web 应用。
- Intranet 应用程序：局域网访问的 Web 应用。
- 单页应用程序：也就是 WebPages，这个现在用得很少。
- 移动应用程序：构建移动端程序。
- Web API：构建 RESTful 服务，轻量级 API 开发框架，微软之前有 Wcf 这一重型框架。

选择了“基本”模板并确定后会创建两个项目，即 FirstMVCApp 和 FirstMVCApp.Tests，如图 3-8 所示。

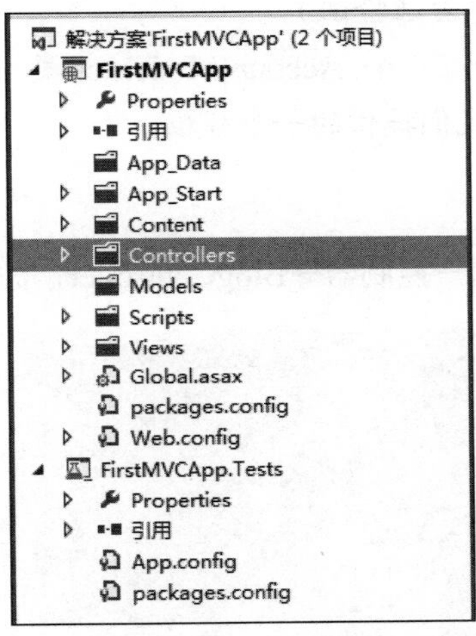

图 3-8

3.3.2　项目框架结构说明

1. 默认项目模板中的内容

- App_Data：用来存储数据库文件，xml 文件或者应用程序需要的一些其他数据
- Content：用来存放应用程序中需要用到的一些静态资源文件，如图片和 CSS 样式文件。Content：目录默认情况下包含了本项目中用到的 css 文件 Site.css，以及一个文件夹 themes ：themes 中主要存放 jQuery UI 组件中要用到的图片和 css 样式。
- Controllers：用于存放所有控制器类，控制器负责处理请求，并决定哪一个 Action 执行，充当一个协调者的角色。
- Models：用于存放应用程序的核心类，数据持久化类，或者视图模型。如果项目比较大，可以把这些类单独放到一个项目中。
- Scripts：用于存放项目中用到的 JavaScript 文件，默认情况下，系统自动添加了一系列的 js 文件，包含 jquery 和 jquery 验证等 js。
- Views：包含了许多用于用户界面展示的模板，这些模板都是使用 Rasor 视图来展示的，子目录对应着控制器相关的视图。

- Global.asax：存放在项目根目录下，代码中包含应用程序第一次启动时的初始化操作，诸如路由注册。
- Web.config：同样存在于项目根目录，包含 ASP.NET MVC 正常运行所需的配置信息。

如果选择的是“Internet 应用程序”模板，那么还会有 favicon.ico 文件：

- favicon.ico：存在于应用程序根目录，应用程序的图标文件显示在浏览器，名称不能更改，可以使用其他图片替换。
- Controllers 目录下面默认会有两个控制器——HomeController（负责主页的请求处理）和 AccountController（身份证验证）。
- Models 目录中默认会含有一个 AccountModels.cs.类，类中包含了一系列和身份验证相关的类，是项目默认为我们提供的一个模板。

2. 添加控制器 Blog

右击文件夹 Controllers→添加→控制器→BlogController，添加如下代码：

```
/// <summary>
/// Action 方法，返回 ActionResult
/// </summary>
/// <returns></returns>
public ActionResult Index()
{
    ViewBag.Message = "First ASP.NET MVC application.";//展现到视图中数据
    //~/Views/Home/Index.cshtml
    return View(); //展现指定的视图，当没有指定视图名称时，默认是指向根目录下 Views 文
     // 件夹中，子文件夹名称为当前控制器名称 Home，视图名称和当前 Action 名称一样
}
```

3. 添加视图

将鼠标移动到 Index()方法体中的任意位置，右击→添加视图→确定，默认会在 Views/Blog 目录下面生成一个 Index.cshtml 文件。

在 Index 视图中添加代码：

```
@ViewBag.Message
```

4. 运行

将 FirstMVCApp 设置为启动项目，然后按 Ctrl+F5 组合键或者选择“调试→开始执行”运

行程序，这时 VS 会自动启动 ASP.NET 开发者服务器，并随机分配一个端口，以便于我们浏览 ASP.NET 程序。

运行会报错，提示“无法找到资源”，这是因为 App_Start 目录下面的 RouteConfig.cs 路由配置类中配置了默认路由，路由规则如下：

```
routes.MapRoute(
            name: "Default",
            url: "{controller}/{action}/{id}",
            defaults: new { controller = "Home", action = "Index", id =
            UrlParameter.Optional }
);
```

可以看到默认控制器是 Home，我们将其修改为 Blog，再按 Ctrl+F5 组合键运行，如图 3-9 所示。

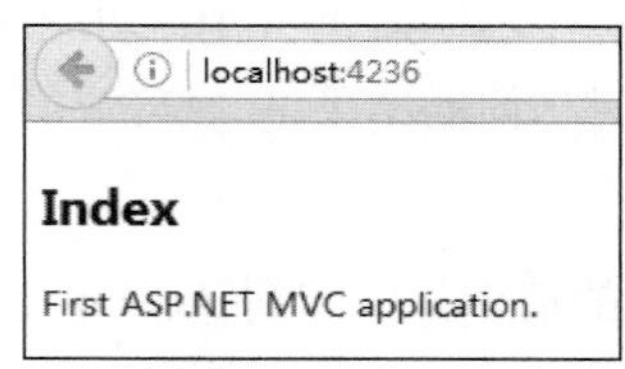

图 3-9

出现这个结果页面表示 OK。

BlogController 通过继承 Controller 来表示它是一个控制器类，类名的后缀和 Controller 一样。控制器通过 URL 来识别执行的是哪一个 Action。

ViewBag 本质上是一个字典，提供了一种 View 可以访问的动态数据存储。这用到了.NET 4.0 的动态语言特性。你可以给 ViewBag 添加任意属性，并且这个属性是动态创建的，不需要修改类的定义就可以从 View 中访问。

3.3.3　路由——映射 URL 到 Action

在 ASP.NET MVC 中是如何将 URL 映射到控制器中指定的 Action 的呢？

在 Global.asax 中，“RouteConfig.RegisterRoutes(RouteTable.Routes);”这行语句进行了路由规则的注册。

打开 App_Start 目录下面的 RouteConfig.cs 文件，代码如下：

```
namespace FirstMVCApp
{
    public class RouteConfig
    {
        public static void RegisterRoutes(RouteCollection routes)
        {
```

```
        routes.IgnoreRoute("{resource}.axd/{*pathInfo}");

        routes.MapRoute(
            name: "Default",
            url: "{controller}/{action}/{id}",
            defaults: new { controller = "Blog", action = "Index", id =
            UrlParameter.Optional }
        );
    }
  }
}
```

从这里可以看出应用程序启动的时候默认调用了 BlogController 控制器中的 Index 方法，而且控制器是不区分大小写的。

所以我们在地址栏中输入“http://localhost:4236/Blog/Index”、“http://localhost:4236/”以及“http://localhost:4236/BLog/Index”时访问的结果是一样的。

Action

我们可以看到 ASP.NET MVC 的请求都归结到 Action 上，所以是 URL 驱动，而且 Action 跟 View 是弱耦合的关系，因为我们可以在 Action 中的 View()方法中指定视图名称。

（1）新建一个视图 Article

右击 Views 目录下面的 Blog 目录，选择“添加→视图”，并将视图名称设置为 Article，在视图中添加文字“我是文章”。

（2）新建一个 Action ShowArticle

在 Blog 视图中添加一个 Action 方法 ShowArticle()，代码如下：

```
public ActionResult ShowArticle()
{
    return View("Article");
}
```

运行程序，在浏览器输入“http:// localhost:4236/Blog/ShowArticle 地址”，结果如图 3-10 所示。

图 3-10

View 模板显示页面的规则是先找对应的 Controller 文件夹，再找对应的 Shared 文件夹中所有继承自 viewpage 的页面。

3.3.4　返回 string 的 MVC 方法

在 Blog 控制器中添加如下代码：

```
public string Say()
{
     return "Hello,World!";
}
```

运行，在地址栏输入“http://localhost:4236/Blog/Say”地址，运行结果如图 3-11 所示。

图 3-11

服务器在接收请求后会解析 URL（根据路由表里配置的 URL 来分析类名和方法名）从中找到请求的类名，并在类名后面加上 Controller 作为真实的类名，创建该类的对象，并调用 URL 中指定的方法。

Blog 控制器里面的 Index 方法返回了一个 View()方法，表示返回加载一个视图，视图文件的路径默认是/View/控制器名称/Action 名称.cshtml。

3.3.5　简单了解 Razor 视图

以 cshtml 为后缀的就是 Razor 视图。在 ASP.NET MVC 中，官方给出了两种默认视图，一种是 ASPX（就是传统的 WebForm），一种就是 Razor。

在视图中，我们可以直接调用 C#代码和代码块，只要在调用之前加一个@符号即可。代码块要用大括号括起来。

修改 Index 视图代码如下：

```
@{
    ViewBag.Title = "Index";
}

<h2>Index</h2>
<body>
    @ViewBag.Message
    <div>
     @for (int i = 0; i < 3; i++)
     {
         <p>@i</p>
```

```
    }
  </div>
</body>
```

注意，修改 View 中的代码不需要重新编译项目，只需要按 Ctrl+S 组合键保存，然后刷新界面即可，运行结果如图 3-12 所示。

图 3-12

3.3.6 ASP.NET MVC 组件之间的关系

ASP.NET MVC 组件之间的关系如图 3-13 所示。

- View 和 Controller 都可以直接请求 Model，但是 Model 不依赖 View 和 Controller;
- Controller 可以直接请求 View 来显示具体页面;
- View 不依赖 Controller，View 可以通过另外的方式来请求 Controller。

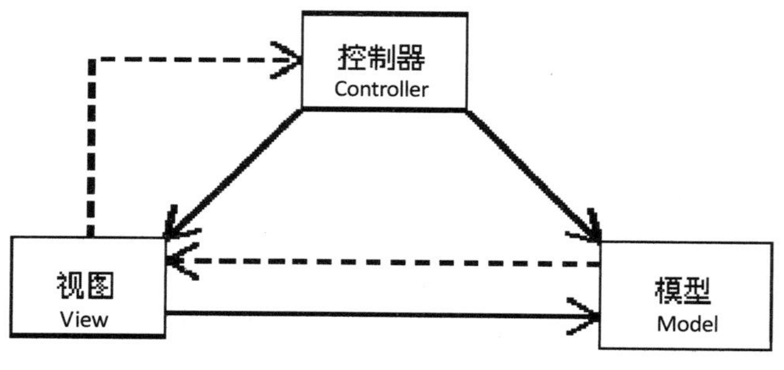

图 3-13

3.4 MVC 的约定

在 ASP.NET MVC 开发中有一个很重要的理念叫“约定大于配置”，也就是说在 ASP.NET MVC 中存在许多潜规则，当然这些潜规则是可以打破的，但是一般而言，我们都会去遵守，因为当开发人员都去默契地遵守这些规则时对开发和维护是非常有好处的。例如，可以一眼就知道哪个目录是干什么的、哪个文件表示什么……这很重要。

3.4.1　控制器的约定大于配置

（1）Controller 放到 controllers 文件夹中，并且命名方式以 Controller 结尾。

（2）每个 Controller 都对应 View 中的一个文件夹，文件夹的名称跟 Controller 名相同。Controller 中的方法名都对应一个 View 视图（非必须，但是建议这么做）而且 View 的名字跟 Action 的名字相同。

（3）控制器必须是非静态类，并且要实现 IController 接口。

（4）Controller 类型可以放到其他项目中。

3.4.2　视图的相关约定

（1）所有的视图必须放到 Views 目录下。

（2）不同控制器的视图用文件夹进行分割，每个控制器都对应一个视图目录。

（3）一般视图名字跟控制器的 Action 相对应（非必须）。

（4）多个控制器公共的视图放到 Shared 目录中。

第 4 章

◂ MVC进阶 ▸

通过上一章的学习，我们对 MVC 已经有了初步的了解和认识，下面我们来继续深入学习 MVC。

4.1 View 详解

View（视图）的职责是向用户提供界面，负责根据提供的模型数据生成准备提供给用户的格式界面，并提供用户和系统交互的入口。

MVC 支持多种视图引擎（Razor 和 ASPX 视图引擎是官方默认给出的，其实还支持 *N* 种视图引擎，甚至都可以是自己写的视图引擎）。

4.1.1 View 和 Action 之间数据传递的方式

View 和 Action 之间前后台数据传递的方式有如下几种。

- 弱类型 ViewData[""]
- 动态型 ViewBag //dynamic
- 强类型 Model
- 临时存储 TempData[""]
- 后台：return View(data); //存入 ViewData.Model
- 前台：Model //其实就是 WebViewPage.Model

下面我们通过一个实例来进行讲解。

打开 VS2012，选择“文件→新建项目→Web→ASP.NET MVC4 Web 应用程序”，并将项目命名为 MvcViewApp，选择“基本”项目模板，然后单击“确定”按钮。

在 Controllers 目录中添加 Home 控制器、Models 目录中添加 User 模型类：

```
public class User
{
    public string Name { get; set; }
}
public ActionResult Index()
```

```
{
        ViewData["One"] = "天机老人孙老头";
        ViewBag.Two = "子母龙凤环上官金虹";
        var _user = new User{Name="小李飞刀李寻欢" };
        TempData["Four"] = "嵩阳铁剑郭嵩阳";

        return View(_user); //等于 ViewData.Model = _user;
}
```

添加 Index 视图，这里使用强类型视图，用法为“@model 类型”，写在 View 最上面：@ViewData.Model（一般直接简写为@Model）、@Html.xxFor(x=>x.**) 。

使用强类型视图的好处：提供智能提示，实现了编译时错误检查，防止写错字段。

```
@model MvcViewApp.Models.User
@{
    ViewBag.Title = "Index";
}

<h2>百晓生兵器谱排名</h2>
<p>第一名：@ViewData["One"]</p>
<p>第二名：@ViewBag.Two</p>
<p>第三名：@Model.Name</p>
<p>第四名：@TempData["Four"] </p>
```

运行结果如图 4-1 所示。

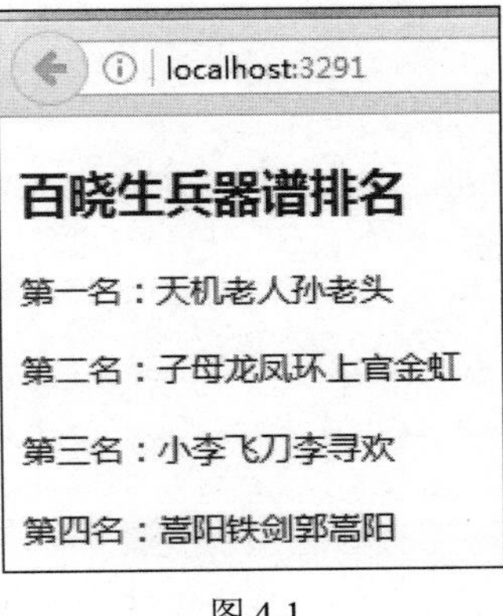

图 4-1

4.1.2　TempData、ViewData 和 ViewBag 的区别

ViewData 是字典型的（Dictionary），ViewBag 不再是字典的键值对结构，而是 dynamic（动态）型，会在程序运行的时候动态解析。ViewData 为 object 型，而 ViewBag 为 dynamic 型。dynamic 型与 object 型的区别是在使用时它会自动根据数据类型转换，而 object 型则需要我

们自己去强制转换。

通过反编译工具查看 System.Web.Mvc.dll 的源码（当然也可以直接下载 System.Web.Mvc4 的源码，本书云盘中提供了 MVC4 的源码 Chapter04 \MVC4SourceCode.zip），代码如下所示。

```
public abstract class WebViewPage : WebPageBase, IViewDataContainer,
IViewStartPageChild
{
    [Dynamic]
    public object ViewBag { [return: Dynamic] get; }
    public ViewContext ViewContext { get; set; }
    public ViewDataDictionary ViewData { get; set; }
    ......
}
```

这里 ViewBag 只有 get 方法，没有 set 方法，但是我们在上面却给 ViewBag 赋值了。通过反编译发现 ViewBag 的代码如下所示。

```
[Dynamic]
public object ViewBag
{
        [return: Dynamic]
        get
        {
            Func<ViewDataDictionary> viewDataThunk = null;
            if (this._dynamicViewDataDictionary == null)
            {
                if (viewDataThunk == null)
                {
                    viewDataThunk = () => this.ViewData;
                    //创建一个委托对象，内部方法返回 ViewData
                }
                this._dynamicViewDataDictionary = new
                DynamicViewDataDictionary(viewDataThunk);
            }
            return this._dynamicViewDataDictionary;
        }
}
```

不难看出 ViewBag 返回的是_dynamicViewDataDictionary，继续跟踪发现_dynamicViewDataDictionary 属于 DynamicViewDataDictionary 类，代码如下：

```
internal sealed class DynamicViewDataDictionary : DynamicObject
{
```

```
    // Fields
    private readonly Func<ViewDataDictionary> _viewDataThunk;

    // Methods
    public DynamicViewDataDictionary(Func<ViewDataDictionary> viewDataThunk);
    public override IEnumerable<string> GetDynamicMemberNames();
    public override bool TryGetMember(GetMemberBinder binder, out object result);
    public override bool TrySetMember(SetMemberBinder binder, object value);

    // Properties
    private ViewDataDictionary ViewData { get; }
}
```

TryGetMember 和 TrySetMember 方法如下：

```
public override bool TryGetMember(GetMemberBinder binder, out object result)
{
    result = this.ViewData[binder.Name];
    return true;
}
public override bool TrySetMember(SetMemberBinder binder, object value)
{
    this.ViewData[binder.Name] = value;
    return true;
}
```

ViewBag 其实就是 ViewData，只是多了一层 Dynamic 控制，可以说它是另一种访问 ViewData 的方式。理论上 ViewBag 要比 ViewData 慢一点点，但是几乎可以忽略，所以使用何种方式完全取决于个人的爱好。

TempData 的使用同 ViewData 和 ViewBag 一样，TempData 也可以用来向视图传递数据，只是 ViewData 和 ViewBag 的生命周期和 View 相同，它们只对当前 View 有用。TempData 则可以在不同的 Action 中进行传值，类似 Webform 里的 Session。有一点需要注意，TempData 的值在取了一次后会自动删除。TempData 用来在一次请求中同时执行的多个 Action 方法之间共享数据。

从源码中可以看到 TempData 是一个 TempDataDictionary 类型。TempDataDictionary 类型中有一个 this 索引器：

```
public TempDataDictionary TempData { get; set; }
public object this[string key]
{
    get
    {
        object obj2;
```

```
        if (this.TryGetValue(key, out obj2))
        {
            this._initialKeys.Remove(key);
            return obj2;
        }
        return null;
    }
    set
    {
        this._data[key] = value;
        this._initialKeys.Add(key);
    }
}
```

每次获取数据之后，就马上将该数据从键值对列表_initialKeys 中移除了。

```
public class TempDataDictionary : IDictionary<string, object>,
ICollection<KeyValuePair<string, object>>, IEnumerable<KeyValuePair<string,
object>>, IEnumerable
{
    // Fields
    private Dictionary<string, object> _data;
    private HashSet<string> _initialKeys; //注意，就是往这里面存数据
    private HashSet<string> _retainedKeys;
    internal const string TempDataSerializationKey = "__tempData";
    ......
}
```

4.2 Razor 视图引擎

4.2.1 什么是 Razor

Razor 是 ASP.NET MVC 3.0 出现的新的视图引擎，为视图提供精简的语法，最大限度减少了语法和额外字符串。Razor 不是编程语言，而是服务器端标记语言，是一种允许用户向网页中嵌入基于服务器代码（Visual Basic 和 C#）的标记语法。

语境 A：在前台声明和使用 C#变量

```
@{
    string schoolName="湖南第一师范";
}
```

```
<span>@schoolName.Models</span> @*错误*@
<span>@(schoolName).Models</span>
<span>我毕业于 @schoolName 信息系</span>
```

说明：声明 C#变量的代码必须写在@{ }代码块中，使用@+C#变量名就能取得 C#变量的值，注意前后必须有空格，否则会被当作普通字符输出。

语境 B：邮件格式中的@符号

```
<span>zouqiongjun@uuch.net</span>@*通过，因为@前面没有空格*@
<span>zouqiongjun @uuch.net</span>@*不通过，因为@前面有空格*@
```

4.2.2　Razor 语法

C# 的主要 Razor 语法规则如下：

- Razor 代码封装于@{ ... } 中。
- 行内表达式（变量和函数）以@开头。
- 代码语句以分号结尾。
- 字符串由引号包围。
- C# 代码对大小写敏感。
- C# 文件的扩展名是 .cshtml。
- Razor 通过理解标记的结构来实现代码和标记之间的顺畅切换。
- @核心转换字符，用来标记代码的转换字符串。
- Razor 表达式自动使用了 HTML 编码。

如果想向浏览器输出 HTML 源代码，那么使用 System.Web.IHtmlString: @Html.Raw（"<b>zouyujie</b>"）会输出结果 zouyujie。

查看 HTML 源码：

```
<b>zouyujie</b>
```

JS 字符串编码：

```
<script>
   alert('@Ajax.JavaScriptStringEncode("小李飞刀")');
</script>
```

运行结果如图 4-2 所示。

图 4-2

@代码块：

```
@{  string s ="zouyujie";
        int age =27;
}
@{Html.RenderPartial("TestPartial");}//调用无返回值方法
```

注释：

```
@* ............*@
```

调用泛型方法：

```
@(Html.SomeMethod<User>());
```

混合代码与文本：

```
@if(1==1){
    <text>我要输出文本在这里！</text>
    @:我要输出文本在这里！
}
```

说明：当前台的 C#代码有多行时，如果这些代码是连续的，中间没有 HTML 代码隔断，那么只需要开头的一个@符号即可；否则 HTML 代码之后被隔断的 C#代码开头需要加上@符号。

@转义：

```
@@
```

操作 Web 内置对象：

```
@Request.RawUrl @Response.Write
```

@作用域与 HTML 标记混合使用：

```
@{
    string userName="刘邦";
    <p>@userName</p>
}
```

在@作用域中输出未转义的 HTML 代码。

（1）使用字符串描述输出：

```
@{
    string strHtm="<p>你好~</p>";
    @strHtml
}
```

（2）使用 HTMLHelper 输出：

```
@{
```

```
    @Html.Raw("<p>哇哈哈哈~</p>");
}
```

（3）使用 HtmlString 类输出：

```
@{
    HtmlString htm =new HtmlString("<p>哈哈</p>");
    @htm
}
```

（4）使用 MvcHtmlString 输出：

```
@{
    var strHtml=MvcHtmlString.Create("<p>哈哈~</p>");
    @strHtml
}
```

数据类型转换：用 As....()方法转换，例如

```
@("120".AsInt())
```

数值类型判断：用 IsInt()方法，例如

```
@(strAge.IsInt()?"是":"否")
```

路径转换用 Href()方法，例如

```
@Href("~/Home/Index");
```

using System.Web.WebPages;//内部为 string 扩展了很多 As..方法

HtmlHelper 重用：相当于是在视图中定义方法。

```
@helper List(List<string> dogs){
   <ul>
      @foreach (string s in dogs)
      {
         <li>@s</li>
      }
   </ul>
   }
@List(new List<string>(){"ruiky","lisa","lucy"})
```

运行效果如图 4-3 所示。

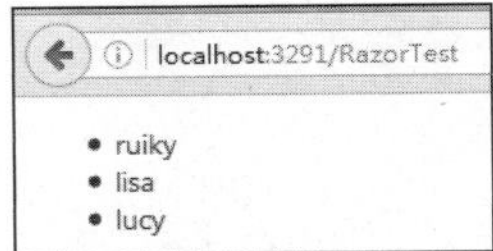

图 4-3

4.2.3 Razor 布局——整体视图模板

这个整体视图模板就和以前 WebForm 中的 Mater 母版页一样。

View 模板显示页面的规则：先找对应的 Controller 文件夹，再找对应的 Shared 文件夹。

为演示应用整体模板视图，在网站根目录下的 Views/Shared 目录新建视图 SiteLayout，代码如下：

```
@{
    ViewBag.Title = "我是模板页";
}
<body>
    <h2>欢迎光临</h2>
      @RenderBody() @*模板页里的占位符*@
</body>
```

然后在 RazorTest 控制器中新建 Action 方法 ViewTemplate，添加视图 ViewTemplate：

```
@{
    Layout = "~/Views/Shared/SiteLayout.cshtml";
    ViewBag.Title = "我是子页面";
}
<h2>今天天气不错</h2>
```

在浏览器中输入地址“http://localhost:3291/RazorTest/ViewTemplate”，效果如图 4-4 所示。

欢迎光临

今天天气不错

图 4-4

我们会发现子页所有 HTML 代码都将替换到模板页的@RenderBody()处应用整体视图模板，我们还可以设置多个“占位符”。

要在模板页设置多个节，可在模板页 SiteLayout.cshtml 添加如下代码：

```
<footer>
    @if(IsSectionDefined("Footer")){
        @RenderSection("Footer");
    }else
    {
        <b>子页面没有设置 Footer 节点</b>
    }
```

```
</footer>
```

在子页面 ViewTemplate.cshtml 定义节点：

```
@section Footer{
    <b>@@ 超级牛逼公司所有</b>
}
```

运行结果如图 4-5 所示。

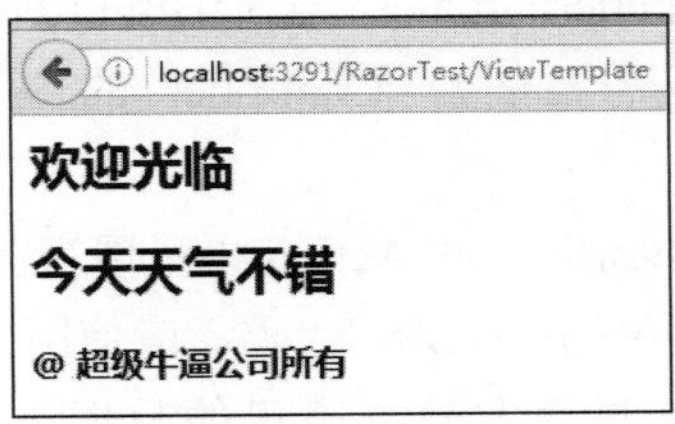

图 4-5

4.2.4 Razor 布局——ViewStart

每个子页面都使用一个 Layout 指定布局。如果多个视图都用同一个布局就会产生冗余，修改维护麻烦。

_ViewStart.cshtml 可解决此问题，此文件代码优先于同目录及子目录下的任何视图代码。执行 View 目录下自动添加的_ViewStart.cshtml：

```
@{
    Layout = "~/Views/Shared/_Layout.cshtml";
}
```

有了它，就可以为某个文件夹下所有的视图添加相同的视图布局了。

因为这个文件代码优先于任何视图，所以任何一个视图都可以重写 Layout 属性来指定自己想要的模板布局页面。

4.2.5 Razor 布局——部分视图

ASP.NET MVC 里的部分视图相当于 Web Form 里的 User Control。我们的页面往往会有许多重用的地方，可以进行封装重用。使用部分视图的好处是既可以简写代码，又可以使页面代码更加清晰、更好维护。

在视图里有多种方法可以加载部分视图，包括 Partial()、Action()、RenderPartial()、RenderAction()、RenderPage()方法。

下面说明一下这些方法的差别。

1. Partial 与 RenderPartial 方法

（1）Razor 语法：@Html.Partial() 与 @{Html.RenderPartial();}。

（2）区别：Partial 可以直接输出内容，在内部将 html 内容转换为 String 字符（MVCHtmlString），然后缓存起来，最后一次性输出到页面。显然，这个转换过程会降低效率，所以通常使用 RenderPartial 代替。

2. RenderPartial 与 RenderAction 方法

（1）Razor 语法：@{Html.RenderPartial();}与@{Html.RenderAction();}。

（2）区别：RenderPartial 不需要创建 Controller 的 Action ，而 RenderAction 需要在 Controller 中创建要加载的 Action。

RenderAction 会先去调用 Contorller 的 Action 再呈现视图，所以这里会再发起一个链接。

如果这个部分视图只是一些简单的 HTML 代码，请使用 RenderPartial。如果这个部分视图除了有 HTML 代码外，还需要通过读取数据库里的数据来渲染，就必须使用 RenderAction 了，因为它可以在 Action 里调用 Model 里的方法读取数据库，渲染视图后再呈现，而 RenderPartial 没有 Action，所以无法做到。

3. RenderAction 与 Action

（1）Razor 语法：@{Html.RenderAction();}与@Html.Action();。

（2）区别：Action 也是直接输出，和 Partial 一样，也存在一个转换的过程，但是不如 RenderAction 直接输出到当前 HttpContext 的效率高。

4. RenderPage 与 RenderPartial 方法

（1）Razor 语法：@{Html.RenderPartial();}与@RenderPage()。

（2）区别：也可以使用 RenderPage 来呈现部分，但它不能使用原来视图的 Model 和 ViewData，只能通过参数来传递；而 RenderPartial 可以使用原来视图的 Model 和 ViewData。

Action 方法可以通过 PartialView 方法以 PartialViewResult 形式返回部分视图。部分视图一般用在 Ajax 请求部分代码中。

在 RazorTest 控制器中添加 Action 方法 PartialViewTest：

```
public ActionResult PartialViewTest()
{
    ViewData["Msg"] = "Hello world!";
    return PartialView();
}
```

然后添加 PartialViewTest 视图，PartialViewTest.cshtml 代码如下：

```
<div>@ViewData["Msg"] </div>
```

最后在 ViewTemplate.cshtml 视图中添加如下代码：

```
<div id="divTest">
    @{Html.RenderAction("PartialViewTest");}
</div>
```

生成项目，运行结果如图 4-6 所示。

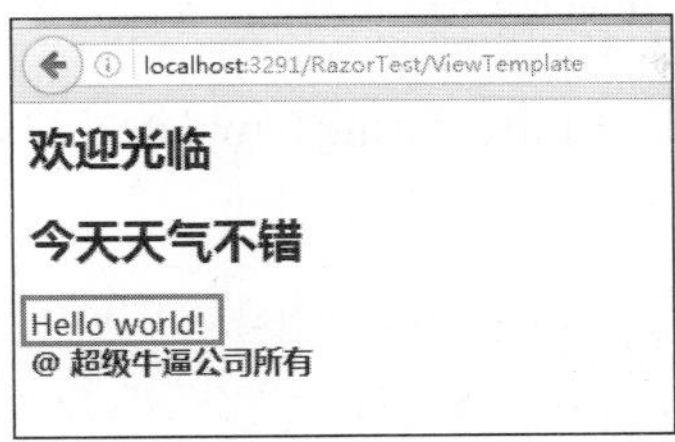

图 4-6

4.2.6　视图引擎

视图引擎仅仅是一个尖括号生成器而已，如图 4-7 所示。

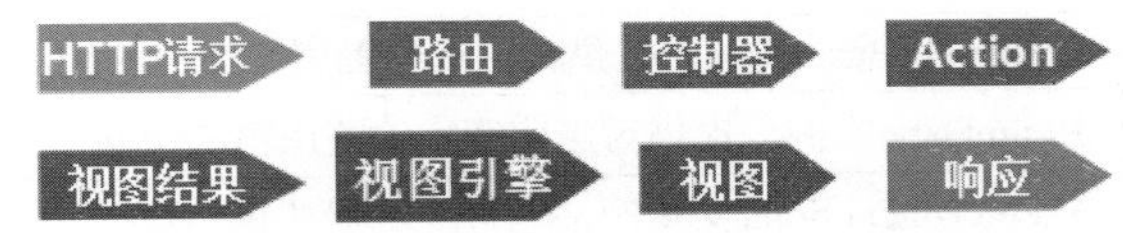

图 4-7

图 4-7 仅仅为了强调如下两点：

（1）视图引擎发挥作用的地方紧跟在 Action 方法执行后，目的是获取从控制器传递的数据，并生成经过格式化的输出。

（2）控制器并不渲染视图，它仅仅准备数据(Model)并通过 ViewResult 实例来决定调用哪个视图。

视图引擎接口 IViewEngine：

```
public interface IViewEngine {
     ViewEngineResult FindPartialView(ControllerContext controllerContext,
     string partialViewName, bool useCache);

     ViewEngineResult FindView(ControllerContext controllerContext, string
     viewName, string masterName, bool useCache);

     void ReleaseView(ControllerContext controllerContext, IView view);
}
```

ViewEngineResult 属性如表 4-1 所示。

表 4-1 ViewEngineResult 属性表

属性名	描述
View	返回与指定视图名称对应的并且可找到的 IView 实例或者 null
ViewEngine	如果能找到一个符合要求的视图，就返回一个 IViewEngine
SearchedLocations	返回一个 IEnumerable<string>，包含引擎搜索到的位置

其他视图引擎：Spart、NHaml、Brail、StringTemplate、NVelocity。

视图 IView：

```
public interface IView {
      void Render(ViewContext viewContext, TextWriter writer);
}
```

ViewPage 类的属性如表 4-2 所示。

表 4-2 ViewPage 类属性表

属性名	描述
HttpContext	HttpContextBase 实例，提供 ASP.NET 内部对象
Controller	ControllerBase 实例，提供对生成试图引擎调用的控制器的访问
RouteData	RouteData 实例，提供对当前请求的路由值的访问
ViewData	ViewDataDictionary 实例，包含控制器传递的数据
TempData	TempDataDictionary 实例，包含单个请求缓存中的数据
View	IView 实例，即将要渲染的视图
Writer	HtmlTextWriter

4.2.7 MVC 视图的“秘密”

其实我们的 cshtml 视图页面在被访问的时候也编译成了页面类，继承于 WebViewPage<T>。在 View 页面 ViewTemplate 中添加代码：

```
<div>@{Response.Write(this.GetType().Assembly.Location);}</div>
```

运行结果（不同机器运行结果会不一致）如下：

```
C:\Users\zouqi\AppData\Local\Temp\Temporary ASP.NET Files\vs\11bac735\53d0ffb7\
App_Web_b5yz5bvn.dll
```

用 reflector 反编译工具查看这个 dll：

```
namespace ASP
{
    [Dynamic(new bool[] { false, true })]
    public class _Page_Views_razortest_Index_cshtml : WebViewPage<object>
```

```
    [Dynamic(new bool[] { false, true })]
    public class _Page_Views_razortest_OutputTest_cshtml : WebViewPage<object>

    [Dynamic(new bool[] { false, true })]
    public class _Page_Views_razortest_PartialViewTest_cshtml :
    WebViewPage<object>

    [Dynamic(new bool[] { false, true })]
    public class _Page_Views_razortest_ViewTemplate_cshtml : WebViewPage<object>
}
```

可以看到 cshtml 页面里的所有代码都编译到这个类的 Execute 方法里了。这里以最后一个类为例：

```
public override void Execute()
{
    ((dynamic) base.ViewBag).Title = "OutputTest";
    base.BeginContext("~/Views/razortest/OutputTest.cshtml", 40, 2, true);
    this.WriteLiteral("\r\n");
    base.EndContext("~/Views/razortest/OutputTest.cshtml", 40, 2, true);
    HtmlString str = new HtmlString("<p>哈哈</p>");
    base.BeginContext("~/Views/razortest/OutputTest.cshtml", 0x66, 3, false);
    this.Write(str);
    base.EndContext("~/Views/razortest/OutputTest.cshtml", 0x66, 3, false);
    base.BeginContext("~/Views/razortest/OutputTest.cshtml", 0x6c, 2, true);
    this.WriteLiteral("\r\n");
    base.EndContext("~/Views/razortest/OutputTest.cshtml", 0x6c, 2, true);
    MvcHtmlString str2 = MvcHtmlString.Create("<p>哈哈~</p>");
    base.BeginContext("~/Views/razortest/OutputTest.cshtml", 0xae, 7, false);
    this.Write(str2);
    base.EndContext("~/Views/razortest/OutputTest.cshtml", 0xae, 7, false);
    base.BeginContext("~/Views/razortest/OutputTest.cshtml", 0xb8, 2, true);
    this.WriteLiteral("\r\n");
    base.EndContext("~/Views/razortest/OutputTest.cshtml", 0xb8, 2, true);
    base.BeginContext("~/Views/razortest/OutputTest.cshtml", 0xc3, 0x18, false);
    this.Write(base.Html.Raw("<p>哇哈哈哈~</p>"));
    base.EndContext("~/Views/razortest/OutputTest.cshtml", 0xc3, 0x18, false);
    base.BeginContext("~/Views/razortest/OutputTest.cshtml", 0xdf, 2, true);
    this.WriteLiteral("\r\n");
    base.EndContext("~/Views/razortest/OutputTest.cshtml", 0xdf, 2, true);
    base.BeginContext("~/Views/razortest/OutputTest.cshtml", 0xea, 0x10, false);
```

```
    this.Write(base.ViewData["Text"]);
    base.EndContext("~/Views/razortest/OutputTest.cshtml", 0xea, 0x10, false);
    base.BeginContext("~/Views/razortest/OutputTest.cshtml", 0xfd, 4, true);
    this.WriteLiteral("\r\n\r\n");
    base.EndContext("~/Views/razortest/OutputTest.cshtml", 0xfd, 4, true);
}
```

4.3 Controller

Controller 主要负责响应应用户的输入，主要关注的是应用程序流、输入数据的处理以及对相关视图（View）输出数据的提供。

Controller 的特征：

- 继承自 System.Web.Mvc.Controller，类 Controller 继承自 ControllerBase，实现了 IController 接口。
- 一个 Controller 可以包含多个 Action。每一个 Action 都是一个方法，返回一个 ActionResult 实例，当然也可以返回其他实例。
- 一个 Controller 对应一个 xxController.cs 控制文件，对应在 View 中有一个 xx 文件夹，一般情况下一个 Action 对应一个 View 页面。

4.3.1 Action 方法参数与返回值

Action 的本质就是类中的公有方法，可以进行重载，但是要求参数不同。Action 还可以通过路由规则传递数据。

Action 方法接收参数：接收浏览器传过来的参数（get、post 两种格式），如果希望某个方法只处理某一种请求，可以在方法前加特性[HttpGet]或[HttpPost]，处理请求时会根据参数进行相应方法的调用。

Post 数据接收方式包括 Request.Form 、FormCollection、同名参数、Model。

1. 通过 Request.Form["name"] 逐个获取表单提交的数据

新建 MVC 项目 MvcControllerApp。为了方便演示，这里使用 DataFirst 的方式连接 Northwind 数据库，右击 Models 文件夹，选择“添加新项→ADO.NET 实体数据模型”，并修改名称为 NorthwindDB.edmx。

添加 Home 控制器，添加 Action 方法 UpdateCustomerInfo：

```
public ActionResult UpdateCustomerInfo()
{
    return View();
}
```

```
[HttpPost]
public string UpdateCustomerInfo(FormCollection form)
{
    return Request.Form["ContactName"];
}
```

添加 UpdateCustomerInfo 视图：

```
@model MvcControllerApp.Models.Customers
@{
    ViewBag.Title = "UpdateCustomerInfo";
}
@using(Html.BeginForm())
{
    @Html.TextBoxFor(x => x.ContactName)
    <input type="submit"  value="修改"/>
}
```

2. 直接使用 FormCollection 来调用

这里在 Action 方法上面添加了一个标记[HttpPost]，表示这个 UpdateCustomerInfo 方法只能处理 Post 请求，并且优先处理 Post 请求。

```
[HttpPost]
public string UpdateCustomerInfo(FormCollection form)
{
   return form["ContactName"];;
}
```

3. 在 Action 中使用同名参数

如果方法的参数名称与表单元素的 name 属性名称一致就会自动填充。

```
[HttpPost]
public string UpdateCustomerInfo(string ContactName)
{
    return ContactName;
}
```

4. 接收 Model

通过实体对象一次性获取表单元素的数据，并设置到实体对象对应的属性中。注意：表单中的表单元素的属性名称必须和实体对象的属性一样，即匹配 input 的 name 与对象的属性相同名称的值。

```
[HttpPost]
```

```
public string UpdateCustomerInfo(Customers model)
{
    return model.ContactName+","+model.CompanyName;
}
```

接收 get 数据：

```
Request.QueryString
```

直接通过请求上下文对象里的 Request 获取 url ?后的参数：

- 浏览器请求路径：/User/UserList/1?kjy=jp。
- 控制器获取：Request.QueryString["kjy"];。

Action 方法输出返回值：ActionResult，返回控制器结果对象，直接或间接继承自 ActionResult 类型。

- ViewResult：使用 View()既可以指定一个页面，也可以指定传递的模型对象，如果没有指定参数就表示返回与 Action 同名的页面。
- ContentResult：使用 Content(string content)返回一个原始字符串。
- RedirectResult：使用 Redirect(string url)将结果转到其他 Action。
- JsonResult：使用 Json(object data)将 data 序列化为 json 数据并返回，推荐加上。
- JsonRequestBehavior.AllowGet：可以处理 Get 请求，一般结合客户端的 ajax 请求进行返回。

ActionResult 包含的内容如图 4-8 所示。

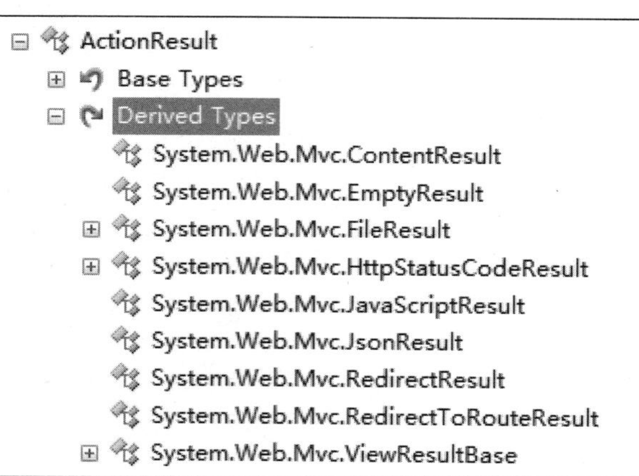

图 4-8

4.3.2 Action 指定使用视图

利用 Action 指定使用视图的方法如下：

```
public ActionResult Index()
{
    return View();//默认情况下不给参数返回和方法同名的视图，即使用视图 Index.cshtml 路径在
    //当前控制器对应的 View 目录下面
    return View("OtherIndex");//使用 OtherIndex.cshtml
    return View("~/Views/Home/Test.cshtml");
}
```

控制器将处理后的数据“传”给视图的方式包括 ViewData、ViewBag、TempData、Model。下面给出一个示例。

控制器代码：

```
public ActionResult Index()
{
        ViewBag.UserName = "小李飞刀";
        ViewData["UserName"] = "陆小凤";
        TempData["UserName"] = "楚留香";//临时数据
        Customers model = new Customers { ContactName = "谢晓峰" };

        return View(model); //这行代码其实就相当于 ViewData.Model=model
}
```

View 代码：

```
@{
   ViewBag.Title = "Index";
}

<div>@ViewBag.UserName </div>
<div>@ViewData["UserName"] </div>
<div>@TempData["UserName"] </div>
<div>@Model.ContactName</div>
```

注意：Model 其实就是 ViewData.Model。

大家可能觉得这十分显而易见，结果肯定是：

```
小李飞刀
陆小凤
楚留香
谢晓峰
```

其实，结果并不是你想象的那样，而是会显示为：

```
陆小凤
陆小凤
```

```
楚留香
谢晓峰
```

为什么呢？因为 ViewData 和 ViewBag 本质上都是 ViewDataDictionary 类型，并且两者之间数据共享，只是提供了不同的语法操作方式而已，所以“陆小凤”覆盖了原先的值“小李飞刀”。

TempData 只是用来临时存储的，存储一次就失效了。

在 4.1.1 小节中，我们已经使用反编译工具 Reflector 查看了源码，这里不再赘述。

Model 强类型参数：在控制器里的 Action 方法最后调用 View 加载视图的时候将数据对象传入。

```
return View(model);
```

在视图中可以通过 Model 属性获取，并且不需要转型就可以直接使用，如图 4-9 所示。

```
@{
    ViewBag.Title = "Index";
}

<div>@ViewBag.UserName </div>
<div>@ViewData["UserName"] </div>
<div>@TempData["UserName"] </div>
<div>@Model.ContactName</div>
```

图 4-9

注意：Model 虽然可以不转型就直接使用，但是因为编译器无法在编译时获取它的类型，所以无法出现智能提示。为了解决这个问题，可以在视图的最上面通过代码指定 Model 类型，如图 4-10 所示。

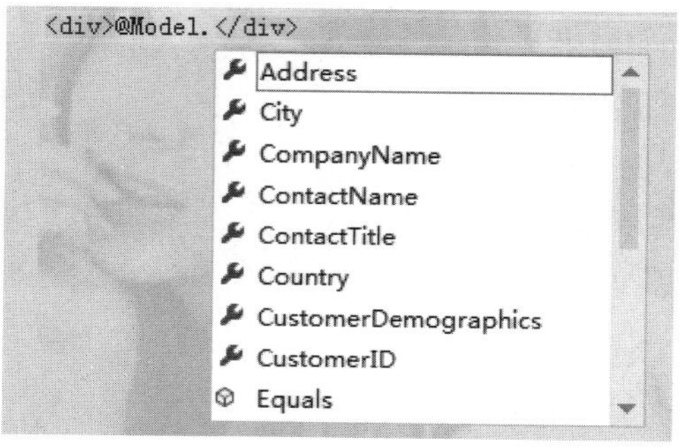

图 4-10

4.3.3 View 和 Controller 之间的关系

单击 View 中的 Model 进行跳转，结果如下：

```
public abstract class WebViewPage<TModel> : WebViewPage
{
```

```
    // Fields
    private ViewDataDictionary<TModel> _viewData;

    // Methods
    protected WebViewPage();
    public override void InitHelpers();
    protected override void SetViewData(ViewDataDictionary viewData);

    // Properties
    public AjaxHelper<TModel> Ajax { get; set; }
    public HtmlHelper<TModel> Html { get; set; }
    public TModel Model { get; }
    public ViewDataDictionary<TModel> ViewData { get; set; }
}
```

当在视图上添加了@model 指令时，当前视图就会继承于 WebViewPage<T> 强类型视图页面类，并且指定 T 为 Customers：

```
@model MvcControllerApp.Models.Customers
```

如果没有添加@model 指令，当前视图就会继承于 WebViewPage<T> 强类型视图页面类，但 T 变成了 dynamic。

这里，Controller 的数据怎么就传到 View 了，明明只给 Controller 中的 ViewData、ViewBag、TempData 赋值了，或者只把对象传给了 View 方法？为什么在 View 中就可以直接调用了呢？我们知道以前的 ASP.NET，aspx 和 aspx.cs 是一个继承关系，子类可以直接调用父类的属性方法。在 ASP.NET MVC 中 View 和 Controller 之间又有什么关系呢？还是使用反编译工具来查看一下源码，这样一目了然。

View()方法返回了一个 ViewResult 类，而这个 ViewResult 对象又继承自 ViewResultBase：

```
public class ViewResult : ViewResultBase
```

点开这个 ViewResultBase 类：

```
public abstract class ViewResultBase : ActionResult
{
    // Fields
    private DynamicViewDataDictionary _dynamicViewData;
    private TempDataDictionary _tempData;
    private ViewDataDictionary _viewData;
    private ViewEngineCollection _viewEngineCollection;
    private string _viewName;

    // Methods
```

```
    protected ViewResultBase();
    public override void ExecuteResult(ControllerContext context);
    protected abstract ViewEngineResult FindView(ControllerContext context);

    // Properties
    public object Model { get; }
    public TempDataDictionary TempData { get; set; }
    public IView View { get; set; }
    [Dynamic]
    public object ViewBag { [return: Dynamic] get; }
    public ViewDataDictionary ViewData { get; set; }
    public ViewEngineCollection ViewEngineCollection { get; set; }
    public string ViewName { get; set; }
}
```

View 和 Controller 中都有 ViewData、ViewBag、TempData、Model 这几个对象，在给 Controller 中的这些对象赋，Controller 会把这些值赋给 View 中对应的对象。

4.4 Model 和验证

Model（模型）就是处理业务想要保存、创建、更新、删除的对象。

4.4.1 Net MVC 请求处理流程

在讲解 Model 和验证之前，我们先来了解一下.NET MVC 的请求处理流程。

MVC 请求流程：客户端请求→IIS→Runtime→Controller→Action → ViewResult(:ActionResult).ExcuteResult()→ RazorView(:IView).RenderView→Response。

对.NET MVC 而言，因为请求都归结到 Action 上，所以是 URL 驱动，Action 跟 View 是松耦合的关系，所有的视图页面继承自 viewpage 类，包括数据、验证规则、数据访问和业务逻辑等应用程序信息，它是我们 MVC Web 应用的主框架。

4.4.2 MVC 模型验证

什么是 Model（模型）？Model 是独立的组件，它不知道 View 和 Controller 的存在，调用的是我们的业务逻辑层。Model 有时候可作为 ViewModel 使用。

对于模型验证，我们可以使用注解，而注解是通过特性来实现的。

以下是一些常用的注解属性：

- DisplayName

- Required
- StringLength(20,MinimumLength=2)
- DataType(System.ComponentModel.DataAnnotations.DataType.MultilineText)
- RegularExpression(@"^\w+([-+.]\w+)*@\w+([-.]\w+)*\.\w+([-.]\w+)*$",ErrorMessage="邮箱格式不对！")

这里通过一个实例来讲解如何使用 MVC 模型验证。

（1）新建一个 MVC 项目 MvcModelApp。

（2）在数据库 Northwind 中新建表 tb_User。

```
create table tb_User(
   ID int identity primary key,
   UserName varchar(20),
   Remark varchar(100),
   Age int not null,
   Pwd varchar(20),
   Email varchar(30)
)
```

（3）添加 ADO.NET 实体模型 ModelTest.edmx，选择从数据库生成，再选择 Northwind 数据进行连接，只勾选新建的 tb_User 表，就会自动生成一个部分类 tb_User。

```
public partial class tb_User
{
      public int ID { get; set; }
      public string UserName { get; set; }
      public string Remark { get; set; }
      public int Age { get; set; }
      public string Pwd { get; set; }
      public string Email { get; set; }
}
```

（4）添加模型注解。使用注解要添加如下两个命名空间的引用：

```
using System.ComponentModel;
using System.ComponentModel.DataAnnotations;
```

MVC 本身内置了一些常用的数据注解，如 Required、StringLength、Range 和[RegularExpression]等，如表 4-3 所示。

表 4-3　模型注解属性表

注解属性名称	注解说明
Required	验证必填
StringLength	验证字段的最大长度
Range	验证字段范围
RegularExpression	自定义验证规则
DisplayName	字段显示名称

由于这个 tb_User 是自动生成的模型类，因此我们不能直接把数据注解添加到这个类上面（自动生成的代码很容易被覆盖）。但是它是一个 partial，所以可以在 Models 文件夹中额外创建一个 partial 类 tb_User 和一个专门针对 tb_User 做数据注解的类 UserMetadata。UserMetadata 是 MetadataTypeAttribute 类型的类，为数据实体的 partial class 添加额外的元数据验证信息。

MetadataTypeAttribute Class 的使用步骤如下：

- 创建 data-model partial class。
- 创建关联的 metadata class。
- 将 metadata class 关联到具体的数据实体类。

MetadataTypeAttribute 使用的注意事项：

标签只能打到一个类上，此标签不能被继承，这里一定要注意，UserMetadata 这个类的命名空间要和自动生成的 tb_User 类的命名空间一致，这里都是 MvcModelApp，此处还使用到了邮箱，我们可以添加一个自定义的注解属性来验证邮箱的格式。要定义自己的定制验证特性，然后应用它们。你可以通过继承自 System.ComponentModel.DataAnnotations 命名空间中的 ValidationAttribute 基类定义完全定制的特性。

```
public class EmailAttribute : RegularExpressionAttribute
{
        public EmailAttribute()
           : base(@"^\w+([-+.]\w+)*@\w+([-.]\w+)*\.\w+([-.]\w+)*$")
        {
        }
}
[MetadataType(typeof(UserMetadata))]
public partial class tb_User
{
        public string RePwd { get; set; }
}
//为实体类添加特性 DisplayName
public class UserMetadata
{
        [DisplayName("用户名")]
```

```
        [Remote("NotExitesUserName", "Home")]
        public string UserName { get; set; }
        /// <summary>
        /// 在实体类中为 Remark 属性设置 DataType 特性，指定为多行文本框
        /// </summary>
        [DataType(DataType.MultilineText)]
        [DisplayName("备注")]
        public string Remark { get; set; }
        [DisplayName("年龄")]
        [Range(1, 120)]
        public int Age { get; set; }
        [PasswordPropertyText]
        [DisplayName("密码")]
        public string Pwd { get; set; }
        [PasswordPropertyText]
        [DisplayName("重输密码")]
        [System.Web.Mvc.Compare("Pwd")]
        public string RePwd { get; set; }
        [Email]
        public string Email { get; set; }
}
```

另外，模型里面我在用户名属性上面添加了一个 Remote 特性，这个是一个异步调用的属性，可以指定触发某个 action，返回值为 bool（true or false），比如注册用户时可用于验证用户名是否存在。

（5）添加 Home 控制器。

```
public class HomeController : Controller
{
        private NorthwindEntities db = new NorthwindEntities();
        //
        // GET: /Home/

        public ActionResult Register()
        {
            return View();
        }
        [HttpPost]
        public ActionResult Register(tb_User model)
        {
            if (ModelState.IsValid)
            {
```

```
                db.tb_User.Add(model);
                db.SaveChanges();
                return RedirectToAction("Index");
            }
            else
            {
                return View();
            }
        }
        [HttpGet]
        public JsonResult NotExitesUserName()
        {
            string UserName = Request.Params["UserName"];

            var user = db.tb_User.Where(x => x.UserName == UserName).FirstOrDefault();
            return user == null ? Json(true, JsonRequestBehavior.AllowGet) :
            Json(false, JsonRequestBehavior.AllowGet);
        }
        public ActionResult Index()
        {
            var user = db.tb_User.ToList();
            return View(user);
        }
}
```

（6）添加相应的视图。

添加 Register 视图。

```
@model MvcModelApp.tb_User
@{
    ViewBag.Title = "用户注册";
}
<script src="~/Scripts/jquery-1.8.2.min.js"></script>
<script src="~/Scripts/jquery.validate.min.js"></script>
<script src="~/Scripts/jquery.validate.unobtrusive.min.js"></script>

<h2>@ViewBag.Title</h2>
@using(Html.BeginForm("Register",null,FormMethod.Post))
{
<table><tr><td>@Html.DisplayNameFor(x=>x.UserName)                          :
</td><td>@Html.TextBoxFor(x=>x.UserName)
@Html.ValidationMessageFor(x=>x.UserName)</td></tr>
```

```
    <tr><td>@Html.DisplayNameFor(x=>x.Age)                                      :
</td><td>@Html.TextBoxFor(x=>x.Age)@Html.ValidationMessageFor(x=>x.Age)</td></tr>
    <tr><td>@Html.DisplayNameFor(x=>x.Pwd)                                      :
</td><td>@Html.PasswordFor(x=>x.Pwd)@Html.ValidationMessageFor(x=>x.Pwd)</td></tr
>
    <tr><td>@Html.DisplayNameFor(x=>x.RePwd)                                    :
</td><td>@Html.PasswordFor(x=>x.RePwd)@Html.ValidationMessageFor(x=>x.RePwd)</td>
</tr>
    <tr><td>@Html.DisplayNameFor(x=>x.Email)                                    :
</td><td>@Html.TextBoxFor(x=>x.Email)@Html.ValidationMessageFor(x=>x.Email)</td><
/tr>
    <tr><td>@Html.DisplayNameFor(x=>x.Remark)                                   :
</td><td>@Html.TextBoxFor(x=>x.Remark)@Html.ValidationMessageFor(x=>x.Remark)</td
></tr>
    <tr><td colspan="2"><input type="submit" value="注册"/></td></tr>
</table>
}
```

添加 Index 视图，在 Index 方法附近右击，选择“添加视图”，这里为了做 Demo 快速演示而选择了“List”支架模板，其实这个 VS 自带的模板功能我们真正在做项目的过程中几乎用不到，因为做项目时的界面一般都是按照美工提供的模板来做的。当然，如果有需要也可以根据美工提供的界面模板自己写一个支架模板，如图 4-11 所示。

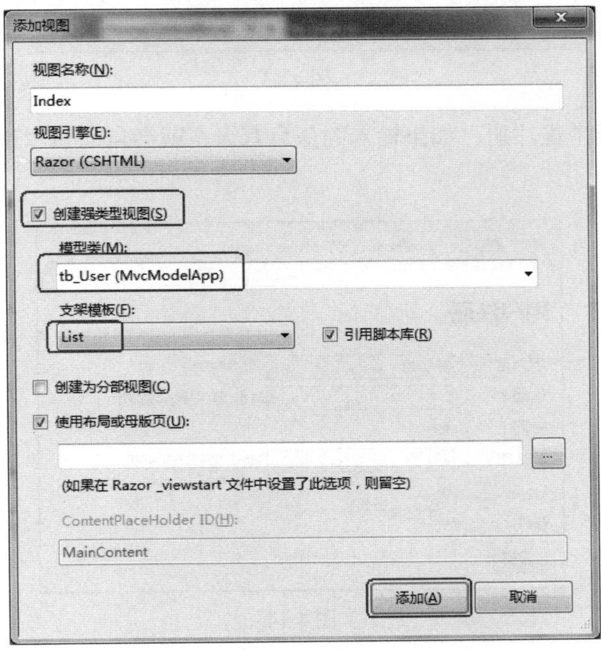

图 4-11

（7）在 Register 视图界面引入 JS 脚本支持。

```
<script src="~/Scripts/jquery-1.8.2.min.js"></script>
<script src="~/Scripts/jquery.validate.min.js"></script>
<script src="~/Scripts/jquery.validate.unobtrusive.min.js"></script>
```

然后生成项目，运行，这里先录入一条数据（按照正确的格式输入），如图 4-12 所示。

图 4-12

这个时候没有任何错误提示，单击“注册”按钮，注册成功，返回 Index 视图，如图 4-13 所示。

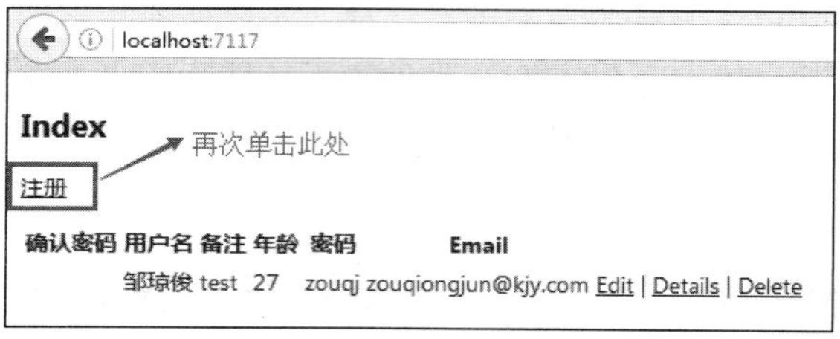

图 4-13

再次录入时不需要单击注册。如果输入的信息有误，就会自动显示错误信息，如图 4-14 所示。

图 4-14

关于邮箱格式验证，要单击“注册”按钮才会显示，如图 4-15 所示。

图 4-15

这是怎么做到的呢？实在是太神奇了，因为从头到尾，我们没有写一行 js 代码。在过去，这样的异步校验都是通过一大串 js 或者 jquery 代码来实现的。

首先，我们在 tb_User 类上面添加各种注解属性，接着在 View 中使用 HtmlHelper 来生成 html 代码，那么我们在生成 html 代码的时候会根据添加的注解属性来生成一些特殊的标签，查看 HTML 源码，如图 4-16 所示。

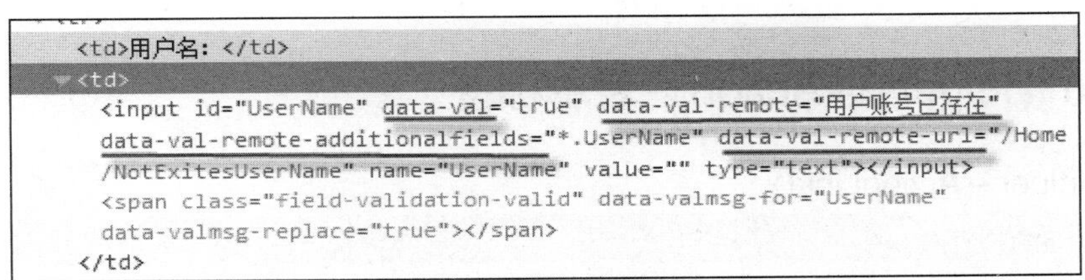

```
<td>用户名: </td>
<td>
  <input id="UserName" data-val="true" data-val-remote="用户账号已存在"
  data-val-remote-additionalfields="*.UserName" data-val-remote-url="/Home
  /NotExitesUserName" name="UserName" value="" type="text"></input>
  <span class="field-validation-valid" data-valmsg-for="UserName"
  data-valmsg-replace="true"></span>
</td>
```

图 4-16

我们可以看到，生成的文本框里面多了许多特殊的标签，而我们在 View 视图中又添加了 <script src="~/Scripts/jquery.validate.unobtrusive.min.js"></script>的 js 文件引用，从名字中我们不难看出，这个 js 实现了隐式验证，也就是说在这个 js 里面它会根据这些特殊的标签来实现相应的验证。

4.5 HtmlHelper

HtmlHelper 是为了方便 View 的开发而产生的。

为什么会出现 HtmlHelper ？我们来看一下 HtmlHelper 的演变。

在 MVC 中，普通首页超级链接为“<a href="/home/index">首页</a>”，当路由改变时可能需要修改为“<a href="/home_ yujie/index">首页</a>”，如果项目里面有很多超级链接就需要改动很多地方。

我们需要的是路由改变也不受影响：<a href="<%=Url.Action("Index","Home")%>">首页</a>。因为没有智能感知，调试不方便，所以应运而生了 Html.Action("Home","Index")。

（1）在 RouteConfig 中修改默认路由：

```
//默认路由信息
routes.MapRoute(
```

```
    name: "Test",
    url: "{controller}_yujie/{action}.html/{id}",
            defaults: new { controller = "Home", action = "Index", id =
            UrlParameter.Optional }
);
```

（2）在视图上创建超链接：

```
<a href="/Home/Index">网站首页</a><br />
@Html.ActionLink("网站首页","Index", "Home")
```

（3）在浏览器中的 HTML 源码：

```
<a href="/Home/Index">网站首页</a><br />
<a href="/Home_yujie/Index.html">网站首页</a>
```

4.5.1 HtmlHelper 的 Action、表单标签

1. HtmlHelper – ActionLink()

```
HtmlHelper - ActionLink()
```

动态生成超链接：根据路由规则，生成对应的 HTML 代码。

2. HtmlHelper – Form

动态生成表单标签。

方式一：{}　　（强烈推荐）

```
@using(Html.BeginForm("HandleForm", "Home"))
{

}
```

方式二：Begin　End

```
@Html.BeginForm()
@{Html.EndForm();}
```

（1）在视图中创建表单

```
@using (Html.BeginForm("Add", "home", FormMethod.Post, new { id="form1" }))
{

}
```

（2）生成的 HTML 代码

```
<form action="/home/Add" id="form1" method="post"> </form>
```

直接在视图中用@调用有返回值的方法，就已经相当于是将返回值写入 Response 了：

```
@Html.Label("UserName")
```

相当于下面的代码：

```
@{
    Response.Write(Html.Label("UserName"));
}
```

4.5.2　HtmlHelper 的弱类型与强类型方法

1. 弱类型方法

Html.xx("xx")，指定 name 和 valueLable()、TextBox()等生成 html 标签方法。

```
@Html.Label("Id");
@Html.TextBox("Id")
DropDownList
var items = new List<SelectListItem>()
{
    (new SelectListItem() {Text = "001", Value = "1", Selected = false}),
    (new SelectListItem() {Text = "002", Value = "2", Selected = false})
};
```

在控制器中将 items 值赋给 ViewData：

```
ViewData["items"] = items;
```

在视图中这样使用：

```
@Html.DropDownList("items")
CheckBox & RadioButton & Hidden & Password
@Html.CheckBox("Role")
```

在 Controller 获取提交的值为：(true,false)未操作、true 选择、false 不选择三种。

RadioButton 第一个参数为控件名字（名字相同说明为同一个 Group），返回值为第二个参数：

```
@Html.RadioButton("Sex","male",true)男<br />
@Html.RadioButton("Sex","female",false)女
@Html.Hidden("Id")
```

```
@Html.Password("pwd")
```

注：所有的方法都默认去视图的 Model 属性所储存的对象中查找匹配属性。

例如，在 Models 文件夹中添加 User 类，引入命名空间：

```
using System.ComponentModel;
```

（1）为实体类添加特性 DisplayName：

```
public class User
{
      [DisplayName("用户名")]
      public string UserName { get; set; }
}
```

（2）在 Action 方法中为视图的 Model 设置值：

```
public ActionResult Index()
{
      return View(new User { UserName = "郭靖" });
}
```

（3）在视图中通过 HTML 的帮助方法，生成 HTML 标签，同时指定要读取的属性名：

```
@Html.Label("UserName")
```

（4）生成对应的 HTML 标签，并自动读取对应属性的 DisplayName 里的文本：

```
<label for="UserName">用户名</label>
```

其他控件方法的使用也一样。

2. 强类型方法

强类型方法直接通过 lambda 表达式，去视图的 Model 属性对象中查找对应的属性数据。以 Html.xxFor 这种方式生成的 HTML 控件的 name 和指定实体对应的属性名一致。

（1）在视图上调用方法：

```
@Html.TextBoxFor(m=>m.UserName);
```

（2） 生成的 HTML 代码：

```
<input type="text" id="UserName" name="UserName" value="郭靖"/>
```

3. 实体属性附加类型

通过属性的 DataType 特性生成 HTML 标签。

（1）在 User 类中，添加命名空间引用：

```
using System.ComponentModel.DataAnnotations;
```

（2）在实体类中为 Remark 属性设置 DataType 特性，指定为多行文本框：

```
[DataType(DataType.MultilineText)]
public string Remark { get; set; }
```

（3）在视图上自动根据 model 对象里属性保存的实体类属性的[DataType] 特性里指定的类型生成对应的 HTML 标签：

```
@Html.EditorFor(a=> a.Remark)
```

（4）生成 HTML 代码：

```
<textarea id="Remark" class="text-box multi-line" name="Remark"></textarea>
HtmlHelper - LabelFor & 模型元数据
```

4. 模型类的元数据

模型类的元数据包括属性（名称和类型）与特性包含的值。

（1）为实体类属性设置 DisplayName 特性：

```
[DisplayName("用户名")]
public string UserName{ get; set; }
```

（2）在“新增/修改”页面上显示某个属性的标签说明：

```
@Html.LabelFor(model => model.UserName)
```

（3）生成 HTML 源码：

```
<label for="UserName">用户名</label>
HtmlHelper - Display / Editor 模型元数据
```

@Html.Editor / @Html.Display 可以通过读取特性值生成 HTML。

（1）在控制器中：

```
public ActionResult Index()
{
    //ViewBag.UserName = "小李飞刀";
    return View(new User { UserName = "郭靖",Remark="武林高手" });
}
```

（2）在“新增/修改”页面上显示某个属性的 input 标签：

```
<div>@Html.DisplayFor(model => model.Remark):@Html.EditorFor(a=> a.Remark)</div>
```

（3）生成 HTML 源码：

```
<div>武林高手:<textarea id="Remark" class="text-box multi-line" name="Remark">武林高手</textarea></div>
```

4.5.3 HtmlHelper 的 RenderPartial

从分部视图里取数据：

```
@Html.RenderPartial( "Top", ViewData.Model );
@Html.RenderPartial( "~/Views/Home/ Top.cshtml ", ViewData.Model );
```

直接将用户控件嵌入到界面上这个方法通过接受分部视图的文件名以及相应的可变化的数据进行的呈现可重用。分部视图和模板页显示到具体的页面中的区别是：布局页面（模板页）子页是把自己的内容填到布局页面上去；分部视图是供某个页面过来取分部视图的内容。

4.5.4 HtmlHelper 的 RenderAction

在视图中请求某个 Action 方法（违反了 MVC 设计）：

```
@{Html.RenderAction("Test");}
```

允许你直接调用某个 Action，并把返回的结果直接显示在当前调用的 View 中。

1. 两者的相同点

- RenderPartial 和 RenderAction 通常都被用来显示一个功能相对独立的“块”，比如说显示菜单或者导航条。
- 两者输出的结果都被作为调用的 View 的一部分显示。

2. 两者的不同点

- RenderPatial 的数据来自于调用的 View，而 RenderAction 来自自己。
- RenderAction 会发起一个新的 Request，而 RenderPatial 不会。

3. 如何选择

根据两者不同点中的第二点，由于 RenderAction 会调用一个新的 Action 方法，而 ASP.NET MVC 中 Action 是最小的缓存单位，因此如果某一个“块”的数据比较固定，不会因为访问者的不同而发生变化，那么这时就是使用 RenderAction 的时候了。

题外话：RenderAction 会发起一个新的 Request，感觉对调用页面的流程有点破坏。一个 View 在显示的时候，自己又发起一个 Request 去获取数据来显示，显然有点破坏作为一个 View 的原则：A View should only know how to render, but not what to render!

4.5.5　HtmlHelper 扩展方法

扩展方法：

- 方法所在的类必须是静态的。
- 方法也必须是静态的。
- 方法的第一个参数必须是要扩展的那个类型，比如要给 int 扩展一个方法，那么第一个参数就必须是 int。
- 在第一个参数前面还需要有一个 this 关键字。

在 MVC 中扩展 HtmlHelper 后，要在使用扩展方法的页面上引用扩展方法所在的名称空间：

```
public static class UserHelper
{
    public static string UserShow(this HtmlHelper<User> helper)
    {
        var user = helper.ViewData.Model;
        if (user.UserName == "郭靖")
        {
            return string.Format("<div>我是{0}</div>", user.UserName);
        }
        else
        {
            return "<div>找不到</div>";
        }
    }
}
```

控制器：

```
public ActionResult Index()
{
    return View(new User { UserName = "郭靖" });
}
```

View 调用：

```
@Html.Raw(Html.UserShow())
```

界面显示：

```
我是郭靖
```

4.6 ASP.NET MVC 分页

说起分页，基本上是我们 Web 开发中遇见得最多的场景，没有之一。即便如此，要做出比较优雅的分页还是需要技巧的。ASP.NET MVC 中很常见的分页实现方式有以下几种。

- HtmlHelper。
- 局部视图。
- MvcPager。
- 第三方 UI 组件。

4.6.1 HtmlHelper

（1）新建 ASP.NET MVC 站点，将项目命名为 MvcAppPager。

（2）在 Models 文件夹中新建类 PageOfList。

在类中新建接口 IPageOfList：

```
public interface IPageOfList
{
        long CurrentStart { get; }
        int PageIndex { get; set; }
        int PageSize { get; set; }
        int PageTotal { get; }
        long RecordTotal { get; set; }
}
```

PageOfList 实现接口 IPageOfList：

```
public interface IPageOfList<T> : IPageOfList, IList<T>
{
}
public class PageOfList<T> : List<T>, IList<T>, IPageOfList, IPageOfList<T>
{
        public PageOfList(IEnumerable<T> items, int pageIndex, int pageSize, long
        recordTotal)
        {
            if (items != null)
                AddRange(items);
            PageIndex = pageIndex;
            PageSize = pageSize;
            RecordTotal = recordTotal;
```

```
        }

        public PageOfList(int pageSize)
        {
            if (pageSize <= 0)
            {
                throw new ArgumentException("pageSize must gart 0", "pageSize");
            }
        }

        public int PageIndex { get; set; }

        public int PageSize { get; set; }

        public int PageTotal
        {
            get
            {
                return (int)RecordTotal / PageSize + (RecordTotal % PageSize > 0 ? 1 : 0);
            }
        }

        public long RecordTotal { get; set; }

        public long CurrentStart
        {
            get
            {
                return PageIndex * PageSize + 1;
            }
        }

        public long CurrentEnd
        {
            get
            {
                return (PageIndex + 1) * PageSize > RecordTotal ? RecordTotal :
(PageIndex + 1) * PageSize;
            }
        }
}
```

（3）新建一个实体类 Order。

```
public class Order
{
        public int ID { get; set; }
        public string OrderNo { get; set; }
        public decimal WayFee { get; set; }
        public string EMS { get; set; }
}
```

（4）新建一个 Home 控制器和一个名为 Index 的 Action 方法：

```
List<Order> list = new List<Order> { new Order { ID = 1, OrderNo =
        "2016050501", WayFee = 20,EMS="C01111" },
            new Order { ID = 2, OrderNo = "2016050502", WayFee = 10,EMS="C01222" },
            new Order { ID = 3, OrderNo = "201605050203", WayFee =
            10,EMS="C01222" }, new Order { ID = 4, OrderNo = "201605050204",
            WayFee = 10,EMS="C01222" },
            new Order { ID = 5, OrderNo = "201605050205", WayFee =
            10,EMS="C01222" }, new Order { ID = 6, OrderNo = "201605050206",
            WayFee = 10,EMS="C01222" }};
private const int PageSize = 2;
private int counts;

public ActionResult Index(int pageIndex = 0)
{
            counts = list.Count;
            list = list.Skip(PageSize * pageIndex).Take(PageSize).ToList();
            PageOfList<Order> _orderList = new PageOfList<Order>(list, pageIndex,
            PageSize, counts);
            return View(_orderList);
}
```

（5）添加一个分页扩展方法。

添加命名空间引用：

```
using System.Text;
using System.Web.Mvc;
using System.Web.Mvc.Html;

public static class ExtHelper
{
        public static MvcHtmlString UlPaging(this HtmlHelper helper, IPageOfList list)
        {
```

```
StringBuilder sb = new StringBuilder();

if (list == null)
{
    return new MvcHtmlString(sb.ToString());
}
sb.AppendLine("<div class=\"fenye\">" + string.Format("<span>共 {0} 条
记录，每页 {1} 条  </span>", list.RecordTotal, list.PageSize));
System.Web.Routing.RouteValueDictionary route = new
System.Web.Routing.RouteValueDictionary();
foreach (var key in helper.ViewContext.RouteData.Values.Keys)
{
    route[key] = helper.ViewContext.RouteData.Values[key];
}

foreach (string key in
helper.ViewContext.RequestContext.HttpContext.Request.QueryString)
{
    route[key]                                                        =
helper.ViewContext.RequestContext.HttpContext.Request.QueryString[key];
}

if (list.PageIndex <= 0)
{
    sb.AppendLine("<a class=\"backpage\" href=\"javascript:void(0);\">上一页</a>");
}
else
{
    route["pageIndex"] = list.PageIndex - 1;

    sb.AppendLine(helper.ActionLink("上一页", route["action"].ToString(),
    route).ToHtmlString());
}

if (list.PageIndex > 3)
{
    route["pageIndex"] = 0;
    sb.AppendLine(helper.ActionLink(@"<b>1</b>",
    route["action"].ToString(), route).ToHtmlString().Replace("&lt;",
    "<").Replace("&gt;", ">"));
    if (list.PageIndex >= 5)
    {
```

```
            sb.AppendLine("<a href='#'>..</a>");
        }
    }

    for (int i = list.PageIndex - 2; i <= list.PageIndex; i++)
    {
        if (i < 1)
            continue;
        route["pageIndex"] = i - 1;

        sb.AppendLine(helper.ActionLink(@"<b>" + i.ToString() + @"</b>",
        route["action"].ToString(), route).ToHtmlString().Replace("&lt;",
        "<").Replace("&gt;", ">"));
    }

    sb.AppendLine(@"<a class='active' href='#'><b>" + (list.PageIndex + 1)
    + @"</b></a>");
    for (var i = list.PageIndex + 2; i <= list.PageIndex + 4; i++)
    {
        if (i > list.PageTotal)
            continue;
        route["pageIndex"] = i - 1;
        sb.AppendLine(helper.ActionLink(@"<b>" + i.ToString() + @"</b>",
        route["action"].ToString(), route).ToHtmlString().Replace("&lt;",
        "<").Replace("&gt;", ">"));
    }

    if (list.PageIndex < list.PageTotal - 4)
    {
        if (list.PageIndex <= list.PageTotal - 6)
        {
            sb.AppendLine("<a href='#'>..</a>");
        }
        route["pageIndex"] = list.PageTotal - 1;

        sb.AppendLine(helper.ActionLink(@"<b>" + list.PageTotal.ToString() +
        "</b>", route["action"].ToString(),
        route).ToHtmlString().Replace("&lt;", "<").Replace("&gt;", ">"));
    }
    if (list.PageIndex < list.PageTotal - 1)
    {
        route["pageIndex"] = list.PageIndex + 1;
```

```
            sb.AppendLine(helper.ActionLink("下一页", route["action"].ToString(),
            route).ToHtmlString());
        }
        else
        {
         sb.AppendLine("<a class=\"nextpage\" href=\"javascript:void(0);\">下一页</a>");
        }
        sb.AppendLine("</div>");
        return new MvcHtmlString(sb.ToString());
    }
}
```

（6）添加 Index 视图。

```
@model MvcAppPager.Models.PageOfList<MvcAppPager.Models.Order>
@{
   ViewBag.Title = "Index";
}
@Styles.Render("~/Content/page.css")
<div id="body" style="width:400px;float:left">
@using (Html.BeginForm("Index", "Home", FormMethod.Get))
{
   <table>
       <tr>
           <th>ID</th>
           <th>订单号</th>
           <th>运单号</th>
           <th>运费</th>
       </tr>
       @if (Model != null && Model.Count > 0)
       {
           foreach (var item in Model.ToList())
           {
           <tr>
               <td>@item.ID</td>
               <td>@item.OrderNo</td>
               <td>@item.EMS</td>
               <td>@item.WayFee</td>
           </tr>
           }
       }
   </table>
   @MvcAppPager.Models.ExtHelper.UlPaging(this.Html, Model)
```

```
}
</div>
```

（7）运行结果如图 4-17 所示。

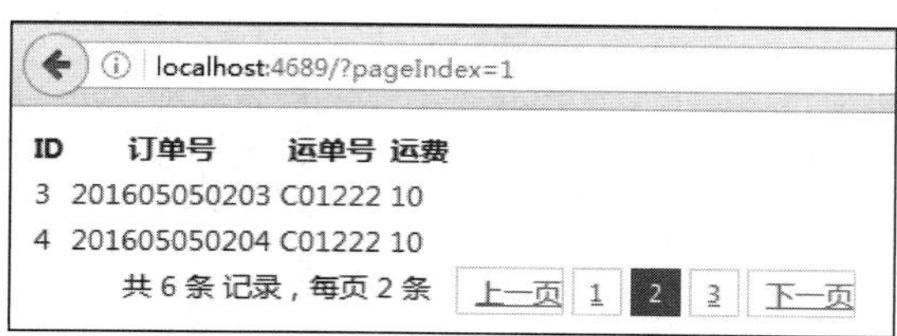

图 4-17

4.6.2 局部视图

如果你觉得这样拼接字符串麻烦，可以改用部分视图来实现。部分视图要建立在根目录 View 下面的文件夹中，因为这样可以达到共用的效果。部分视图就跟我们过去做 WebForm 开发里面的用户控件一样。

右击 Shared 文件夹，选择“添加→视图”，如图 4-18 所示。

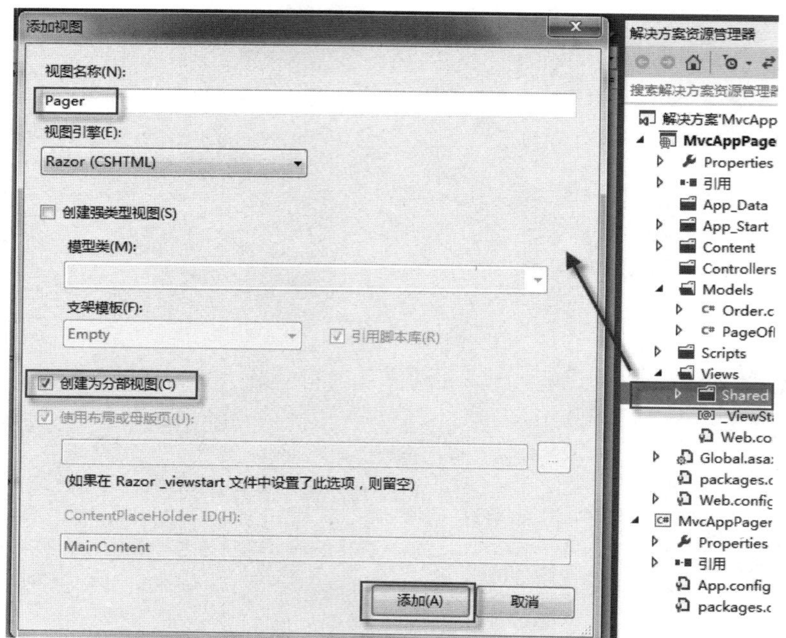

图 4-18

```
@model MvcAppPager.Models.IPageOfList
@Styles.Render("~/Content/page.css")
<div class="fenye"><span>共 @Model.RecordTotal 条 记录， 每页 @Model.PageSize 条
 </span>
```

```
@{
    System.Web.Routing.RouteValueDictionary route = new
    System.Web.Routing.RouteValueDictionary();
    foreach (var key in Html.ViewContext.RouteData.Values.Keys)
    {
        route[key] = Html.ViewContext.RouteData.Values[key];
    }

    foreach (string key in
    Html.ViewContext.RequestContext.HttpContext.Request.QueryString)
    {
        route[key] = Html.ViewContext.RequestContext.HttpContext.Request.QueryString[key];
    }
    if (Model.PageIndex <= 0)
    {
        <a class="backpage" href="javascript:void(0);">上一页</a>
    }
    else
    {
        route["pageIndex"] = Model.PageIndex - 1;
        Html.ActionLink("上一页", route["action"].ToString(),
        route).ToHtmlString();
    }

    if (Model.PageIndex > 3)
    {
        route["pageIndex"] = 0;
        Html.ActionLink("<b>1</b>", route["action"].ToString(),
        route).ToHtmlString().Replace("&lt;", "<").Replace("&gt;", ">");

        if (Model.PageIndex >= 5)
        {
            <a href='#'>..</a>;
        }
    }

    for (int i = Model.PageIndex - 2; i <= Model.PageIndex; i++)
    {
        if (i < 1)
        {continue;}
        route["pageIndex"] = i - 1;
        @Html.ActionLink(i.ToString(), route["action"].ToString(),
```

```
        route["controller"]);

    }

    <a class='active' href='#'><b> @(Model.PageIndex+1) </b></a>
    for (var i = Model.PageIndex + 2; i <= Model.PageIndex + 4; i++)
    {
        if (i > Model.PageTotal)
        {continue;}
        else{
        route["pageIndex"] = i - 1;
        @Html.ActionLink(i.ToString(), route["action"].ToString(), route);
        }
    }

    if (Model.PageIndex < Model.PageTotal - 4)
    {
        if (Model.PageIndex <= Model.PageTotal - 6)
        {
            <a href='#'>..</a>
        }
        route["pageIndex"] = Model.PageTotal - 1;
        @Html.ActionLink(Model.PageTotal.ToString(), route["action"].ToString(),
        route).ToHtmlString();
    }
    if (Model.PageIndex < Model.PageTotal - 1)
    {
        route["pageIndex"] = Model.PageIndex + 1;
        Html.ActionLink("下一页", route["action"].ToString(),
        route).ToHtmlString();
    }
    else
    {
        <a class="nextpage" href="javascript:void(0);">下一页</a>
    }
  }
</div>
```

其他地方和 HtmlHelper 方式一样，只需要在 Index 视图中修改分页调用方式，将如下代码：

```
@MvcAppPager.Models.ExtHelper.UlPaging(this.Html, Model)
```

修改为：

```
Html.RenderPartial("pager", Model);
```

即可。

在以上两种方式中，我们在 Index 视图上面调用的时候，每次单击页码，都会修改浏览器 URL 上的 pageIndex 参数值，也就是说每次分页都刷新整个界面，其实我们只需要改为异步表单调用就可以实现局部刷新了，如@Ajax.BeginForm。

后面会单独讲解 MVC 中的 Ajax 应用。

4.6.3 MvcPager

除了上述两种闭门造车的方式，我们还可以学习拿来主义，使用现成的分页控件。

MvcPager 分页控件是在 ASP.NET MVC Web 应用程序中实现分页功能的一系列扩展方法，该分页控件的最初如下实现方法借鉴了网上流行的部分源代。

MvcPager 的主要功能如下：

- 实现最基本的 url route 分页功能。
- 支持手工输入或选择页索引并对输入的页索引进行有效性验证。
- 支持使用 jQuery 实现 Ajax 分页，生成的 HTML 代码更精简。
- Ajax 分页模式下支持在分页过程中通过 GET 或 POST 方法提交表单数据，实现查询功能。
- Ajax 分页模式下支持浏览器历史记录功能（暂不支持 IE 7 及早期版本和 Opera 浏览器）。
- 在 Ajax 分页模式下，当客户端浏览器不支持或禁用 Javascript 功能时优雅降级为普通分页。
- 搜索引擎友好，无论是普通分页还是 Ajax 分页，搜索引擎都可以直接搜索到所有页面。
- 支持最新的 ASP.NET MVC 4.0 或更高版本。
- 支持 IE、Firefox、Opera、Chrome 及 Safari 等常用浏览器。
- 支持调用客户端 Javascript API 跳转到指定页（3.0 版新增）。
- 多语言支持（简体中文、繁体中文和英文）（3.0 版新增）。

下面我们看一个演示。

（1）添加 MvcPager.dll 的引用。本书云盘中有提供（Chapter04\Chapter04），当然也可以从网上下载。

（2）Controller 中的方法 MvcPager 代码如下：

```
public ActionResult MvcPager(int id = 1)
{
    counts = list.Count;
    PagedList<Order> lst = list.AsQueryable().ToPagedList(id, PageSize);
    lst.TotalItemCount = counts;
```

```
    lst.CurrentPageIndex = id;
    return View(lst);
}
```

（3）在 View 中使用 MvcPager：

```
@model PagedList<MvcAppPager.Models.Order>
 @using Webdiyer.WebControls.Mvc;
 <table>
        <tr>
            <th>ID</th>
            <th>订单号</th>
            <th>运单号</th>
            <th>运费</th>
        </tr>
        @if (Model != null && Model.Count > 0)
        {
            foreach (var item in Model.ToList())
             {
            <tr>
                <td>@item.ID</td>
                <td>@item.OrderNo</td>
                <td>@item.EMS</td>
                <td>@item.WayFee</td>
            </tr>
            }
        }
   </table>
@Html.Pager(Model, new PagerOptions
{
    PageIndexParameterName = "id",
    ShowPageIndexBox = true,
    FirstPageText = "首页",
    PrevPageText = "上一页",
    NextPageText = "下一页",
    LastPageText = "末页",
    PageIndexBoxType = PageIndexBoxType.TextBox,
    PageIndexBoxWrapperFormatString = "请输入页数{0}",
    GoButtonText = "转到"
})
    <br />
    >>分页 共有 @Model.TotalItemCount 条订单 @Model.CurrentPageIndex/@Model.TotalPageCount
```

（4）运行结果如图 4-19 所示。

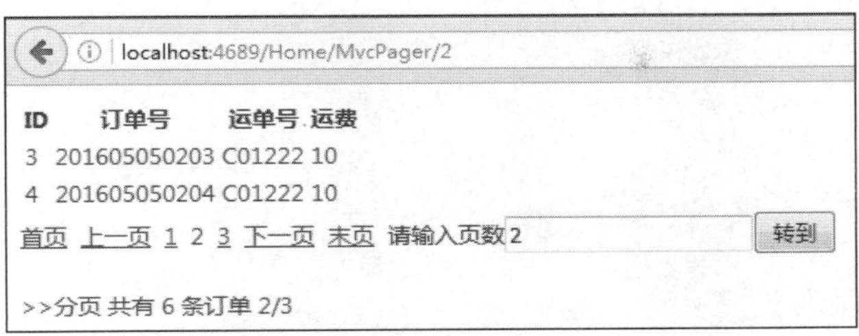

图 4-19

MvcPager 无刷新分页

在某一些场景下，我们不想每次单击分页都刷新整个界面，可以使用 ajax 进行局部更新。使用 MvcPager 进行局部更新的操作如下：

（1）在根目录 View/Shared 中，新建一个局部视图 OrderList.cshtml。右击“Shared”文件夹，选择“添加→视图”，选中“创建为分部视图”。

```
@model PagedList<MvcAppPager.Models.Order>
@using Webdiyer.WebControls.Mvc;
<div id='divList'>
<table>
        <tr>
            <th>ID</th>
            <th>订单号</th>
            <th>运单号</th>
            <th>运费</th>
        </tr>
        @if (Model != null && Model.Count > 0)
        {
            foreach (var item in Model.ToList())
            {
            <tr>
                <td>@item.ID</td>
                <td>@item.OrderNo</td>
                <td>@item.EMS</td>
                <td>@item.WayFee</td>
            </tr>
            }
        }
</table>

@Ajax.Pager(Model, new PagerOptions
```

```
{
    PageIndexParameterName = "id",
    ShowPageIndexBox = true,
    FirstPageText = "首页",
    PrevPageText = "上一页",
    NextPageText = "下一页",
    LastPageText = "末页",
    PageIndexBoxType = PageIndexBoxType.TextBox,
    PageIndexBoxWrapperFormatString = "请输入页数{0}",
    GoButtonText = "转到"
 }, new AjaxOptions { UpdateTargetId = "divList" })
    <br />
    >>分页 共有 @Model.TotalItemCount 条订单
    @Model.CurrentPageIndex/@Model.TotalPageCount
</div>
```

我们看到@Ajax.Pager...这里的代码在前面使用的是@Html.Pager，而且在这里，多了一句代码：new AjaxOptions { UpdateTargetId = "divList" }。这句代码的意思是，指定要更新的容器 ID，这里指定的更新容器是<div id='divList'>。一定要注意的是@Ajax.Pager...的代码必须放置在 UpdateTargetId 所指向的容器之内，这里就是放在<div id='divList'>之内。

（2）在 Home 控制器中添加 Action 方法 MvcPagerAjax。

```
public ActionResult MvcPagerAjax(int id = 1)
{
        counts = list.Count;
        PagedList<Order> lst = list.AsQueryable().ToPagedList(id, PageSize);
        lst.TotalItemCount = counts;
        lst.CurrentPageIndex = id;
        if (Request.IsAjaxRequest())//如果是 Ajax 请求
        {
            return PartialView("OrderList", lst);
        }
        else
        {
            return View(lst);
        }
}
```

（3）添加视图 MvcPagerAjax，为了方便展示局部更新，这里特意加了一行代码显示当前时间。

还有就是，这里要添加 js 引用，因为要用到 ajax。

```
@model PagedList<MvcAppPager.Models.Order>
```

```
@using Webdiyer.WebControls.Mvc;
@Scripts.Render("~/Scripts/jquery-1.8.2.min.js")
@Scripts.Render("~/Scripts/jquery.unobtrusive-ajax.min.js")
@{
    Html.RenderPartial("OrderList", Model);
}
<div>@DateTime.Now.ToString()</div>
```

（4）运行结果如图 4-20 所示。

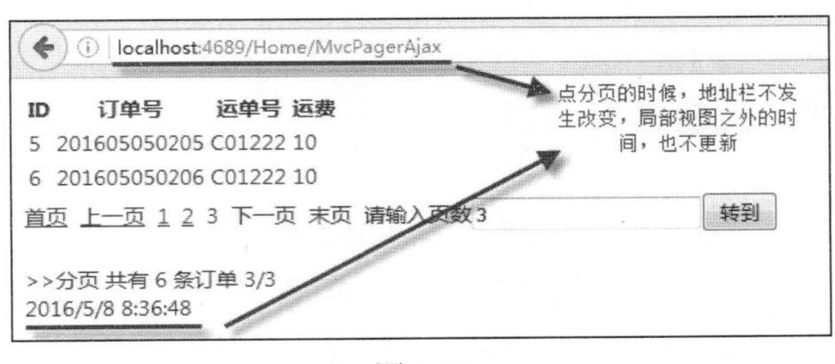

图 4-20

4.6.4　第三方 UI 组件

如果我们在 MVC 项目中使用了一些第三方的 UI 框架或者 UI 组件，可能就用不到 HtmlHelper 和局部视图了，因为许多第三方的 UI 框架和 UI 组件都自带分页功能，而且十分强大，诸如 EasyUI、ExtJS、MiniUI 等。用过第三方 UI 控件的都知道，Web 后台开发人员可以节省许多调整样式和用原生 js 或者 jQuery 实现一些功能的时间，也基本上用不着去考虑各种兼容性问题，因为这些都由 UI 组件替我们实现了。

在后面的项目中会基于 jQuery 表格组件 dataTables 来进行分页和展示，本章暂时不讲。dataTables 的更多资料请参考 http://dt.thxopen.com/example/。站在一个使用者的角度来说，这些第三方的 UI 组件和框架非常简单，因为通常成熟的 UI 框架和组件都有十分详细的文档和各种各样的 Demo，网上也有许多人发表的相关文章。

第 5 章

MVC核心透析

本章主要介绍 MVC 的一些高级特性和核心技术。

5.1 MVC Routing

在项目要使用 Routing（路由），需要添加程序集引用 System.Web.Routing.dll，而在 ASP.NET MVC 项目中则自动添加了此引用。严格来说，System.Web.Routing.dll 并不属于 ASP.NET MVC 的一部分，微软把此项目单独列出来了，并没有开放源代码，但是未来微软也许会把 Routing 开源，毕竟开源是大势所趋，而且微软也在不断地拥抱开源。

MapRoute()有如下重载，它们都是扩展方法。

```
public class AreaRegistrationContext
{
    // Fields
    private readonly HashSet<string> _namespaces;
    // Methods
    public AreaRegistrationContext(string areaName, RouteCollection routes);
    public AreaRegistrationContext(string areaName, RouteCollection routes,
    object state);
    public Route MapRoute(string name, string url);
    public Route MapRoute(string name, string url, object defaults);
    public Route MapRoute(string name, string url, string[] namespaces);
    public Route MapRoute(string name, string url, object defaults, object
    constraints);
    public Route MapRoute(string name, string url, object defaults, string[]
    namespaces);
    public Route MapRoute(string name, string url, object defaults, object
    onstraints, string[] namespaces);
    // Properties
    public string AreaName { get; private set; }
    public ICollection<string> Namespaces { get; }
    public RouteCollection Routes { get; private set; }
    public object State { get; private set; }
}
```

MapRoute 参数介绍：

- name：规则名称，可以随意起名，但是不可以重名，否则会发生错误（路由集合中已存在名为“xx”的路由）。路由名必须是唯一的。
- url：url 获取数据的规则，这里不是正则表达式，将要识别的参数括起来即可，如“{controller}/{action};”。只需要传递 name 和 url 参数就可以建立一条 Routing（路由）规则，比如“routes.MapRoute("New","{controller}/{action}");”。
- constraints：用来限定每个参数的规则或 HTTP 请求的类型。constraints 属性是一个 RouteValueDictionary 对象，也就是一个字典表，但是这个字典表的值可以有两种：
  - 用于定义正则表达式的字符串，正则表达式不区分大小写。
  - 一个用于实现 IRouteConstraint 接口且包含 Mathc 方法的对象。

通过使用正则表达式可以规定参数格式化，比如 Controller 参数只能为 6 位数字：

```
new{controller=@"\d{6}"}
```

- namespaces：string[]类型，指定在哪些命名空间下面查找控制器进行匹配。如果不设置此属性，默认是在 bin 目录下面遍历所有的 dll，依次查找这些 dll 中继承自 Controller 的类进行匹配。

5.1.1　Routing——URL

作为广泛使用的 Web 用户接口，URL 需要被重视，因为好的 URL 有利于 SEO（Search Engine Optimization）优化。那么什么是好的 URL 呢？

好的 URL 应该满足如下条件：

- URL 应为获取某种资源提供信息，不一定是物理文件路径。
- 简短易于记忆和拼写输入。
- 可以反映出站点结构。
- 应该是“可拆分”，用户移除末尾，进而获得更高层次信息。
- 持久、不应改变。

对一个网站而言，为了 SEO 友好，一个网址的 URL 层次最好不要超过三层：http://localhost/{分类}/{具体页}；

其中，域名作为第一层，分类作为第二层，最后的网页就是最后一层了。使用默认路由中的"{controller}/{action}/{id}"形式会影响网站的 SEO。

下面的两种 URL 你更喜欢哪种呢？

```
http://www.cnblogs.com/jiekzou/1.html
http://www.cnblogs.com/jiekzou/1
```

如果是后者，那么服务器要怎么识别呢？ASP.NET MVC 使用路由机制完成由 URL 到具体调用方法的映射过程。

注：在以前传统的 ASP.NET WebForm 开发中，Url 代表服务器磁盘上的物理文件。

5.1.2 Routing 的作用

Routing 的作用是：

- 匹配传入的请求(不匹配服务器物理文件)，并将请求映射到“控制器”的具体操作“Action 方法”和“参数”中。
- 调用并执行对应控制器类的 Action 方法。

Global.asax.cs 文件中的 Application_Start()方法里面进行了路由注册：

```
RouteConfig.RegisterRoutes(RouteTable.Routes);
```

App_Start 文件夹中的 RouteConfig.cs 定义了路由的识别规则：

```
public static void RegisterRoutes(RouteCollection routes)
{
        routes.IgnoreRoute("{resource}.axd/{*pathInfo}");

        routes.MapRoute(
                name: "Default",// 1.路由名称
                url: "{controller}/{action}/{id}",// 2.带有参数的URL
                defaults: new { controller = "Home", action = "Index", id =
                UrlParameter.Optional }
                // 3.参数默认值，(UrlParameter.Optional 表示可选的意思)
        );
}
```

注：{controller}和{action}是特定参数名，不能改。RegisterRoutes 方法中可以注册多个路由，通常越详细的路由配置放置到越前面，但是路由名称不能重复，路由按照先后顺序与传入的 URL 匹配，直到匹配成功为止，一旦有路由匹配成功，就不会再继续匹配，如果所有的路由都没有匹配上就会报错。

5.1.3 Routing 包含字面值的 URL

路由 URL 在段中也允许包含“字面值”，如/jiekzou/{controller}/{action}/{id}规定第一段必须以 jiekzou 开头才能与该路由匹配，如/jiekzou/home/index/1URL 中可以将字面量和参数混合在一起，如图 5-1 所示。

```
{controller}-{action}-{id} 匹配：/Home-Index-1
{controller}.{action}.{id} 匹配：/Home.Index.1
{controller}/{action}-{id} 匹配：/Home/Index-1
{controller}/{action}-{month}/{day} 匹配：/Home/Index-1/24
```

图 5-1

不能有两个连续的 URL 参数，参数之间必须有字符隔开，如{controller}{action}-{id}就是错误的。

（1）路由约束

允许 URL 段使用正则表达式来限制路由是否匹配请求，通过 constraints 属性来配置。

```
routes.MapRoute(
        name: "Default",// 1.路由名称
        url: "{controller}/{action}/{id}",// 2.带有参数的 URL
        defaults: new { controller = "Home", action = "Index", id =
        UrlParameter.Optional },
        // 3.参数默认值，(UrlParameter.Optional 表示可选的意思)
        constraints: new { controller = @"Home", action =@"Index|Test"} //正则
);
```

在 Home 控制器中添加 3 个 Action 方法和对应的视图，即 Index、Test、Test1，然后在 URL 地址栏访问，发现 Test1 的 Action 无法访问到，因为我们在路由里面进行的正则校验规则不包括名为 Test1 的 Action。

（2）命名路由

生成指定路由名的 URL 超链接：

```
@Html.RouteLink("链接文本", "blog",new {contorller="home",action ="index",id=1 } )
```

5.1.4 Routing 测试

在项目上右击，选择“管理 NuGet 程序包(N)...→联机”→搜索“RouteDebugger”，单击“安装”按钮，如图 5-2 所示。

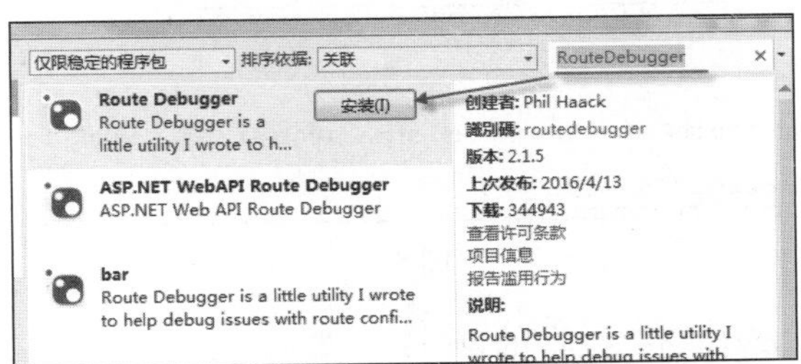

图 5-2

安装完成之后，就会发现在项目中自动添加了 RouteDebugger 引用，如图 5-3 所示。

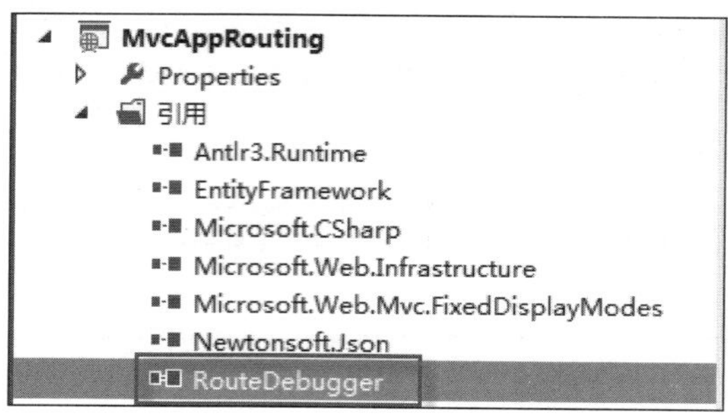

图 5-3

而且在 Web.config 中，<appSettings>节点里面自动添加了如下配置代码：

```
<add key="RouteDebugger:Enabled" value="true" /></appSettings>
```

由于我们使用的是 MVC 4，因此只需要按 F5 键运行程序即可看到效果。浏览器地址为 http://localhost:18901/，页面效果如图 5-4 所示。

链接文本

Route Debugger

Type in a url in the address bar to see which defined routes match it. A {*catchall} route is added to the list of routes automatically in case none of your routes match.

To generate URLs using routing, supply route values via the query string. example: http://localhost:14230/?id=123

Matched Route: {controller}/{action}/{id}

Route Data

Key	Value
controller	Home
action	Index

Data Tokens

Key	Value

All Routes

Matches Current Request	Url	Defaults	Constraints	DataTokens
False	api/{controller}/{id}	id =	(empty)	(null)
False	{resource}.axd/{*pathInfo}	(null)	(empty)	(null)
True	{controller}/{action}/{id}	controller = Home, action = Index, id = UrlParameter.Optional	controller = Home, action = Index|Test	(empty)
True	{*catchall}	(null)	(null)	(null)

Current Request Info

AppRelativeCurrentExecutionFilePath is the portion of the request that Routing acts on.

AppRelativeCurrentExecutionFilePath: ~/

图 5-4

对于.NET 3.5 和 MVC 3 之前的项目，如果要使用 RouteDebugger，还需要在 Application_Start 中注册：

```
protected void Application_Start()
{
        AreaRegistration.RegisterAllAreas();
        RegisterRoutes(RouteTable.Routes);
        //注册 RouteDebug
        RouteDebug.RouteDebugger.RewriteRoutesForTesting(RouteTable.Routes);
}
```

要禁用 RouteDebugger，只需在 Web.config 中修改配置文件即可（将 value 值修改为 false）：

```
!--禁用-->
<add key="RouteDebugger:Enabled" value="false" />
```

5.2 异步 Ajax

Ajax 异步刷新应用在 Web 开发中经常用到；在过去 WebForm 中通常是使用 jQuery 和一般处理程序或者 aspx 页面来实现的；在 MVC 中，虽然依旧可以使用一般处理程序，但是一般还是通过在 Controller 中新建 Action 方法来实现。

5.2.1 传统 Ajax 实现方式

1. jQuery+一般处理程序

（1）右击项目，选择“添加→HTML 页”，命名为 ShowDateTime.html，引入 jquery 文件，添加如下代码：

```
<!DOCTYPE html>
<html xmlns="http://www.w3.org/1999/xhtml">
<head>
<meta http-equiv="Content-Type" content="text/html; charset=utf-8"/>
   <title></title>
   <script src="Scripts/jquery-1.8.2.min.js"></script>
   <script type="text/javascript">
       $(function () {
          $("#btnGetDateTime").click(function () {
             $.post("GetDateTimeHandler.ashx", {}, function (data)
             { $("#divDateTime").html(data); });
          });
       });
```

```
    </script>
</head>
<body>
    <input type="button"  value="获取时间" id="btnGetDateTime"/>
    <div id="divDateTime"></div>
</body>
</html>
```

（2）右击项目，选择“添加→新建项→一般处理程序”，并设置名称，这里设为“GetDateTimeHandler.ashx”。

```
public class GetDateTimeHandler : IHttpHandler
{

        public void ProcessRequest(HttpContext context)
        {
            context.Response.ContentType = "text/plain";
            context.Response.Write(DateTime.Now.ToString());
        }

        public bool IsReusable
        {
            get
            {
                return false;
            }
        }
}
```

（3）运行 ShowDateTime.html（右击该文件，在浏览器中查看），如图 5-5 所示。

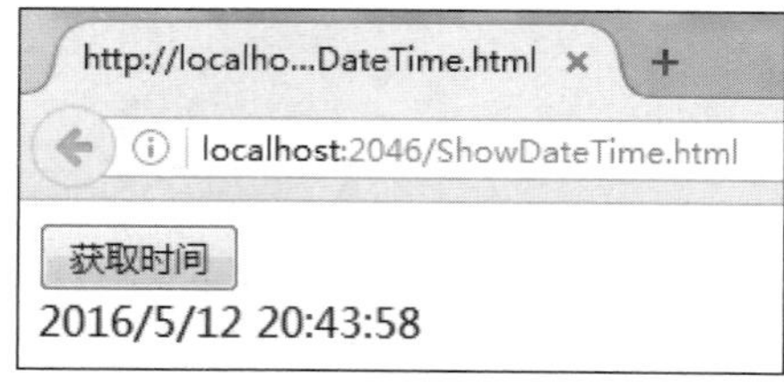

图 5-5

2. jQuery+Action

（1）在控制器中添加 Action 方法 ShowEmployeesList。这里用到了对象序列化，我们在新建 ASP.NET MVC 项目的时候，VS 自动帮我们引入了 System.Web.Extensions.dll，此程序集中的 JavaScriptSerializer 类提供了序列化和反序列化的方法。

```
[HttpGet]
public ActionResult ShowEmployeesList()
{
        return View();
}
 [HttpPost]
public ActionResult GetEmployeesList()
{
        using (NorthwindEntities db=new NorthwindEntities())
        {
           var list = db.Employees.Select(x => new {
            FirstName=x.FirstName,LastName=x.LastName,City=x.City
            }).ToList();
           System.Web.Script.Serialization.JavaScriptSerializer jss = new
           System.Web.Script.Serialization.JavaScriptSerializer();
           return Content(jss.Serialize(list));
           //return  Content(JsonConvert.SerializeObject(list));

        }
}
```

同时，VS 也自动添加了 Newtonsoft.Json.dll 这个第三方的 dll 程序集，功能比 System.Web.Extensions.dll 更加强大，性能更好。使用方式一是添加命名空间引用“using Newtonsoft.Json;”，二是调用代码“return Content(JsonConvert.SerializeObject(list));”。

（2）添加视图 ShowEmployeesList：

```
@{
   ViewBag.Title = "ShowEmployeesList";
}
@Scripts.Render("~/Scripts/jquery-1.8.2.min.js")
<script type="text/javascript">
   $(function () {
      $("#btnEmployeesList").click(function () {
         $.post("/Home/GetEmployeesList", {}, function (data) {
            var result = $.parseJSON(data);
            $.each(result, function (key, value) {
               $("#divEmployeesList").append(result[key].FirstName + "\t\b" +
               result[key].LastName + "\t\b" + result[key].City + "<br/>");
            });
         });
      });
   });
```

```
</script>
<div style="margin-bottom:30px;"><input type="button" id="btnEmployeesList"
value="获取员工信息列表"/></div> <div id="divEmployeesList"></div>
```

（3）运行 http://localhost:2046/Home/ShowEmployeesList，结果如图 5-6 所示。

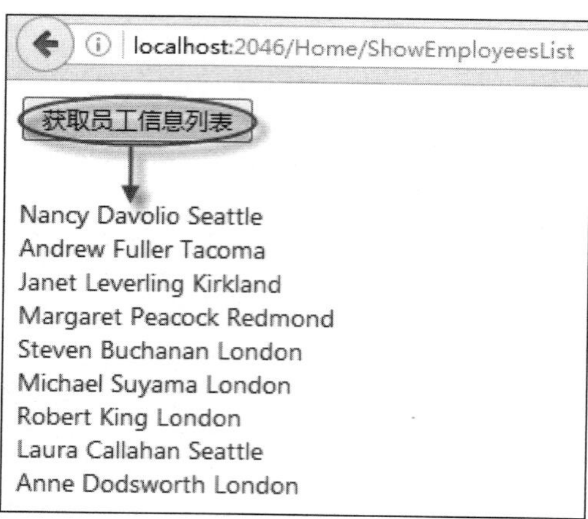

图 5-6

5.2.2 Unobtrusive Ajax 使用方式

通俗来讲，非入侵式就是将嵌入 HTML 中的 JavaScript 全部取出来，放在单独的 js 文件中，HTML 标签中不要出现任何 onclick、onload 等。

Unobtrusive Ajax 方便程序员编写简单易于维护的 Ajax 代码（Code is cleaner and easier to maintain），基本特点如下：

- 网页内容和表单使用纯 HTML。
- 无须借助 JavaScript，表单和超链接也能正常使用。
- 页面外观完全由 CSS 控制，而不是 HTML（不要用 table 来布局）或 JavaScript。
- 任何人都能通过任何设备（考虑不支持 JavaScript 的设备）访问。

ASP.NET MVC 全局开启非入侵 Ajax

ASP.NET MVC 4 已经默认开启客户端验证和非侵入式 js。在 Web.config 中可以看到如下配置信息：

```
<add key="ClientValidationEnabled" value="true" />
<add key="UnobtrusiveJavaScriptEnabled" value="true" />
```

如果只想在指定页面使用此功能，可以在页面中单独添加非入侵 js 文件：

```
@Scripts.Render("~/Scripts/jquery.unobtrusive-ajax.min.js")
```

或者

```
<script                src="@Url.Content("~/Scripts/jquery.unobtrusive-ajax.min.js")"
type="text/javascript"> </script>
```

这两种方法都可以，推荐使用@Scripts.Render 的形式，因为它支持压缩合并和服务器缓存（后面会讲）。

我们还可以在单个视图页面（View）上关闭非入侵 Ajax 功能，如下：

```
@{Html.EnableUnobtrusiveJavaScript(false);}
@{Html.EnableClientValidation(false);}
```

注意 js 的引用顺序：

```
@Scripts.Render( "~/Scripts/jquery.min.js")
@Scripts.Render("~/Scripts/jquery.validate.min.js")
@Scripts.Render("~/Scripts/jquery.validate.unobtrusive.min.js")
@Scripts.Render("~/Scripts/jquery.unobtrusive-ajax.min.js")
```

其中两个 validate 的 js 文件是用于异步验证的。

5.2.3　AjaxHelper

1. 异步链接按钮

使用异步链接按钮时，必须开启非入侵式 Ajax（导入 jQuery 和 unobtrusive 相关的 Ajax 文件）。

（1）在 View 中，@Ajax.ActionLink 创建 Ajax 超链接按钮，一般用来请求动态生成的部分 HTML 代码（分部视图）或者普通的字符串文本。

```
@Ajax.ActionLink("链接文本", "PartialViewTest", new AjaxOptions() {
                    UpdateTargetId="divMsg",//数据显示的 html 容器 id
                    InsertionMode= InsertionMode.Replace, //替换容器内容
                    HttpMethod="Post" })
```

（2）在 Controller 中：

```
public PartialViewResult PartialViewTest()
{
          ViewData["Msg"] = "Hello world!";
          return PartialView();
}
```

生成的 HTML 代码如下：

```
<a data-ajax="true" data-ajax-method="Post" data-ajax-mode="replace" data-ajax-
update="#divMsg" href="/Home/PartialViewTest">链接文本</a>
```

2. 异步表单

（1）在 Controller 中添加如下 Action 方法：

```
public ActionResult AjaxForm()
{
        return View();
}
public ActionResult BaseInfo(string txtName)
{
        using (NorthwindEntities db = new NorthwindEntities())
         {
            var result = db.Employees.Where(x => x.LastName ==
            txtName).FirstOrDefault();
            return Content("姓: "+result.FirstName+"名: "+result.LastName);
        }
}
```

（2）添加名为 AjaxForm 的 View，在该 View 中通过@Ajax.BeginForm 创建异步表单：

```
@{
    ViewBag.Title = "AjaxForm";
}
@Scripts.Render( "~/Scripts/jquery-1.8.2.min.js")
@Scripts.Render("~/Scripts/jquery.unobtrusive-ajax.min.js")
@using (Ajax.BeginForm("BaseInfo", "Home", new AjaxOptions
{
        UpdateTargetId="msgDiv" , InsertionMode= InsertionMode.Replace,
        HttpMethod="post" , OnFailure= "fail", OnSuccess="success" ,
        LoadingElementId="lodeingmsg"}))
{
    <input type="text" name="txtName" />
    <input type="submit" />
}
<div id="lodeingmsg" style="display:none;">加载中...</div>
<div id="msgDiv"></div>
<script type="text/javascript">function fail(txt) { alert("查询失败, 失败信息: " +
txt.responseText); }
    function success(txt) { alert("查询成功, 查询信息: " + txt); }
</script>
```

运行结果如图 5-7~图 5-9 所示。

图 5-7

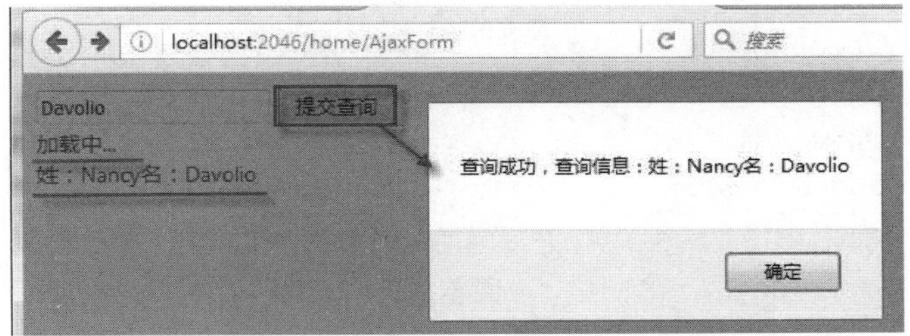

图 5-8

图 5-9

View 中的代码分析如下：

- UpdateTargetId：目标元素 id，获取服务器响应后，将获取的响应报文体显示到目标元素的 innerHTML 中。
- InsertionMode：InsertAfter 插入目标元素原有内容之后，InsertBefore 插入目标元素原有内容之前，Replace 替换目标元素原有内容。
- LoadingElementId：异步对象 readyState==4 之前显示“正在加载”状态的元素 id。
- OnSuccess：请求成功之后，后续执行的 js 回调函数，可以根据返回的参数确定服务器端处理情况。
- OnFailure：请求失败之后，后续执行的 js 回调函数。
- HttpMethod：设置请求方式（get、post）。
- Confirm：提交前的回调函数，指定为一个 js 的 function。

生成的 HTML 代码如下：

```
<form action="/Home/BaseInfo" data-ajax="true" data-ajax-failure="fail" data-ajax-loading="#lodeingmsg" data-ajax-method="post" data-ajax-mode="replace" data-ajax-success="success" data-ajax-update="#msgDiv" id="form0" method="post">
<input name="txtName" type="text"><inputtype="submit"></form>
```

AjaxOptions 对象生成“对应”触发 Ajax 请求标签的属性如图 5-10 所示。

AjaxOptions	HTML attribute
Confirm	data-ajax-confirm
HttpMethod	data-ajax-method
InsertionMode	data-ajax-mode *
LoadingElementDuration	data-ajax-loading-duration **
LoadingElementId	data-ajax-loading
OnBegin	data-ajax-begin
OnComplete	data-ajax-complete
OnFailure	data-ajax-failure
OnSuccess	data-ajax-success
UpdateTargetId	data-ajax-update
Url	data-ajax-url

图 5-10

5.2.4 请求 Json 数据

添加控制器 Json，在 Controller 中使用 Json 方法返回一个 JsonResult：

```
NorthwindEntities db = new NorthwindEntities();
public ActionResult CustomersList() //返回类型也可写 JsonResult
{
        var dogList = db.Customers.Select(x => new { CustomerID = x.CustomerID,
        ContactName = x.ContactName, Phone = x.Phone }).ToList();
        return Json(dogList, JsonRequestBehavior.AllowGet);
}
```

MVC 框架默认不允许使用 Json 响应 Get 请求，需要开启，开启代码为 JsonRequestBehavior.Allow。在 GetView 中添加代码：

```
@Ajax.ActionLink("click here", "CustomersList", new AjaxOptions() {
                    UpdateTargetId="divMsg",
                    InsertionMode= InsertionMode.Replace,
                    HttpMethod="Get",OnSuccess="ajaxFinish" })
<div id="divMsg"></div>
   <table id="tbList"><tr>
      <th>ID</th>
      <th>姓名</th>
      <th>电话</th>
</tr></table>
```

1. jQuery 请求控制器 Action

除了 URL 指向控制器的 Action 方法不同外，其他和以前一样，具体参见 jQuery 帮助文档 ajax 方法：

```
$.ajax
$.post
$.load
$.get
```

在 View 中如果不使用@Ajax.ActionLink，那么也可以使用：

```
$(function(){
      $('#btnAjax').click(function () {
         $.get('Json/CustomersList', function (json) {
            $("#temp").tmpl(json).appendTo("#tbList");
         });
});
<input type="button" id="btnAjax" value="获取客户信息列表" />
```

2. jQuery 模板插件

从云盘中获取 jQuery 模板插件 jquery.tmpl.min.js，然后在 View 中导入脚本：

```
@Scripts.Render("~/Scripts/jjquery-1.8.2.min.js")
@Scripts.Render("~/Scripts/jquery.tmpl.min.js")
```

添加模板-占位符格式，即$（json 对象属性名）：

```
<script id="temp" type="text/x-jquery-tmpl">
         <tr>
            <td>${CustomerID}</td><td>${ContactName}</td><td>${Phone}</td>
         </tr>
</script>
```

为模板装载数据并最终生成 HTML，添加到表格中：

```
function ajaxFinish(jsonObjArray)
{ //[{CustomerID:"1",ContactName:"zouqj",Phone:"15243641131"},....{}]
      $("#temp").tmpl(jsonObjArray).appendTo("#tbList");
}
```

运行结果如图 5-11 所示。

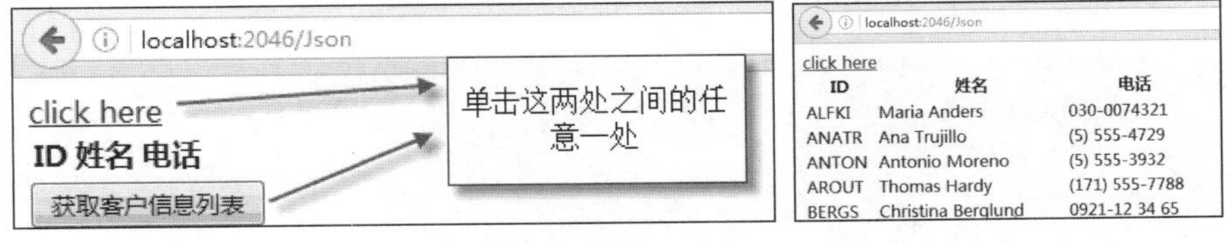

图 5-11

5.3 MVC Areas

为什么会出现 Areas（区域）？

随着业务的需要，结构需求越来越多，Views 目录下面的文件夹也越来越多，又或者需要更改结构的页面路径，另外可能是多个人合作开发，有多个 Web 项目需要合并在一个网站中访问。

MVC 项目目录结构缺点：

- 不利于分功能协作开发（购物车、商品管理、用户权限管理……）。
- 代码结构臃肿。

用 Areas 可以解决上述问题。在项目中新建一个区域，就像引用另外一个项目一样，可以理解为当前项目中的子项目，其本质就是一个子文件夹。就好比使用 Areas 之前是把所有文件放到一个文件夹中，使用了 Areas 就相当于使用了子文件夹来进行管理。这样，Areas 就可以将复杂的网站模块化，有利于页面的规范。

5.3.1 Area 使用入门

操作步骤：

（1）在项目上右击，选择“添加→Area”，填写名称为 Admin，然后确定，就会在这个项目上创建一组文件夹，就像是一个 MVC 子项目，如图 5-12 所示。

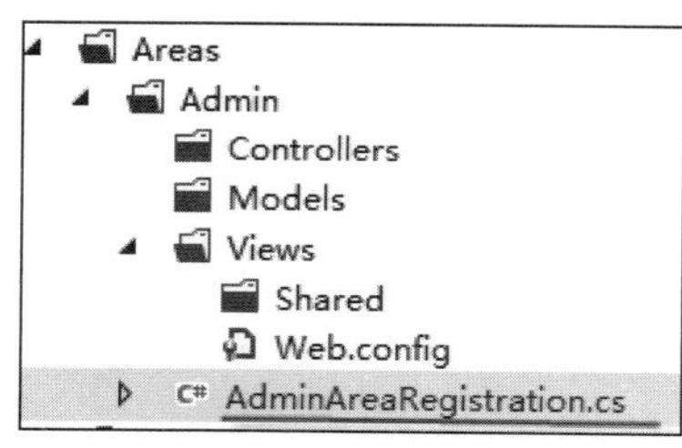

图 5-12

（2）修改文件夹下的 XXAreaRegistration 代码，可以在这个文件中进行路由注册，这个 XX 就是刚才填写的区域名称 Admin。

```
/// <summary>
/// 区域注册类
/// </summary>
public class AdminAreaRegistration : AreaRegistration
{
/// <summary>
```

```
        /// 区域名称，很明显可以重写，默认名称就是我们新建区域时取的名称，在某个区域中控制器的
        /// Action 方法加载视图时会作为路径的一部分，说白了就是路径中的一个子文件夹名称。
        /// </summary>
        public override string AreaName
        {
            get
            {
                return "Admin";
            }
        }
/// <summary>
        /// 这里是进行区域路由注册，和我们之前看到的普通路由注册不一样的地方是：路由规则在最前面
        /// 添加了区域名称“Admin”，少了一个默认的 controller = "Home"，其他地方一模一样，用
        /// 法也一样
        /// </summary>
        /// <param name="context"></param>
        public override void RegisterArea(AreaRegistrationContext context)
        {
            context.MapRoute(
                "Admin_default",
                "Admin/{controller}/{action}/{id}",
                new { action = "Index", id = UrlParameter.Optional }
            );
        }
}
```

（3）我们在新建 MVC 项目的时候会自动在 Global 中 Application_Start()里面的第一行进行区域路由注册：

```
AreaRegistration.RegisterAllAreas();
```

也就是说区域路由注册优先于普通路由注册。

（4）在 Admin 区域下面添加控制器 Home。

（5）添加 Index 视图：

```
@{
    ViewBag.Title = "Index";
}
<h2>这是我的 Admin 区域首页</h2>
```

（6）运行项目，在浏览器地址栏中输入“http://localhost:14359/Admin/Home”，结果如图 5-13 所示。

图 5-13

5.3.2 Area 注册类放到单独程序集

在 5.3.1 小节中，Area（区域）展示在同一项目中，也就是说新建的区域还是处于 MVC 站点之内，类似于在 MVC 站点中通过新建文件夹的形式存在。

（1）新建类库 AdminArea。

（2）将原来 MvcAppArea 项目中的 AdminAreaRegistration.cs 文件复制到 AdminArea 类库项目中。

（3）添加 System.Web.Mvc 和 System.Web 的程序集引用。

（4）删除原来 MvcAppArea 项目中的 AdminAreaRegistration.cs 文件。

（5）在 MvcAppArea 项目中添加对类库项目 AdminArea 的引用。

（6）运行项目，在浏览器中输入地址"http://localhost:14359/Admin/Home"，就会发现它和 5.3.1 小节中的运行结果一模一样。

为什么现在依旧能够正常运行？区域路由又是如何工作的？

推测：首先，我们新建一个 AdminArea 类库，又把区域路由注册类 AdminAreaRegistration 放到了这个类库中，这样看起来就在物理层面上进行了分离。但是 AdminAreaRegistration 类的命名空间没有改变，也就是说我们把区域路由注册类包装到了 AdminArea 类库中。

然后，MvcAppArea 项目添加了对类库项目 AdminArea 的引用，这样的话 MvcAppArea 项目的 bin 目录就必然会存在 AdminArea.dll 程序集。当一个请求过来的时候，首先会遍历 bin 目录下面的所有程序集，然后筛选出命名空间是 MvcAppArea.Areas.Admin（默认情况下是当前区域注册类所在的命名空间）并且继承了 AreaRegistration 类的所有类，最后进行路由规则匹配。

接下来，通过 Reflector 反编译工具来证明推测。先从"AreaRegistration.RegisterAllAreas();"这行代码入手，查看 RegisterAllAreas 类的源码，源码上面有相应的注释：

```
// Generated by .NET Reflector from
//C:\Windows\Microsoft.Net\assembly\GAC_MSIL\System.Web.Mvc\v4.0_4.0.0.0__31bf385
//6ad364e35\System.Web.Mvc.dll
namespace System.Web.Mvc
{
   using System;
   using System.Web.Routing;

   public abstract class AreaRegistration
   {
```

```
private const string TypeCacheName = "MVC-AreaRegistrationTypeCache.xml";
//在这里声明了一个缓存配置文件名称的常量，可是在 RegisterAllAreas 方法中又硬编码了一个
//同名字符串，可见微软的牛人并不是神，也是人，是人就不可能把代码写得完美无瑕，我们要学会“找茬”
//无参构造函数
protected AreaRegistration()
{
}

internal void CreateContextAndRegister(RouteCollection routes, object state)
{
    AreaRegistrationContext context = new
    AreaRegistrationContext(this.AreaName, routes, state);
    string str = base.GetType().Namespace; //当前区域注册类所在的命名空间
    if (str != null)
    {
        context.Namespaces.Add(str + ".*");
    }
    this.RegisterArea(context);
}
/// <summary>
/// 判断是否是一个区域路由注册类
/// </summary>
/// <param name="type"></param>
/// <returns></returns>
private static bool IsAreaRegistrationType(Type type)
{
    //继承自类 AreaRegistration 并且包含一个无参构造函数
    return (typeof(AreaRegistration).IsAssignableFrom(type) &&
    (type.GetConstructor(Type.EmptyTypes) != null));
}
//1
public static void RegisterAllAreas()
{
    RegisterAllAreas(null);
}
//2
public static void RegisterAllAreas(object state)
{
    RegisterAllAreas(RouteTable.Routes, new BuildManagerWrapper(), state);
}
/// <summary>
/// 3 可以看到最终我们调用的其实就是这个静态方法
```

```
        /// </summary>
        /// <param name="routes">静态路由表</param>
        /// <param name="buildManager"></param>
        /// <param name="state"></param>
        internal static void RegisterAllAreas(RouteCollection routes,
        IBuildManager buildManager, object state)
        {
            //MVC-AreaRegistrationTypeCache.xml 这个文件是在运行时生成的，它存在于内存当中
            foreach (Type type in
            TypeCacheUtil.GetFilteredTypesFromAssemblies("MVC-
            AreaRegistrationTypeCache.xml", new
            Predicate<Type>(AreaRegistration.IsAreaRegistrationType), buildManager))
            {
                //如果是区域注册类，就创建一个区域注册类对象，然后调用
                //CreateContextAndRegister 方法创建一个区域注册上下文对象，并进行区域路由注册
                ((AreaRegistration)Activator.CreateInstance(type)).
                CreateContextAndRegister(routes, state);
            }
        }
        //这个区域注册方法，在 AdminAreaRegistration 这个实体类中来实现
        public abstract void RegisterArea(AreaRegistrationContext context);
        //抽象属性，区域名称，同样在 AdminAreaRegistration 实体类中实现
        public abstract string AreaName { get; }
    }
}
```

5.3.3 Area 注册控制器放到单独程序集

在项目比较复杂、开发人员比较多的情况下，我们在同一个项目中使用区域的方式会出现一些弊端，最常见的就是不好进行版本控制。前面已经说了，在同一项目中使用区域就好比是在文件夹中使用子文件夹，那么就会有一个问题，大家其实还是都在同一个总文件夹下面操作文件，并没有做到物理上面的分离，这样的话大家都各自在区域下面添加文件，然后提交代码，各种冲突就出现了。要知道，解决代码冲突是一件很痛苦的事情，尤其是面临 N 个冲突的时候。为了更好地分工合作，减少代码冲突，我们可以把 Area 注册控制器放到单独的程序集中，从物理上进行分离。

（1）新建一个空的 MVC 项目 AdminAreaController，把其他文件删除，只保留 Controllers 文件夹和 Web.config 文件。

（2）新建 User 控制器：

```
namespace AdminAreaController.Controllers
```

```
{
    public class UserController : Controller
    {
        public ActionResult Index()
        {
            return View();
        }


    }
}
```

（3）从项目 AdminArea 中复制 AdminAreaRegistration.cs，修改命名空间为 AdminAreaController，修改路由名称为“Admin”，添加路由约束：

```
new string[1] { "AdminAreaController.Controllers" }
```

这里添加了一个命名空间约束，这个命名空间就是我们刚才新建的 User 控制器所在的命名空间。

（4）在项目 MvcAppArea 中，Areas 文件夹下面添加文件夹 User，然后在 User 文件夹中添加 Index 视图：

```
@{
    ViewBag.Title = "Index";
}
<h2>控制器作为单独程序集/h2>
```

（5）运行项目 MvcAppArea，在浏览器中输入地址“http://localhost:14359/Admin/User/Index”，运行结果如图 5-14 所示。

图 5-14

5.4 MVC Filter

MVC Filter 是典型的 AOP（面向切面编程）应用，在 ASP.NET MVC 中有 4 种过滤器类型，如图 5-15 所示。

过滤器类型	接口	默认实现	描述
Action	IActionFilter	ActionFilterAttribute	在动作方法之前及之后运行
Result	IResultFilter	ActionFilterAttribute	在动作结果被执行之前和之后运行
AuthorizationFilter	IAuthorizationFilter	AuthorizeAttribute	首先运行，在任何其它过滤器或动作方法之前
Exception	IExceptionFilter	HandleErrorAttribute	只在另一个过滤器、动作方法、动作结果弹出异常时运行

图 5-15

5.4.1 Action

在 ASP.NET MVC 项目 MvcAppAjax 中，新建文件夹 Filter，然后新建类 MyActionFilterAttribute（为了遵循默认的约定，名称以 Attribute 结尾），继承自 ActionFilterAttribute 类。ActionFilterAttribute 类有如下 4 个方法，从命名就可以看出它们的执行时机。

```
[AttributeUsage(AttributeTargets.Method | AttributeTargets.Class,
Inherited=true, AllowMultiple=false)]
public abstract class ActionFilterAttribute : FilterAttribute, IActionFilter,
IResultFilter
{
    // Methods
    protected ActionFilterAttribute();
    public virtual void OnActionExecuted(ActionExecutedContext filterContext);
    public virtual void OnActionExecuting(ActionExecutingContext
    filterContext);
    public virtual void OnResultExecuted(ResultExecutedContext filterContext);
    public virtual void OnResultExecuting(ResultExecutingContext
    filterContext);
}
```

接下来是 ActionFilterAttribute 类：

```
public class MyActionFilterAttribute: ActionFilterAttribute
{
      public override void OnActionExecuting(ActionExecutingContext filterContext)
      {
           filterContext.HttpContext.Response.Write("Action 执行前: " +
           DateTime.Now.ToString("yyyy-MM-dd hh:mm:ss fff")+"<br/>");
           base.OnActionExecuting(filterContext);
      }
      public override void OnActionExecuted(ActionExecutedContext filterContext)
      {
```

```
        filterContext.HttpContext.Response.Write("Action 执行后: " +
        DateTime.Now.ToString("yyyy-MM-dd hh:mm:ss fff"));
        base.OnActionExecuted(filterContext);
    }
}
```

对于过滤器，我们可以把它们加在 3 个地方，一个是控制器上面（控制器下面的所有 Action），一个是 Action 上面（指定标识的 Action），另一个就是全局位置（所有控制器中的 Action)。这里只演示在 Action 上面和 Home 控制器中：

```
[MyActionFilter]
public ActionResult Index()
{
        return View();
}
public ActionResult Index1()
{
        return View();
}
```

创建项目 MvcAppAjax，然后运行，如图 5-16、图 5-17 所示。

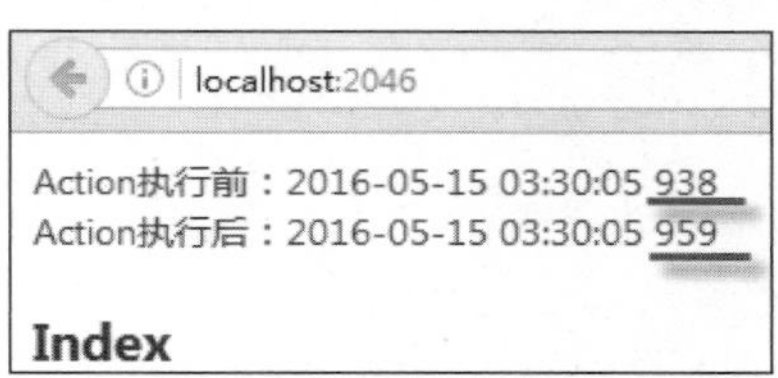

图 5-16

图 5-17

5.4.2　Result

新建类 MyResultFilterAttribute，继承 ActionFilterAttribute：

```
public class MyResultFilterAttribute : ActionFilterAttribute
{
    /// <summary>
    /// 加载 "视图" 前执行
    /// </summary>
    /// <param name="filterContext"></param>
    public override void
    OnResultExecuting(System.Web.Mvc.ResultExecutingContext filterContext)
    {
        filterContext.HttpContext.Response.Write("加载视图前执行 OnResultExecuting
        " + DateTime.Now.ToString("yyyy-MM-dd hh:mm:ss fff") + "<br/>");
```

```
            base.OnResultExecuting(filterContext);
        }

        /// <summary>
        /// 加载"视图" 后执行
        /// </summary>
        /// <param name="filterContext"></param>
        public override void OnResultExecuted(System.Web.Mvc.ResultExecutedContext
        filterContext)
        {
            filterContext.HttpContext.Response.Write("加载视图后执行 OnResultExecuted
            " + DateTime.Now.ToString("yyyy-MM-dd hh:mm:ss fff") + "<br/>");
            base.OnResultExecuted(filterContext);
        }
}
```

这里把 MyResultFilter 过滤器加在控制器上面，相当于给 Home 控制器中所有的 Action 方法添加了 MyResultFilter 过滤器。

```
 [MyResultFilter]
public class HomeController : Controller
{
        //
        // GET: /Home/

        [MyActionFilter]
        public ActionResult Index()
        {
            return View();
        }
        public ActionResult Index1()
        {
            return View();
        }
}
```

运行结果如图 5-18、图 5-19 所示。

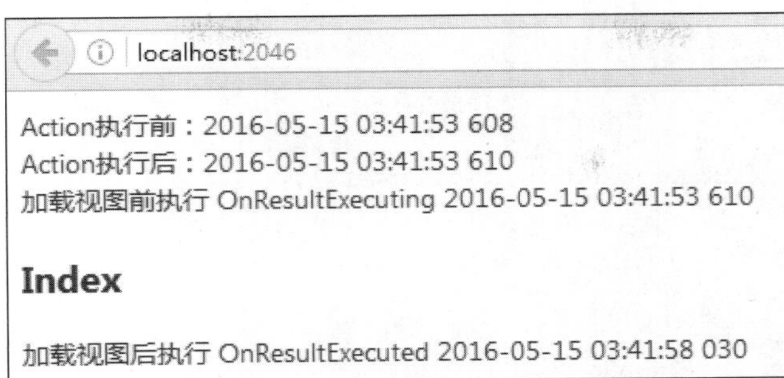

图 5-18

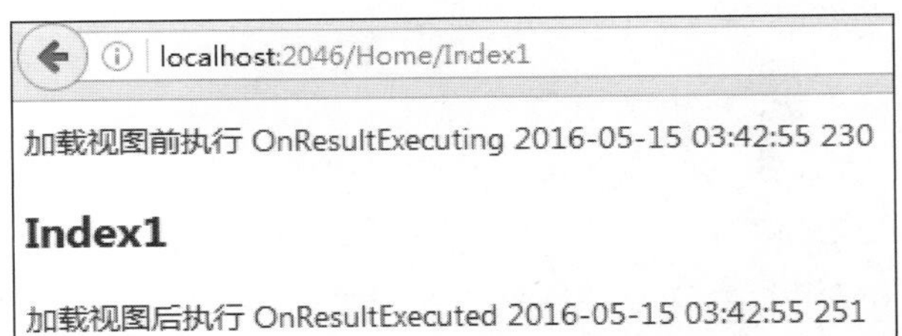

图 5-19

注意：无论 Action 的返回类型是什么（甚至 void），Result 过滤器都将执行。

RouteData 中保存了当前请求匹配的路由信息和路由对象，添加 MyCustormFilterAttribute.cs：

```
public class MyCustormFilterAttribute:ActionFilterAttribute
{
    public override void OnActionExecuting(ActionExecutingContext filterContext)
    {
        //1.获取请求的类名和方法名
        string strController =
        filterContext.RouteData.Values["controller"].ToString();
        string strAction = filterContext.RouteData.Values["action"].ToString();
        //2.用另一种方式 获取请求的类名和方法名
        string strAction2 = filterContext.ActionDescriptor.ActionName;
        string strController2 =
        filterContext.ActionDescriptor.ControllerDescriptor.ControllerName;

        filterContext.HttpContext.Response.Write("控制器:" + strController + "</br>");
        filterContext.HttpContext.Response.Write("控制器:" + strController2 + "</br>");
        filterContext.HttpContext.Response.Write("Action:" + strAction + "</br>");
        filterContext.HttpContext.Response.Write("Action:" + strAction2 + "</br>");
        filterContext.HttpContext.Response.Write("Action 执行前: " +
        DateTime.Now.ToString("yyyy-MM-dd hh:mm:ss fff") + "<br/>");
        base.OnActionExecuting(filterContext);
    }
    public override void OnActionExecuted(ActionExecutedContext filterContext)
```

```
        {
            filterContext.HttpContext.Response.Write("Action 执行后: " +
            DateTime.Now.ToString("yyyy-MM-dd hh:mm:ss fff") + "<br/>");
            base.OnActionExecuted(filterContext);
        }
        /// <summary>
        /// 加载 "视图" 前执行
        /// </summary>
        /// <param name="filterContext"></param>
        public override void
        OnResultExecuting(System.Web.Mvc.ResultExecutingContext filterContext)
        {
            filterContext.HttpContext.Response.Write("加载视图前执行 OnResultExecuting
            " + DateTime.Now.ToString("yyyy-MM-dd hh:mm:ss fff") + "<br/>");
            base.OnResultExecuting(filterContext);
        }
        /// <summary>
        /// 加载"视图" 后执行
        /// </summary>
        /// <param name="filterContext"></param>
        public override void OnResultExecuted(System.Web.Mvc.ResultExecutedContext
        filterContext)
        {
            filterContext.HttpContext.Response.Write("加载视图后执行 OnResultExecuted
            " + DateTime.Now.ToString("yyyy-MM-dd hh:mm:ss fff") + "<br/>");
            base.OnResultExecuted(filterContext);
        }
}
```

添加 FilterTest 控制器和 Index 视图：

```
[MyCustormFilter]
        public ActionResult Index()
        {
            return View();
}
```

运行结果如图 5-20 所示。

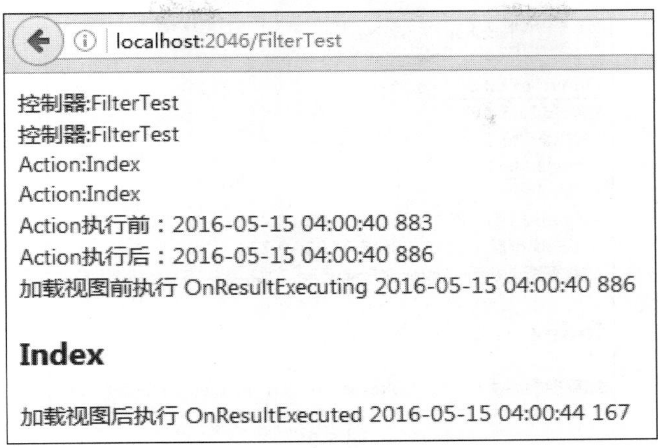

图 5-20

5.4.3　AuthorizeAttribute

新建类 TestAuthorizeAttribute.cs：

```
/// <summary>
/// 授权过滤器——在 Action 过滤器前执行
/// </summary>
public class TestAuthorizeAttribute : AuthorizeAttribute
{
        public override void OnAuthorization(AuthorizationContext filterContext)
        {
            filterContext.HttpContext.Response.Write("OnAuthorization<br/>");
            //注释掉父类方法，因为父类里的 OnAuthorization 方法会调用 ASP.NET 的授权验证机制！
            //base.OnAuthorization(filterContext);
        }
}
```

在控制器 FilterTest 中的 Index 上添加 TestAuthorize 标记：

```
[MyCustormFilter]
[TestAuthorize]
public ActionResult Index()
{
        return View();
}
```

运行结果如图 5-21 所示。

图 5-21

通常 Authorize 过滤器也是在全局过滤器上面的，主要用来做登录验证或者权限验证，在 FilterConfig 类的 RegisterGlobalFilters 方法中添加：

```
filters.Add(new TestAuthorizeAttribute());
```

在全局中注册过滤器，则所有控制器的所有行为（Action）都会执行这个过滤器。

5.4.4 Exception

新建 TestHandleErrorAttribute.cs：

```
/// <summary>
/// 异常处理 过滤器
/// </summary>
public class TestHandleErrorAttribute: HandleErrorAttribute
{
       public override void OnException(ExceptionContext filterContext)
        {
           //1.获取异常对象
           Exception ex = filterContext.Exception;
           //2.记录异常日志
           //3.重定向友好页面
           filterContext.Result = new RedirectResult("~/error.html");
           //4.标记异常已经处理完毕
           filterContext.ExceptionHandled = true;

           base.OnException(filterContext);
       }
}
```

在 Action 上面加 TestHandleError：

```
[TestHandleError]
public ActionResult GetErr()
{
        int a = 0;
        int b = 1 / a;
        return View();
}
```

在浏览器输入“http://localhost:2046/FilterTest/GetErr”，运行，会自动跳转到 error.html 页面，如图 5-22 所示。

图 5-22

如果没有跳转，就需要去 Web.config 配置文件中的 <system.web>节点下面添加如下配置节点，开启自定义错误：

```
<customErrors mode="On"></customErrors>
```

通常这样的异常处理是放在全局过滤器上面的，只要任意 Action 方法报错就会执行 TestHandleError 过滤器中的代码。

修改 App_Start 目录下面的 FilterConfig 代码：

```
public class FilterConfig
{
    public static void RegisterGlobalFilters(GlobalFilterCollection filters)
    {
        //filters.Add(new HandleErrorAttribute());
        //添加全局过滤器
        filters.Add(new TestHandleErrorAttribute ());
    }
}
```

5.5 MVC 整体运行流程

5.5.1 进入管道

（1）当我们在浏览器端输入一个 URL“地址 http://www.cnblogs.com/jiekzou/”时，客户端

会发送一个基于 HTTP 的 Web 请求给服务器，在浏览器中看到请求报文信息，如图 5-23 所示。

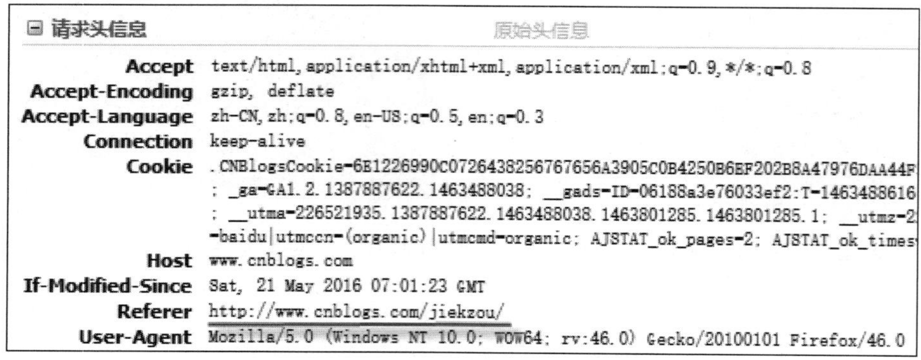

图 5-23

在请求到达 Web 服务器的那一刻，Web 服务器 Windows 内核中的 HTTP.SYS 组件就会捕获到请求。当 HTTP.SYS 组件分析到这是一个需要交给 IIS 服务器处理的 HTTP 请求时，HTTP.SYS 组件就会将 Request 请求交给 IIS 服务器来处理。

HTTP.SYS 是一个位于 Windows Server 和 Windows XP SP2 中的操作系统核心组件（内核模式中），能够让任何应用程序通过它提供的接口以 HTTP 进行信息通信。

关于内核模式与用户模式。

在 Windows Server 操作系统中，一个进程既可以运行于内核模式，也可以运行于用户模式。如果一个进程运行于内核模式，那么这个进程就可以访问所有硬件和系统数据；如果一个进程运行于用户模式，那么这个进程不能直接访问硬件，而且访问系统数据时也会受到限制。在 Intel 处理器架构中一共有 0~3 四个特权级，内核模式运行于 0 级之内，而用户模式运行于 3 级。通过在内核模式运行 HTTP.SYS，侦听器可以直接访问 TCP/IP 协议栈，但是又能够位于 WWW 服务之外，这样就不会受到应用程序中代码缺陷的影响，也不会因为应用程序崩溃而出现问题。

（2）如果该请求有缓存内容就直接响应，如图 5-24 所示。

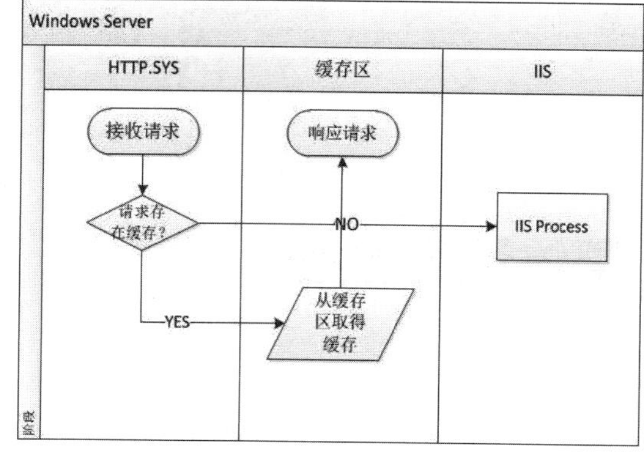

图 5-24

HTTP.SYS 组件的一个重要作用就在于它有一个缓存区，会将近期处理的响应结果放入缓存区之中，如果再次请求这个内容，就会从缓存区中取得内容并进行响应，提高了响应速度。而且，静态的内容现在被缓存于内核模式下，使服务响应速度更快。

（3）IIS 服务器会分析 Request 请求的 context-type 类型，然后从处理程序映射表中去匹配，当在处理程序映射表中能够匹配到 Request 请求的 context-type 类型时，IIS 服务器就将请求交给映射表中所对应的程序处理。当 IIS 发现在处理程序映射表中没有能匹配的项的时候（当没有匹配项的时候，一般情况下是请求“静态文件”），就直接去下载 Request 请求所对应路径的文件，如.jpg、html、xml.css 文件等，如图 5-25 所示。

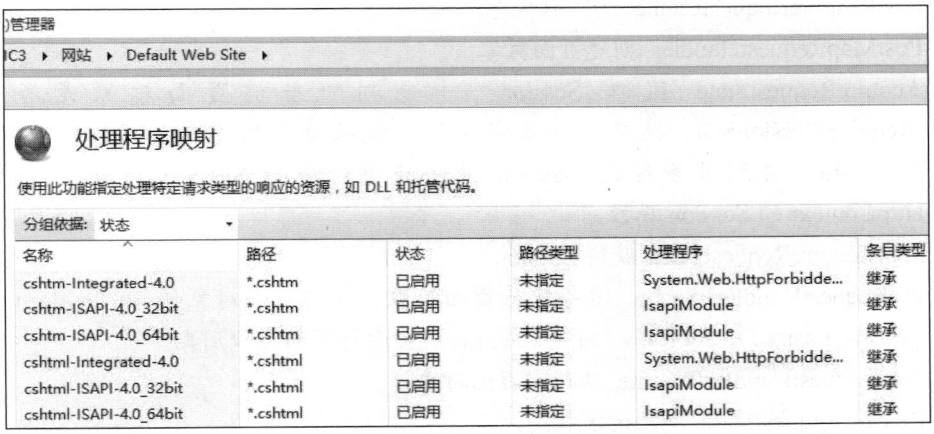

)管理器

IC3 ▸ 网站 ▸ Default Web Site ▸

处理程序映射

使用此功能指定处理特定请求类型的响应的资源，如 DLL 和托管代码。

分组依据: 状态

名称	路径	状态	路径类型	处理程序	条目类型
cshtm-Integrated-4.0	*.cshtm	已启用	未指定	System.Web.HttpForbidde...	继承
cshtm-ISAPI-4.0_32bit	*.cshtm	已启用	未指定	IsapiModule	继承
cshtm-ISAPI-4.0_64bit	*.cshtm	已启用	未指定	IsapiModule	继承
cshtml-Integrated-4.0	*.cshtml	已启用	未指定	System.Web.HttpForbidde...	继承
cshtml-ISAPI-4.0_32bit	*.cshtml	已启用	未指定	IsapiModule	继承
cshtml-ISAPI-4.0_64bit	*.cshtml	已启用	未指定	IsapiModule	继承

图 5-25

（4）对于动态资源的处理，IIS 首先会通过一个工作进程去加载具体的处理组件 dll。以 IIS 6.0 为例，如果 IIS 判断它自己无法处理 asp.net 的请求，就会由 W3WP.exe 所维护的工作进程来加载 aspnet_isapi.dll。

（5）如果 Web 应用程序是第一次加载，那么首先会由 aspnet_isapi.dll 加载.NET 运行时（主要是调用服务器上的.Net Framework 创建 CLR 运行时）。IIS 工作进程里有一个应用程序池，其中可以承载多个应用程序域 AppDomain。

（6）HTTP.SYS 接收请求，通过应用程序域工厂 AppDomainFactory 创建应用程序域 AppDomain。

（7）一个 IsapiRuntime 被加载，并创建一个 IsapiWorkerRequest 对象封装当前的 HTTP 请求，并把该 IsapiWorkerRequest 对象传递给 ASP.NET 的 HttpRuntime 运行时，此时，HTTP 请求开始进入 ASP.NET 请求管道。HttpRuntime 是 ASP.NET 请求管道的入口。当请求进来时，首先进入 HttpRuntime，由 HttpRuntime 来决定如何处理请求。

（8）HttpRuntime 维护着一个 HttpApplication 池，当有 HTTP 请求过来时就从池中选取可用的 HttpApplication 处理请求。

（9）HttpRuntime 根据 IsapiWorkerRequest 对象创建 HttpContext 对象。

（10）HttpApplicationFactory 创建新的或者从 HttpApplication 池获取现有的、可用的 HttpApplication 对象。

（11）HttpApplication 调用 ProcessRequest 方法，内部执行 19 个管道事件。MVC 的 19 个管道事件如下：

- BeginRequest 开始处理请求。
- AuthenticateRequest 授权验证请求，获取用户授权信息。
- PostAuthenticateRequest 获取成功。
- AunthorizeRequest 授权，一般用来检查用户是否获得权限。
- PostAuthorizeRequest 获得授权。
- ResolveRequestCache 获取页面缓存结果(如果没有则执行)。
- PostResolveRequestCache 已获取缓存。
- PostMapRequestHandler 创建页面对象。
- AcquireRequestState 获取 Session——先判断当前页面对象是否实现了 IRequiresSessionState 接口，如果实现了，就从浏览器发来的请求报文头中获得 SessionId，并到服务器的 Session 池中获得对应的 Session 对象，最后赋值给 HttpContext 的 Session 属性。
- PostAcquireRequestState 获得 Session。
- PreRequestHandlerExecute 准备执行页面对象。执行页面对象的 ProcessRequest 方法（1.ashx,1.aspx）。如果请求的是 1.aspx，就会运行页面生命周期。
- PostRequestHandlerExecute 执行完页面对象了。
- ReleaseRequestState 释放请求状态。
- PostReleaseRequestState 已释放请求状态。
- UpdateReuqestCache 更新缓存。
- PostUpdateRequestCache 已更新缓存。
- LogRequest 日志记录。
- PostLogRequest 已完成日志。
- EndRequest 完成。

5.5.2 路由注册

应用程序一启动就会调用 Global.asax 中的 Application_Start()方法。这个方法中的 RouteConfig.RegisterRoutes(RouteTable.Routes);进行了路由注册。

首先看一下 RouteTable.Routes 这个静态属性，它取的就是_instance 静态变量，代码如下：

```
public class RouteTable
{
    private static RouteCollection _instance = new RouteCollection();

    public static RouteCollection Routes
    {
```

```
            get
            {
                return _instance;
            }
        }
}
```

（2）查看 RouteConfig.RegisterRoutes 方法：

```
public static void RegisterRoutes(RouteCollection routes)
{
            routes.IgnoreRoute("{resource}.axd/{*pathInfo}");

            routes.MapRoute(
                name: "Default",
                url: "{controller}/{action}/{id}",
                defaults: new { controller = "Home", action = "Index", id =
                UrlParameter.Optional }
            );
}
```

（3）接着看 routes.MapRoute 方法，很显然它就是一个扩展方法，通过这个扩展方法把路由对象都添加到 RouteCollection routes 这个路由集合对象中去，而这个 routes 对象就是 RouteTable 类中的静态变量_instance。从下面的代码中可以看到 routes 是一个键值对列表，路由名称作为 key，其他的属性封装成一个 Route 对象作为 Value。

```
public static Route MapRoute(this RouteCollection routes, string name, string url,
object defaults, object constraints, string[] namespaces)
{
            if (routes == null)
            {
                throw new ArgumentNullException("routes");
            }
            if (url == null)
            {
                throw new ArgumentNullException("url");
            }
            Route item = new Route(url, new MvcRouteHandler()) {
                Defaults = CreateRouteValueDictionary(defaults),
                Constraints = CreateRouteValueDictionary(constraints),
                DataTokens = new RouteValueDictionary()
            };
            if ((namespaces != null) && (namespaces.Length > 0))
```

```
        {
            item.DataTokens["Namespaces"] = namespaces;
        }
        routes.Add(name, item);
        return item;
}
```

至此，路由注册就完成了。

注意：整个 MVC 只有一个路由集合，因为它是静态变量。

5.5.3 创建 MvcHandler 对象

在管道第 7 个事件的时候创建一个 MvcHandler 对象，然后赋值给 HttpContext 对象的 remapHanlder 属性，存入 HttpContext 对象中。

（1）当浏览器 URL 请求是非静态资源时，根据 URL 到路由表里查找匹配 URL 规则的路由。

（2）遍历路由表（RouteCollection）里的所有路由，元素类型为 Route，并调用每个 Route 对象的 GetRouteData 方法。UrlRoutingModule 类的 PostResolveRequestCache 方法中有如下代码：

```
RouteData routeData = this.RouteCollection.GetRouteData(context);
//context 是 HttpContextBase 类型
```

（3）Route 类中的 GetRouteData 方法创建了一个 RouteData 对象，并将 Route 里的 RouteHandler 对象传进去，如下代码所示：

```
RouteData data = new RouteData(this, this.RouteHandler); //这里的 this 是 Route 对象
```

这个 this.RouteHandler 对应着 MvcRouteHandler 对象。

```
Route item = new Route(url, new MvcRouteHandler()) {
            Defaults = CreateRouteValueDictionary(defaults),
            Constraints = CreateRouteValueDictionary(constraints),
            DataTokens = new RouteValueDictionary()
};
```

调用 ProcessConstraints 检查约束，如果 URL 不满足约束就直接返回 null，停止方法执行，最后将 Route 对象里的 URL 默认值和命名空间存入 RouteData，最后返回 RouteData 对象。

（4）创建 MvcHandler，并将 RequestContext 对象存进来。

```
RequestContext requestContext = new RequestContext(context, routeData);
context.Request.RequestContext = requestContext;
```

```
IHttpHandler httpHandler = routeHandler.GetHttpHandler(requestContext);
```

MvcRouteHandler 对象中有一个 GetHttpHandler 方法，这里新建了一个 MvcHandler 对象：

```
protected virtual IHttpHandler GetHttpHandler(RequestContext requestContext)  {
requestContext.HttpContext.SetSessionStateBehavior(this.GetSessionStateBehavior(r
equestContext));
          return new MvcHandler(requestContext);
}
```

PostResolveRequestCache 方法的完整代码如下：

```
public virtual void PostResolveRequestCache(HttpContextBase context)
{
   RouteData routeData = this.RouteCollection.GetRouteData(context);//获取路由数据
   if (routeData != null)
   {
      IRouteHandler routeHandler = routeData.RouteHandler;
      if (routeHandler == null)
      {
         throw new InvalidOperationException(string.Format(CultureInfo.CurrentCulture,
         SR.GetString("UrlRoutingModule_NoRouteHandler"), new object[0]));
      }
      if (!(routeHandler is StopRoutingHandler))
      {
         RequestContext requestContext = new RequestContext(context, routeData);
         context.Request.RequestContext = requestContext;
         IHttpHandler httpHandler = routeHandler.GetHttpHandler(requestContext);
         if (httpHandler == null)
         {
            object[] args = new object[] { routeHandler.GetType() };
            throw newInvalidOperationException(string.Format(CultureInfo.
            CurrentUICulture, SR.GetString("UrlRoutingModule_NoHttpHandler"), args));
         }
         if (httpHandler is UrlAuthFailureHandler)
         {
            if (!FormsAuthenticationModule.FormsAuthRequired)
            {
               throw new HttpException(0x191, SR.GetString("Assess_Denied_Description3"));
            }

UrlAuthorizationModule.ReportUrlAuthorizationFailure(HttpContext.Current, this);
         }
```

```
            else
            {
                context.RemapHandler(httpHandler);
            }
        }
    }
}
```

5.5.4 执行 MvcHandler ProcessRequest 方法

第 8 个事件会判断 HttpContext 中的 remapHandler 对象是否为空，如果为空，就根据 URL 创建页面对象，否则不创建页面对象，直接跳过，执行后面的事件，如图 5-26 所示。

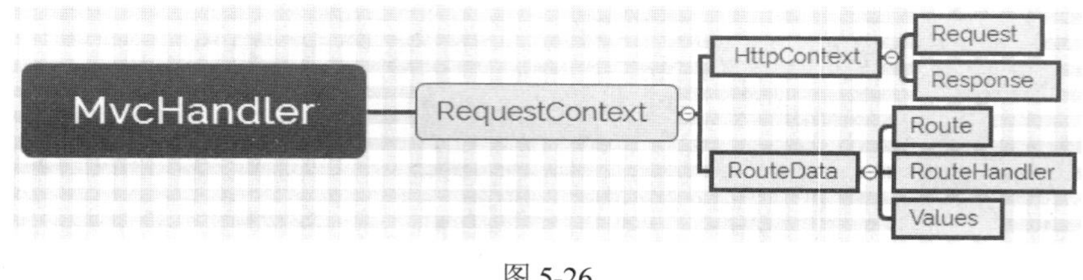

图 5-26

源码分析

（1）首先我们来看 MvcHandler 这个类，位于 System.Web.Mvc 命名空间下，并有如下属性：

```
public RequestContext RequestContext { get; private set; }
public class RequestContext
{
    // Methods
    public RequestContext();
    public RequestContext(HttpContextBase httpContext, RouteData routeData);

    // Properties
    public virtual HttpContextBase HttpContext { get; set; }
    public virtual RouteData RouteData { get; set; }
}
```

再来看一下 RouteData 类：

```
public class RouteData
{
    // Fields
    private RouteValueDictionary _dataTokens;
    private IRouteHandler _routeHandler;
```

```
    private RouteValueDictionary _values;

    // Methods
    public RouteData();
    public RouteData(RouteBase route, IRouteHandler routeHandler);
    public string GetRequiredString(string valueName);

    // Properties
    public RouteValueDictionary DataTokens { get; }
    public RouteBase Route { get; set; }
    public IRouteHandler RouteHandler { get; set; }
    public RouteValueDictionary Values { get; }
}
```

HttpContextBase 类中有如下属性：

```
public virtual HttpRequestBase Request { get; }
public virtual HttpResponseBase Response { get; }
public virtual HttpServerUtilityBase Server { get; }
public virtual HttpSessionStateBase Session { get; }
```

RouteTable 类中有一个静态路由集合对象 Routes：

```
public class RouteTable
{
    // Fields
    private static RouteCollection _instance;

    // Methods
    static RouteTable();
    public RouteTable();

    // Properties
    public static RouteCollection Routes { get; }
}
```

RouteCollection 类中有一个 MapPageRoute 方法：

```
public Route MapPageRoute(string routeName, string routeUrl, string physicalFile,
bool checkPhysicalUrlAccess, RouteValueDictionary defaults, RouteValueDictionary
constraints, RouteValueDictionary dataTokens)
{
    if (routeUrl == null)
    {
```

```
        throw new ArgumentNullException("routeUrl");
    }
    Route item = new Route(routeUrl, defaults, constraints, dataTokens, new
    PageRouteHandler(physicalFile, checkPhysicalUrlAccess));
    this.Add(routeName, item);
    return item;
}
```

（2）MvcHandler 中有 ProcessRequest 方法，代码如下：

```
protected internal virtual void ProcessRequest(HttpContextBase httpContext)
{
        IController controller;
        IControllerFactory factory;
        this.ProcessRequestInit(httpContext, out controller, out factory);
        try
        {
            controller.Execute(this.RequestContext);//调用控制器父类 ControllerBase
            的 Execute 方法
        }
        finally
        {
            factory.ReleaseController(controller);
        }
}
```

在管道的第 11、12 事件之间执行 MvcHandler 的 ProcessRequest 方法，创建控制器类对象。在这个方法里面又调用了方法 ProcessRequestInit。ProcessRequestInit 方法的代码如下：

```
private void ProcessRequestInit(HttpContextBase httpContext, out IController
controller, out IControllerFactory factory)
{
    HttpContext current = HttpContext.Current;
    if ((current != null) && (ValidationUtility.IsValidationEnabled(current) ==
    true))
    {
        ValidationUtility.EnableDynamicValidation(current);
    }
    this.AddVersionHeader(httpContext);
    this.RemoveOptionalRoutingParameters();
    string requiredString = this.RequestContext.RouteData.GetRequiredString
    ("controller"); //获取控制器名字
    factory = this.ControllerBuilder.GetControllerFactory(); //获取控制器工厂对象
```

```
    controller = factory.CreateController(this.RequestContext, requiredString);
    //通过工厂创建被请求的控制器类对象
    if (controller == null)
    {
        throw new InvalidOperationException(string.Format(CultureInfo.CurrentCulture,
        MvcResources.ControllerBuilder_FactoryReturnedNull, new object[]
        { factory.GetType(), requiredString }));
    }
}
```

5.5.5　调用控制器里面的 Action 方法

（1）ControllerBase 类中有如下 Initialize 方法：

```
protected virtual void Initialize(RequestContext requestContext)
{
        this.ControllerContext = new ControllerContext(requestContext, this);
        //将上下文对象赋值给 ControllerContext，requestContext 就是 RequestContext，this
        //表示控制器本身
}
```

（2）ControllerBase 类中的 Execute 方法代码如下：

```
protected virtual void Execute(RequestContext requestContext)
{
        if (requestContext == null)
        {
            throw new ArgumentNullException("requestContext");
        }
        if (requestContext.HttpContext == null)
        {
            throw new ArgumentException(MvcResources.ControllerBase_
             CannotExecuteWithNullHttpContext, "requestContext");
        }
        this.VerifyExecuteCalledOnce(); //验证这个方法是否只调用一次
        this.Initialize(requestContext);
        using (ScopeStorage.CreateTransientScope())
        {
            this.ExecuteCore(); //执行控制器中的核心处理方法
        }
}
```

（3）这里的 this.ExecuteCore()方法是由 ControllerBase 的子类 Controller 来实现的：

```
protected override void ExecuteCore()
{
    this.PossiblyLoadTempData();
    try
    {
        string requiredString = this.RouteData.GetRequiredString("action");
        //获取 Action 方法名称
        //调用控制器里面的 Action 方法
        if(!this.ActionInvoker.InvokeAction(base.ControllerContext, requiredString))
        {
            this.HandleUnknownAction(requiredString);
        }
    }
    finally
    {
        this.PossiblySaveTempData();
    }
}
```

（4）在调用 Action 方法之前会执行过滤器。ActionInvoker 是一个 IActionInvoker 类型的接口，可以找到它的实现类 ControllerActionInvoker：

```
public class ControllerActionInvoker : IActionInvoker
```

找到 InvokeAction 方法：

```
public virtual bool InvokeAction(ControllerContext controllerContext, string
actionName)
{
    if (controllerContext == null)
    {
        throw new ArgumentNullException("controllerContext");
    }
    if (string.IsNullOrEmpty(actionName))
    {
        throw new ArgumentException(MvcResources.Common_NullOrEmpty, "actionName");
    }
    ControllerDescriptor controllerDescriptor =
    this.GetControllerDescriptor(controllerContext);
    //获取控制器描述对象（控制器名称、类型..）
    ActionDescriptor actionDescriptor = this.FindAction(controllerContext,
    controllerDescriptor, actionName); //获取 Action 描述对象（Action 名称，控制器描述）
    if (actionDescriptor == null)
```

```
{
   return false;
}
FilterInfo filters = this.GetFilters(controllerContext,
actionDescriptor);//获取所有的过滤器
try
{
  //最先执行授权过滤器
   AuthorizationContext context =
   this.InvokeAuthorizationFilters(controllerContext,
   filters.AuthorizationFilters, actionDescriptor);
   if (context.Result != null)
   {
      this.InvokeActionResult(controllerContext, context.Result);
   }
   else
   {
      if (controllerContext.Controller.ValidateRequest)
      {
         ValidateRequest(controllerContext); //验证文本输入数据
      }
      IDictionary<string, object> parameterValues =
      this.GetParameterValues(controllerContext, actionDescriptor);
      //获取浏览器传过来的参数
      ActionExecutedContext context2 = this.InvokeActionMethodWithFilters
      (controllerContext, filters.ActionFilters, actionDescriptor,
      parameterValues); //获得 Action 方法执行结果上下文
      this.InvokeActionResultWithFilters(controllerContext,
      filters.ResultFilters, context2.Result);
      // context2.Result，控制器中 Action 方法中 View 方法返回对象 ViewResult
   }
}
catch (ThreadAbortException)
{
   throw;
}
catch (Exception exception)
{
   ExceptionContext context3 = this.InvokeExceptionFilters(controllerContext,
   filters.ExceptionFilters, exception);
   if (!context3.ExceptionHandled)
   {
```

```
            throw;
        }
        this.InvokeActionResult(controllerContext, context3.Result);
    }
    return true;
}
```

5.5.6 根据 Action 方法返回的 ActionResult 加载 View

（1）InvokeActionResultWithFilters 方法如下：

```
protected virtual ResultExecutedContext InvokeActionResultWithFilters
(ControllerContext controllerContext, IList<IResultFilter> filters,
ActionResult actionResult)
{
    ResultExecutingContext preContext = new ResultExecutingContext
    (controllerContext, actionResult);
    Func<ResultExecutedContext> seed = delegate {
        this.InvokeActionResult(controllerContext, actionResult);
        return new ResultExecutedContext(controllerContext, actionResult, false, null);
    };
    return filters.Reverse<IResultFilter>().Aggregate<IResultFilter,
    Func<ResultExecutedContext>>(seed, (next, filter) => () =>
    InvokeActionResultFilter(filter, preContext, next))();
}
```

（2）这个方法里面又调用了 this.InvokeActionResult 方法：

```
protected virtual void InvokeActionResult(ControllerContext controllerContext,
ActionResult actionResult)
{
    actionResult.ExecuteResult(controllerContext);
}
```

（3）InvokeActionResult 方法中调用了 ActionResult 抽象类中的 ExecuteResult 方法。子类 ViewResultBase 的具体实现如下：

```
public override void ExecuteResult(ControllerContext context)
{
    if (context == null)
    {
        throw new ArgumentNullException("context");
    }
```

```
    //先检查是否传入指定视图的名称，如果没有传入，就取 Action 方法的名字作为要读取的视图名称
    if (string.IsNullOrEmpty(this.ViewName))
    {
        this.ViewName = context.RouteData.GetRequiredString("action");
    }
    ViewEngineResult result = null;
    if (this.View == null)
    {
        result = this.FindView(context);
        //找到对应的视图引擎，this 是 ViewResultBase 的某个子类（ViewResult）
        this.View = result.View;
    }
    TextWriter output = context.HttpContext.Response.Output;
    ViewContext viewContext = new ViewContext(context, this.View, this.ViewData,
    this.TempData, output);
    this.View.Render(viewContext, output);
    if (result != null)
    {
        result.ViewEngine.ReleaseView(context, this.View);
    }
}
```

（4）ViewResult 类：

```
protected override ViewEngineResult FindView(ControllerContext context)
{
    //调用方法找到合适的视图引擎（遍历视图引擎集合，哪个视图引擎可以找到对应的视图就返回哪个视
    //图引擎的结果，此结果中包含视图接口对象）
    ViewEngineResult result = base.ViewEngineCollection.FindView(context,
    base.ViewName, this.MasterName);
    if (result.View != null)
    {
        return result;
    }
    StringBuilder builder = new StringBuilder();
    foreach (string str in result.SearchedLocations)
    {
        builder.AppendLine();
        builder.Append(str);
    }
    throw new InvalidOperationException(string.Format(CultureInfo.CurrentCulture,
    MvcResources.Common_ViewNotFound, new object[] { base.ViewName, builder }));
}
```

此方法返回了一个 ViewEngineResult 对象，获取返回的 ViewEngineResult 里的 View 对象，然后调用 Render 方法来生成 HTML 代码并写入 Response 中。

```
public abstract class BuildManagerCompiledView : IView
public void Render(ViewContext viewContext, TextWriter writer)
{
    if (viewContext == null)
    {
        throw new ArgumentNullException("viewContext");
    }
    object instance = null;
    Type compiledType = this.BuildManager.GetCompiledType(this.ViewPath);
    if (compiledType != null)
    {
        instance = this.ViewPageActivator.Create(this._controllerContext,
        compiledType);
    }
    if (instance == null)
    {
        throw   new   InvalidOperationException(string.Format(CultureInfo.CurrentCulture,
MvcResources.CshtmlView_ViewCouldNotBeCreated, new object[] { this.ViewPath }));
    }
    this.RenderView(viewContext, writer, instance);
}
```

至此，整个 MVC 运行流程就结束了。

第 6 章

网站性能和安全优化

只要做 Web 开发，就不得不考虑性能和安全。因为这一块内容是和公司的利益紧密联系的。网站性能好，用户体验就好，系统所能承受的压力就大，所能提供的服务也就越好。就拿电商网站天猫来说，如果网站性能不给力，遇到双 11 这种情况，客户下单时，系统直接卡死或者崩溃了，天猫一天要亏多少钱啊！

安全比性能更重要，如果一个网站连最基本的安全都无法保障，那么一旦网站被入侵，后果就非常严重了。举个例子，如果淘宝被人黑了，导致系统停运一天，或者支付宝被人入侵了，许多客户里面的钱被直接转走了，那么给公司造成的损失将会是巨大的。

6.1 缓存

一提到性能，我们第一时间想到的就是缓存。

1. 缓存的分类

- ASP.NET Identity。
- 客户端缓存（Client Caching）。
- 代理缓存（Proxy Caching）。
- 反向代理缓存（Reverse Proxy Caching）。
- 服务器缓存（Web Server Caching）。

2. 缓存的好处及应用

- 可以让 css、js、image、aspx 等资源文件在第二次访问时读取本地而不用再次请求服务器端，减少客户端对服务器资源请求的压力，加快客户端响应速度。
- 对于经常使用的数据源，将其存储在数据缓存或者内存中，这样来减少数据库请求，缓解数据库压力。
- 将网站部署在多台计算机上，采用分布式方式处理，可以有效解决多个用户请求对一台服务器所造成的压力，加快客户端请求响应（分布式部署）。
- 对经常访问但数据经常不更新的页面进行静态化处理，有效减少服务器压力和客户端响应速度。

3. OutputCache 属性

- Duration：页或用户控件进行缓存的时间（以秒计）。在页或用户控件上设置该特性为来自对象的 HTTP 响应建立一个过期策略，并将自动缓存页或用户控件输出。
- Location：OutputCacheLocation 枚举值之一，默认值为 Any。
- Any：输出缓存可位于产生请求的浏览器客户端、参与请求的代理服务器（或任何其他服务器）或处理请求的服务器上。此值对应于 HttpCacheability.Public 枚举值。
- Client：输出缓存位于产生请求的浏览器客户端上。此值对应于 HttpCacheability.Private 枚举值。
- Downstream：输出缓存可存储在任何 HTTP 1.1 可缓存设备中，源服务器除外。这包括代理服务器和发出请求的客户端。
- None：对于请求的页，禁用输出缓存。此值对应于 HttpCacheability.NoCache 枚举值。
- Server：输出缓存位于处理请求的 Web 服务器上。此值对应于 HttpCacheability.Server 枚举值。
- ServerAndClient：输出缓存只能存储在源服务器或发出请求的客户端中。代理服务器不能缓存响应。此值对应于 HttpCacheability.Private 和 HttpCacheability.Server 枚举值的组合。
- CacheProfile：与该页关联的缓存设置的名称。这是可选特性，默认值为空字符串("")。
- NoStore：一个布尔值，它决定了是否阻止敏感信息的二级存储。
- ProviderName：一个字符串值，标识要使用的自定义输出缓存提供程序。
- Shared：一个布尔值，确定用户控件输出是否可以由多个页共享，默认值为 false。
- SqlDependency：标识一组数据库/表名称对的字符串值，页或控件的输出缓存依赖于这些名称对。注意，SqlCacheDependency 类监视输出缓存所依赖的数据库中的表，因此当更新表中的项时，使用基于表的轮询将从缓存中移除这些项。
- CommandNotification：使用通知（在 Microsoft SQL Server 2005 中），最终将会使用 SqlDependency 类向 SQL Server 2005 服务器注册查询通知。
- VaryByCustom：任何表示自定义输出缓存要求的文本。如果特性的赋值为 browser，那么缓存将随浏览器名称和主要版本信息的不同而异。如果输入自定义字符串，就必须在应用程序的 Global.asax 文件中重写 GetVaryByCustomString 方法。
- VaryByHeader：分号分隔的 HTTP 标头列表，用于使输出缓存发生变化。将该特性设为多标头时，对于每个指定标头组合，输出缓存都包含一个不同版本的请求文档。
- VaryByParam：分号分隔的字符串列表，用于使输出缓存发生变化。默认情况下，这些字符串对应于使用 GET 方法特性发送的查询字符串值，或者使用 POST 方法发送的参数。将该特性设置为多个参数时，对于每个指定参数组合，输出缓存都包含一个不同版本的请求文档。可能的值包括 none、星号 (*) 以及任何有效的查询字符串或 POST 参数名称。

- VaryByControl：分号分隔的字符串列表，用于改变用户控件的输出缓存。这些字符串代表用户控件中声明的 ASP.NET 服务器控件的 ID 属性值。
- VaryByContentEncodings：：以分号分隔的字符串列表，用于更改输出缓存。将 VaryByContentEncodings 特性用于 Accept-Encoding 标头，可确定不同内容编码获得缓存响应的方式。有关如何指定 Accept-Encoding 标头的更多信息，可参见 W3C 网站上的 Hypertext Transfer Protocol——HTTP/1.1。

4. 示例 1

（1）添加 Action 方法 Index：

```
[OutputCache(Duration=5,VaryByParam="none")]
public ActionResult Index()
{
    ViewBag.Now = DateTime.Now.ToString();
    Response.Cache.SetOmitVaryStar(true); //解决一个隐蔽的 bug
    return View();
}
```

（2）添加 Index 视图：

```
@{
    ViewBag.Title = "Index";
}
<h2>@ViewBag.Now</h2>
```

（3）运行结果如图 6-1 所示。

2016/5/21 17:19:07

图 6-1

按 F12 键打开 Firebug，然后按 F5 键刷新浏览器界面。我们会发现网络请求状态为 304 Not Modified，表示取的是缓存数据，如图 6-2 所示。

图 6-2

5. 示例 2

（1）添加 Action 方法 GetName：

```
[OutputCache(Duration=5,VaryByParam="name")]
public ActionResult GetName(string name)
{
    ViewData["Name"] = name;
    Response.Cache.SetOmitVaryStar(true); //解决一个隐蔽的bug
    return View();
}
```

（2）添加 GetName 视图：

```
@{
    ViewBag.Title = "GetName";
}
<h2>@ViewData["Name"]</h2>
```

（3）在浏览器中输入地址“http://localhost:13043/Home/GetName?name=a”，依旧使用 Firebug 工具进行监听，只要 name 后面的参数名称不变，在 5 秒内，状态都是 304，而只要 name 后面的参数改变，状态就是 200。

6.2 压缩合并 css 和 js

在 APS.NET MVC4 中，App_Start 文件夹下面多了一个 BundleConfig.cs 类，专门用于压缩合并文件的，默认情况下压缩合并功能是开启的，当然我们也可以使用“BundleTable.EnableOptimizations = true;”来显示设置开启。

但是，注意要在 Web.config 中将调试设置为 false，压缩才会生效，Web.config 中配置如下：

```
<compilation debug="false" targetFramework="4.5" />
```

BundleConfig 类配置如下：

```
bundles.Add(new ScriptBundle("~/bundles/jquery").Include("~/Scripts/jquery.min.js",
"~/Scripts/jquery.easyui.min.js"));// "~/Scripts/jquery-{version}.js",
bundles.Add(new StyleBundle("~/Content/css").Include("~/Content/themes/default/easyui.css",
"~/Content/themes/icon.css"));//"~/Content/site.css",
```

我们来看下压缩合并前和压缩合并后的对比，压缩合并前如图 6-3、图 6-4 所示。

```
6        <title>Index</title>
7        <link href="/Content/themes/bootstrap/easyui.css" rel="stylesheet" />
8        <link href="/Content/themes/icon.css" rel="stylesheet" />
9        <script src="/Scripts/jquery.min.js"></script>
10       <script src="/Scripts/jquery.easyui.min.js"></script>
```

图 6-3

Name	Status	Type	Initiator	Size	Time	Timeline
localhost	200	document	Other	1.8 KB	412 ms	
easyui.css	200	stylesheet	(index):7	13.0 KB	146 ms	
icon.css	200	stylesheet	(index):8	912 B	145 ms	
jquery.min.js	200	script	(index):9	94.1 KB	26 ms	
jquery.easyui.min.js	200	script	(index):10	368 KB	177 ms	
blank.gif	200	gif	jquery.min.js:4	418 B	17 ms	
panel_tools.png	200	png	jquery.min.js:4	577 B	19 ms	
layout_arrows.png	200	png	jquery.min.js:4	703 B	15 ms	
ok.png	200	png	jquery.min.js:4	1.2 KB	5 ms	
loading.gif	200	gif	jquery.min.js:4	2.1 KB	84 ms	
datagrid_data1.json	404	xhr	jquery.min.js:5	5.0 KB	10 ms	
accordion_arrows.png	200	png	jquery.min.js:4	571 B	5 ms	

图 6-4

压缩合并后如图 6-5、图 6-6 所示。

```
<link href="/Content/css?v=grHCwXdL-yNcirS5Ut9H4v6ZrEAyJcHV_S_WTf0cxUk1" rel="stylesheet"/>

<script src="/bundles/jquery?v=VLIumFmvwODR2qX9hE4z1CaO3hVi83asiFRjFMX_TsA1"></script>
```

图 6-5

Name	Status	Type	Initiator	Size	Time	Timeline
localhost	200	document	Other	1.9 KB	394 ms	
css?v=grHCwXdL-yNcirS5...	304	stylesheet	(index):8	258 B	13 ms	
jquery?v=VLIumFmvwOD...	304	script	(index):10	262 B	13 ms	

图 6-6

很明显，我们看到文件被合并了，减少了网络请求数，同时文件的大小也减小了，说明被压缩处理了。

注意：我们只能合并同一类型的文件，也就是说不能把 js 和 css 文件合并到一起，只能单独合并 js 文件和 css 文件。

6.3 删除无用的视图引擎

默认情况下，ASP.NET MVC 同时支持 WebForm 和 Razor 引擎，而我们通常在同一个项目中只用到了一种视图引擎，如 Razor，那么，我们就可以移除掉没有使用的视图引擎，提高 View 视图的检索效率。在没有删除 WebForm 引擎之前，检索控制器中不存在的视图时，可以看到视图的检索顺序先是 Home 目录，再是 Shared 目录下面的 aspx、 ascx 文件，如图 6-7 所示。

图 6-7

在 Global.asax 中添加如下代码：

```
void RemoveWebFormEngines()
{
        var viewEngines = ViewEngines.Engines;
        var webFormEngines = viewEngines.OfType<WebFormViewEngine>().
 FirstOrDefault();
        if (webFormEngines != null)
        {
            viewEngines.Remove(webFormEngines);
        }
}

protected void Application_Start()
{
        RemoveWebFormEngines(); //移除 WebForm 视图引擎
        AreaRegistration.RegisterAllAreas();

        WebApiConfig.Register(GlobalConfiguration.Configuration);
        FilterConfig.RegisterGlobalFilters(GlobalFilters.Filters);
        RouteConfig.RegisterRoutes(RouteTable.Routes);
        BundleConfig.RegisterBundles(BundleTable.Bundles);
        AuthConfig.RegisterAuth();
}
```

运行结果如图 6-8 所示。

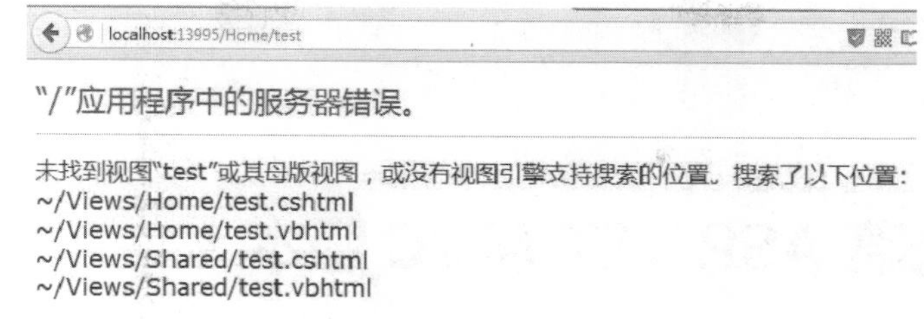
localhost:13995/Home/test

"/"应用程序中的服务器错误。

未找到视图"test"或其母版视图，或没有视图引擎支持搜索的位置。搜索了以下位置:
~/Views/Home/test.cshtml
~/Views/Home/test.vbhtml
~/Views/Shared/test.cshtml
~/Views/Shared/test.vbhtml

图 6-8

6.4 使用防伪造令牌来避免 CSRF 攻击

对表达提交来说，要关注的就是安全问题。ASP.NET MVC 提供了探测某种攻击类型的机制，其中一种措施就是防伪造令牌。这种令牌包含服务器端和客户端组件，代码会在表单中插入一个隐藏域以保存用户特定的令牌 @Html.AntiForgeryToken():

```
@using (Html.BeginForm(new { ReturnUrl = ViewBag.ReturnUrl })) {
   @Html.AntiForgeryToken()
```

@Html.AntiForgeryToken()只能添加在 Html.BeginForm()形式声明的表单中，纯 HTML 的<form>标签表单是不行的。

Html.AntiForgeryToken 辅助方法会写入一个加密过的数据到用户端浏览器的 Cookie 里，然后在表单内插入一个名为_RequestVerificationToken 的隐藏字段，该隐藏字段的内容在每次刷新页面时都会不一样，每次执行 Action 动作方法时都会让这个隐藏字段的值与 Cookie 的加密数据进行验证比对，符合验证才允许执行这个 Action 方法，而且服务器端会优先在数据处理之前执行这些令牌验证代码，具体如下:

```
[ValidateAntiForgeryToken]
[HttpPost]
[AllowAnonymous]
[ValidateAntiForgeryToken]
public ActionResult Login(LoginModel model, string returnUrl)
{
   if (ModelState.IsValid && WebSecurity.Login(model.UserName, model.Password,
   persistCookie: model.RememberMe))
   {
      return RedirectToLocal(returnUrl);
   }

   // 如果我们进行到这一步时某个地方出错，就重新显示表单
   ModelState.AddModelError("", "提供的用户名或密码不正确。");
```

```
    return View(model);
}
```

6.5 隐藏 ASP.NET MVC 版本

默认情况下，ASP.NET MVC 网站会把版本号提供给浏览器，如图 6-9 所示。

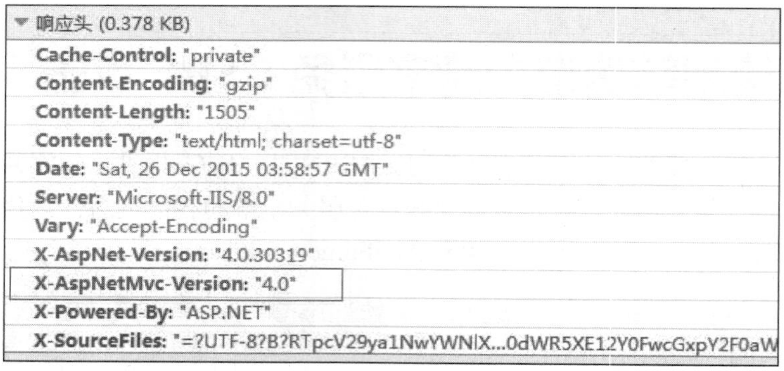

```
▼ 响应头 (0.378 KB)
Cache-Control: "private"
Content-Encoding: "gzip"
Content-Length: "1505"
Content-Type: "text/html; charset=utf-8"
Date: "Sat, 26 Dec 2015 03:58:57 GMT"
Server: "Microsoft-IIS/8.0"
Vary: "Accept-Encoding"
X-AspNet-Version: "4.0.30319"
X-AspNetMvc-Version: "4.0"
X-Powered-By: "ASP.NET"
X-SourceFiles: "=?UTF-8?B?RTpcV29ya1NwYWNlX...0dWR5XE12Y0FwcGxpY2F0aW
```

图 6-9

在 Global.asax 中添加“MvcHandler.DisableMvcResponseHeader = true;”：

```
protected void Application_Start()
{
    MvcHandler.DisableMvcResponseHeader = true;
    AreaRegistration.RegisterAllAreas();

    WebApiConfig.Register(GlobalConfiguration.Configuration);
    FilterConfig.RegisterGlobalFilters(GlobalFilters.Filters);
    RouteConfig.RegisterRoutes(RouteTable.Routes);
    BundleConfig.RegisterBundles(BundleTable.Bundles);
}
```

判断客户端请求是否为 Ajax：Request.IsAjaxRequest。

6.6 Nginx 服务器集群

6.6.1 Nginx 是什么

在学习 Nginx 之前，先来了解几个概念。

代理服务器：一般是指局域网内部的机器通过代理服务器发送请求到互联网上的服务器，

代理服务器一般作用在客户端。应用比如 GoAgent 和 FQ 神器。

完整的代理请求过程：客户端首先与代理服务器创建连接，接着根据代理服务器所使用的代理协议请求对目标服务器创建连接或者获得目标服务器的指定资源。Web 代理（proxy）服务器是网络的中间实体。代理位于 Web 客户端和 Web 服务器之间，扮演“中间人”的角色。HTTP 的代理服务器既是 Web 服务器又是 Web 客户端。代理服务器是介于客户端和 Web 服务器之间的另一台服务器，有了它之后，浏览器不是直接到 Web 服务器去取回网页而是向代理服务器发出请求，信号会先送到代理服务器，由代理服务器来取回浏览器所需要的信息并传送给你的浏览器，如图 6-10 所示。

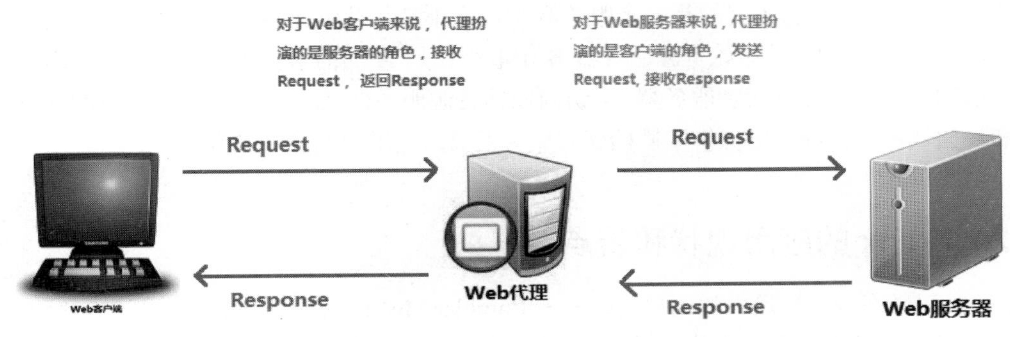

图 6-10

正向代理：是一个位于客户端和原始服务器（origin server）之间的服务器，为了从原始服务器取得内容，客户端向代理发送一个请求并指定目标（原始服务器），然后代理向原始服务器转交请求并将获得的内容返回给客户端。客户端必须要进行一些特别的设置才能使用正向代理。

反向代理服务器：在服务器端接受客户端的请求，然后把请求分发给具体的服务器进行处理，再将服务器的响应结果反馈给客户端。Nginx 就是其中的一种反向代理服务器软件。

Nginx（读为“engine x”），是俄罗斯人 Igor Sysoev（塞索耶夫）编写的一款高性能的 HTTP 和反向代理服务器。也是一个 IMAP/POP3/SMTP 代理服务器；也就是说，Nginx 本身就可以托管网站，进行 HTTP 服务处理，也可以作为反向代理服务器使用。

Nginx 客户端必须设置正向代理服务器，当然前提是要知道正向代理服务器的 IP 地址，还有代理程序的端口。

反向代理正好与正向代理相反。对于客户端而言，代理服务器就像是原始服务器，并且客户端不需要进行任何特别的设置。客户端向反向代理的命名空间（name-space）中的内容发送普通请求，接着反向代理将判断向何处（原始服务器）转交请求，并将获得的内容返回给客户端，如图 6-11 所示。

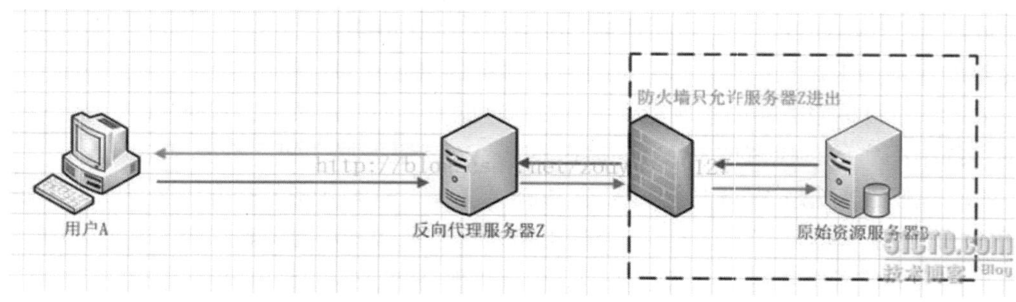

图 6-11

用户 A 始终认为它访问的是原始服务器 B，而不是代理服务器 Z，但实际上反向代理服务器接受用户 A 的应答，从原始资源服务器 B 中取得用户 A 的需求资源，然后发送给用户 A。由于防火墙的作用，只允许代理服务器 Z 访问原始资源服务器 B。尽管在这个虚拟的环境下，防火墙和反向代理的共同作用保护了原始资源服务器 B，但用户 A 并不知情。

6.6.2 Nginx 的应用现状和特点

Nginx 已经在俄罗斯最大的门户网站—— Rambler Media（www.rambler.ru）上运行了 3 年时间，同时俄罗斯超过 20%的虚拟主机平台采用 Nginx 作为反向代理服务器。在国内，已经有淘宝、新浪博客、新浪播客、网易新闻、六间房、56.com、Discuz!、水木社区、豆瓣、YUPOO、海内、迅雷在线等多家网站使用 Nginx 作为 Web 服务器或反向代理服务器。

Nginx 有以下特点：

- 跨平台：Nginx 可以在大多数 UNIX like OS 编译运行，而且也有 Windows 的移植版本。
- 配置异常简单：非常容易上手。配置风格跟程序开发一样，神一般的配置。
- 非阻塞、高并发连接：复制数据时，磁盘 I/O 的第一阶段是非阻塞的。官方测试能够支撑 5 万并发连接，在实际生产环境中跑到 2 万～3 万并发连接数。（这得益于 Nginx 使用了最新的 epoll 模型。）
- 事件驱动：通信机制采用 epoll 模型，支持更大的并发连接。

6.6.3 Nginx 的事件处理机制

对于一个基本的 Web 服务器来说，事件通常有 3 种类型，即网络事件、信号、定时器。

首先看一个请求的基本过程：建立连接→接收数据→发送数据。

再看系统底层的操作：上述过程（建立连接→接收数据→发送数据）在系统底层就是读写事件。

（1）如果采用阻塞调用的方式，当读写事件没有准备好时，必然不能够进行读写事件，所以只好等待，等事件准备好了才能进行读写事件，所以请求将会被耽搁。阻塞调用会进入内核等待，CPU 就会让出去给别人使用。对单线程的 worker 来说显然不合适，当网络事件较多时，大家都在等待，CPU 空闲下来没人用，CPU 利用率自然上不去，更别谈高并发了。

（2）既然没有准备好阻塞调用不行，那么采用非阻塞方式。非阻塞就是事件马上返回 EAGAIN，告诉你事件还没准备好呢，过会儿再来。好吧，过一会儿再来检查事件，直到事件准备好了为止，在这期间你可以先去做其他事情，然后再来看看事件好了没有。虽然不阻塞了，但你需要时不时地过来检查一下事件的状态，虽然可以做更多的事情了，但是带来的开销也是不小的。

非阻塞通过不断检查事件的状态来判断是否进行读写操作，这样带来的开销很大。

（3）因此才有了异步非阻塞的事件处理机制，具体到系统调用就是 select/poll/epoll/kqueue 等。它们提供了一种机制，让你可以同时监控多个事件，调用它们是阻塞的，但可以设置超时时间，在超时时间之内，如果有事件准备好了就返回。这种机制解决了我们上面的两个问题。

以 epoll 为例：当事件没有准备好时就放入 epoll（队列）里面。如果事件准备好了就去处理；如果事件返回的是 EAGAIN，那么继续将其放入 epoll 里面。只要有事件准备好了，我们就去处理它，只有当所有时间都没有准备好时才在 epoll 里面等着。这样，我们就可以并发处理大量的并发了。当然，这里的并发请求是指未处理完的请求，线程只有一个，所以同时能处理的请求也只有一个，只是在请求间进行不断切换而已，切换也是因为异步事件未准备好而主动让出的。这里的切换没有任何代价，可以理解为循环处理多个准备好的事件，事实上就是这样的。

（4）与多线程相比，这种事件处理方式是有很大优势的，不需要创建线程，每个请求占用的内存也很少，没有上下文切换，事件处理非常轻量级。并发数再多也不会导致无谓的资源浪费（上下文切换）。

通过异步非阻塞的事件处理机制，Nginx 实现由进程循环处理多个准备好的事件，从而实现高并发和轻量级。

- master/worker 结构：一个 master 进程，生成一个或多个 worker 进程。
- 内存消耗小：处理大并发的请求，内存消耗非常小。在 3 万并发连接下，开启的 10 个 Nginx 进程才消耗 150MB 内存（15MB × 10=150MB）。
- 成本低廉：Nginx 为开源软件，可以免费使用；而购买 F5 BIG-IP、NetScaler 等硬件负载均衡交换机则需要十多万至几十万人民币。
- 内置的健康检查功能：如果 Nginx Proxy 后端的某台 Web 服务器宕机了，就不会影响前端访问。
- 节省带宽：支持 GZIP 压缩，可以添加浏览器本地缓存的 Header 头。
- 稳定性高：用于反向代理，宕机的概率微乎其微。

6.6.4 Nginx 不为人知的特点

（1）Nginx 代理和后端 Web 服务器间无须长连接。

（2）接收用户请求是异步的，即先将用户请求全部接收下来，再一次性发送后端 Web 服务器，极大地减轻后端 Web 服务器的压力。

（3）发送响应报文时，一边接收来自后端 Web 服务器的数据，一边发送给客户端。

（4）网络依赖性低。Nginx 对网络的依赖程度非常低，从理论上讲，只要能够 ping 通就可以实施负载均衡，而且可以有效区分内网和外网流量。

（5）支持服务器检测。Nginx 能够根据应用服务器处理页面返回的状态码、超时信息等检测服务器是否出现故障，并及时返回错误的请求，重新提交到其他节点上。

6.6.5 Nginx 的内部模型

Nginx 的内部模型如图 6-12 所示。

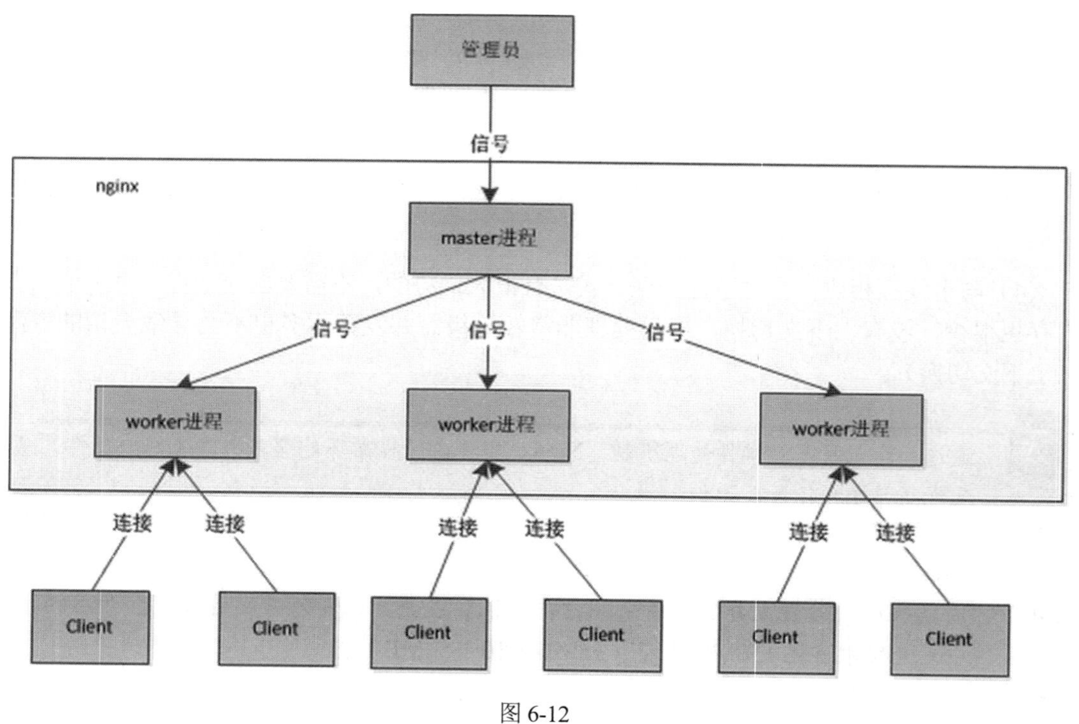

图 6-12

Nginx 是以多进程的方式来工作的，当然 Nginx 也是支持多线程方式的，只是主流方式还是多进程方式，也是 Nginx 的默认方式。Nginx 采用多进程的方式有诸多好处。

（1）Nginx 在启动后会有一个 master 进程和多个 worker 进程。master 进程主要用来管理 worker 进程，包含接收来自外界的信号、向各 worker 进程发送信号、监控 worker 进程的运行状态、当 worker 进程退出后（异常情况下）自动重新启动新的 worker 进程。而基本的网络事件，则是放在 worker 进程中来处理。多个 worker 进程之间是对等的，它们同等竞争来自客户端的请求，各进程互相之间是独立的。一个请求只可能在一个 worker 进程中处理，一个 worker 进程不可能处理其他进程的请求。worker 进程的个数是可以设置的，一般我们会设置与机器 CPU 核数一致，这与 Nginx 的进程模型以及事件处理模型是分不开的。

（2）Master 接收到信号以后怎样进行处理（./nginx -s reload）?首先 master 进程在接到信号后会先重新加载配置文件，再启动新的进程，并向所有老的进程发送信号，告诉它们可以光荣退休了。新的进程在启动后会开始接收新的请求，而老的进程在收到来自 master 的信号后则不再接收新的请求，并且在当前进程中所有未处理完的请求处理完成后再退出。

（3）worker 进程又是如何处理请求的呢？我们前面有提到，worker 进程之间是平等的，每个进程处理请求的机会也是一样的。当我们提供 80 端口的 HTTP 服务时，一个连接请求过来，每个进程都有可能处理这个连接，怎么做到的呢？首先，每个 worker 进程都是从 master 进程 fork 过来，在 master 进程里面先建立好需要 listen 的 socket，再 fork 出多个 worker 进程，这样每个 worker 进程都可以去 accept 这个 socket（当然不是同一个 socket，只是每个进程的这个 socket 会监控在同一个 IP 地址与端口，这个在网络协议里面是允许的）。一般来说，当一个连接进来后，所有在 accept 这个 socket 上面的进程都会收到通知，而只有一个进程可以 accept 这个连接，其他的则 accept 失败，这是所谓的惊群现象。当然，Nginx 也不会视而不见，所以 Nginx 提供了一个 accept_mutex，从名字上我们可以看出这是一个加在 accept 上的一把共享锁。有了这把锁之后，同一时刻就只会有一个进程在 accept 连接，这样就不会有惊群问题了。accept_mutex 是一个可控选项，我们可以显示地关掉，默认是打开的。当一个 worker 进程在 accept 这个连接之后就开始读取请求、解析请求、处理请求，产生数据后再返回给客户端，最后才断开连接。一个完整的请求就是这样的。我们可以看到，一个请求完全由 worker 进程来处理，而且只在一个 worker 进程中处理。

（4）Nginx 采用这种进程模型有什么好处呢？采用独立的进程，可以让互相之间不受影响，一个进程退出后，其他进程还在工作，服务不会中断，master 进程则很快重新启动新的 worker 进程。当然，worker 进程的异常退出肯定是程序有 bug 了，异常退出会导致当前 worker 上的所有请求失败，不过不会影响到所有请求，所以降低了风险。当然，好处还有很多，大家可以慢慢体会。

（5）有人可能要问了，Nginx 采用多 worker 的方式来处理请求，每个 worker 里面只有一个主线程，那能够处理的并发数很有限啊，多少个 worker 就能处理多少个并发，何来高并发呢？非也，这就是 Nginx 的高明之处，Nginx 采用了异步非阻塞的方式来处理请求，也就是说，Nginx 是可以同时处理成千上万个请求的。对于 IIS 服务器每个请求会独占一个工作线程，当并发数上到几千时，就同时有几千的线程在处理请求了。这对操作系统来说是一个不小的挑战，线程带来的内存占用非常大，线程的上下文切换带来的 CPU 开销很大，自然性能就上不去了，而这些开销完全是没有意义的。我们之前说过，推荐设置 worker 的个数为 CPU 的核数，在这里就很容易理解了，更多的 worker 数只会导致进程竞争 CPU 资源，从而带来不必要的上下文切换。而且，Nginx 为了更好地利用多核特性，提供了 CPU 亲缘性的绑定选项，我们可以将某一个进程绑定在某一个核上，这样就不会因为进程的切换带来 cache 的失效。

6.6.6 Nginx 如何处理请求

首先，Nginx 在启动时会解析配置文件，得到需要监听的端口与 IP 地址，然后在 Nginx 的 master 进程里面先初始化好这个监控的 socket（创建 socket，设置 addrreuse 等选项，绑定到指定的 ip 地址端口，再 listen），再 fork（一个现有进程可以调用 fork 函数创建一个新进程。由 fork 创建的新进程被称为子进程）出多个子进程，然后子进程会竞争 accept 新的连接。此时，客户端就可以向 Nginx 发起连接了。当客户端与 Nginx 进行三次握手、与 Nginx 建立好一个连接后，某一个子进程会 accept 成功，得到这个建立好的连接的 socket，然后创建 Nginx 对连接的封装，即 ngx_connection_t 结构体。接着，设置读写事件处理函数并添加读写事件来与客户端进行数据的交换。最后，Nginx 或客户端主动关掉连接，到此，一个连接就结束了。

当然，Nginx 也是可以作为客户端来请求其他 server 数据的（如 upstream 模块），此时，与其他 server 创建的连接也封装在 ngx_connection_t 中。作为客户端，Nginx 先获取一个 ngx_connection_t 结构体，然后创建 socket，并设置 socket 的属性（比如非阻塞）。然后通过添加读写事件、调用 connect/read/write 来调用连接，最后关掉连接，并释放 ngx_connection_t。

在实现时 Nginx 是通过连接池来管理的。每个 worker 进程都有一个独立的连接池，连接池的大小是 worker_connections。这里的连接池里面保存的其实不是真实的连接，只是一个 worker_connections 大小的 ngx_connection_t 结构数组。并且，Nginx 会通过一个链表 free_connections 来保存所有的空闲 ngx_connection_t，每次获取一个连接时就从空闲连接链表中获取一个，用完后再放回空闲连接链表里面。

在这里，很多人会误解 worker_connections 这个参数的意思，认为这个值就是 Nginx 所能建立连接的最大值。其实不然，这个值表示每个 worker 进程所能建立连接的最大值，所以一个 Nginx 能建立的最大连接数应该是 worker_connections * worker_processes。当然，这里说的是最大连接数，对于 HTTP 请求本地资源来说，能够支持的最大并发数量是 worker_connections * worker_processes，如果是 HTTP 作为反向代理，那么最大并发数量应该是 worker_connections * worker_processes/2。因为作为反向代理服务器，每个并发会建立与客户端的连接和与后端服务的连接，即占用两个连接。

6.6.7 Nginx 典型的应用场景

负载均衡技术在现有网络结构之上提供了一种廉价、有效、透明的方法来扩展网络设备和服务器的带宽、增加吞吐量、加强网络数据处理能力、提高网络的灵活性和可用性。它有两方面的含义：首先，大量的并发访问或数据流量分担到多台节点设备上分别处理，减少用户等待响应的时间；其次，单个重负载的运算分担到多台节点设备上做并行处理，每个节点设备处理结束后将结果汇总，返回给用户，系统处理能力得到大幅度提高，如图 6-13 所示。

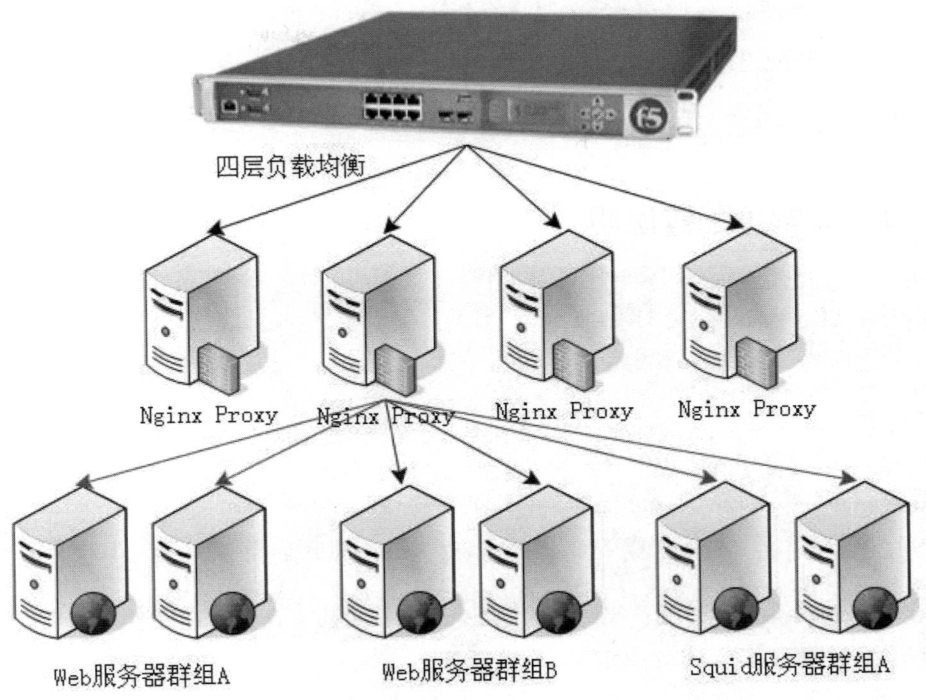

图 6-13

6.6.8　Nginx 的应用

（1）到官方网站下载 Windows 版本。下载地址为 http://nginx.org/en/download.html。

（2）解压到磁盘任一目录，这里解压到目录 D:\WorkSpace\nginx-1.11.1，如图 6-14 所示。

此电脑 › 本地磁盘 (D:) › WorkSpace › nginx-1.11.1

名称	修改日期	类型	大小
conf	2016/6/4 15:13	文件夹	
contrib	2016/6/4 15:13	文件夹	
docs	2016/6/4 15:13	文件夹	
html	2016/6/4 15:13	文件夹	
logs	2016/5/31 21:44	文件夹	
temp	2016/5/31 21:44	文件夹	
nginx.exe	2016/5/31 7:47	应用程序	2,833 KB

图 6-14

（3）修改配置文件。打开目录 Conf，找到 nginx 核心的配置文件 nginx.conf 进行修改。

（4）启动服务：直接运行 nginx.exe（这种方式启动的缺点是：控制台窗口关闭，服务关闭）。可以考虑守护进程的方式启动：start nginx.exe。

（5）停止服务：

```
nginx -s stop
```

（6）重新加载配置：

```
nginx -s  reload
```

6.6.9 Nginx 常见配置说明

```
worker_processes 8;
#nginx 进程数，建议设置为等于 CPU 总核心数
worker_connections 65535;
#单个进程最大连接数（最大连接数=连接数*进程数）
client_header_buffer_size 32k; #上传文件大小限制
large_client_header_buffers 4 64k; #设定请求缓冲
client_max_body_size 8m; #允许客户端请求的最大单文件字数
autoindex on; #开启目录列表访问，合适下载服务器，默认关闭
tcp_nopush on; #防止网络阻塞
tcp_nodelay on; #防止网络阻塞
keepalive_timeout 120; #长连接超时时间，单位是秒
gzip on; #开启 gzip 压缩输出
gzip_min_length 1k; #最小压缩文件大小
gzip_buffers 4 16k; #压缩缓冲区
gzip_http_version 1.0; #压缩版本（默认1.1，前端如果是 squid2.5就请使用1.0）
gzip_comp_level 2; #压缩等级
upstream blog.cnblogs.com {
#upstream 的负载均衡，weight 是权重，可以根据机器配置定义权重；weigth 参数表示权值，权值越高被#
分配到的概率越大
server 192.168.80.121:80 weight=3;
server 192.168.80.122:80 weight=2;
server 192.168.80.123:80 weight=3;
}
#虚拟主机的配置
server
{
#监听端口
listen 80;
#域名可以有多个，用空格隔开
server_name www.cnblogs.com cnblogs.com;
index index.html index.htm index.php;
root /data/www/ha97;
location ~ .*.(php|php5)?$
{
```

```
fastcgi_pass 127.0.0.1:9000;
fastcgi_index index.php;
include fastcgi.conf;
}
```

模块参数：

```
#定义 Nginx 运行的用户和用户组
   user www www;
   #nginx 进程数，建议设置为等于 CPU 总核心数
   worker_processes 8;

   #全局错误日志定义类型，[ debug | info | notice | warn | error | crit ]

   error_log ar/loginx/error.log info;

   #进程文件

   pid ar/runinx.pid;

   #一个 Nginx 进程打开的最多文件描述符数目，理论值应该是最多打开文件数（系统的值 ulimit -n）与
   #Nginx 进程数相除，但是 Nginx 分配请求并不均匀，所以建议与 ulimit -n 的值保持一致

   worker_rlimit_nofile 65535;

   #工作模式与连接数上限
   events
   {
   #参考事件模型，use [ kqueue | rtsig | epoll | /dev/poll | select | poll ]; epoll
   #模型是 Linux 2.6以上版本内核中的高性能网络 I/O 模型，如果跑在 FreeBSD 上面，就用 kqueue 模型

   use epoll;
   #单个进程最大连接数（最大连接数=连接数*进程数）
   worker_connections 65535;
   }

   #设定 http 服务器
   http
   {
   include mime.types; #文件扩展名与文件类型映射表
   default_type application/octet-stream; #默认文件类型
   #charset utf-8; #默认编码
   server_names_hash_bucket_size 128; #服务器名字的 hash 表大小
```

```
client_header_buffer_size 32k; #上传文件大小限制
large_client_header_buffers 4 64k; #设定请求缓冲
client_max_body_size 8m; #允许客户端请求的最大单文件字节数
sendfile on; #开启高效文件传输模式，sendfile 指令指定 Nginx 是否调用 sendfile 函数来输出
#文件，对于普通应用设为 on，如果用来进行下载等应用磁盘 IO 重负载应用，#可设置为 off，以平衡磁
#盘与网络 I/O 处理速度，降低系统的负载。注意：如果图片显示不正常就把这个改成 off
autoindex on; #开启目录列表访问，合适下载服务器，默认关闭
tcp_nopush on;
#数据包不会马上传送出去，等到数据包最大时，一次性传输出去，这样有助于解决网络堵塞
tcp_nodelay on; #马上发送数据
keepalive_timeout 120; #长连接超时时间，单位是秒

#FastCGI 相关参数是为了改善网站的性能：减少资源占用，提高访问速度。下面参数看字面意思都能理解
fastcgi_connect_timeout 300;
fastcgi_send_timeout 300;
fastcgi_read_timeout 300;
fastcgi_buffer_size 64k;
fastcgi_buffers 4 64k;
fastcgi_busy_buffers_size 128k;
fastcgi_temp_file_write_size 128k;

#gzip 模块设置
gzip on; #开启 gzip 压缩输出
gzip_min_length 1k; #最小压缩文件大小
gzip_buffers 4 16k; #压缩缓冲区
gzip_http_version 1.0; #压缩版本（默认1.1，前端如果是 squid2.5请使用1.0）
gzip_comp_level 2; #压缩等级
gzip_types text/plain application/x-javascript text/css application/xml;

#压缩类型，默认已包含 textml，所以就不用再写了，写上去也不会有问题，但是会有一个 warn
gzip_vary on;
#limit_zone crawler $binary_remote_addr 10m; #开启限制 IP 连接数的时候需要使用


upstream blog.cnblogs.com {
#upstream 的负载均衡，weight 是权重，可以根据机器配置定义权重；weigth 参数表示权值，权值越
#高被分配到的概率越大
server 192.168.80.121:80 weight=3;
server 192.168.80.122:80 weight=2;
server 192.168.80.123:80 weight=3;
}
```

```
#虚拟主机的配置
server
{
#监听端口
listen 80;
#域名可以有多个，用空格隔开
server_name www.cnblogs.com cnblogs.com;
index index.html index.htm index.php;
root /data/www/ha97;
location ~ .*.(php|php5)?$
{
fastcgi_pass 127.0.0.1:9000;
fastcgi_index index.php;
include fastcgi.conf;
}
#图片缓冲时间设置
location ~ .*.(gif|jpg|jpeg|png|bmp|swf)$
{
expires 10d;
}
#JS 和 CSS 缓冲时间设置
location ~ .*.(js|css)?$
{
expires 1h;
}
#日志格式设定
log_format access '$remote_addr - $remote_user [$time_local] "$request" '
'$status $body_bytes_sent "$http_referer" '
'"$http_user_agent" $http_x_forwarded_for';

#定义本虚拟主机的访问日志
access_log ar/loginx/ha97access.log access;

#对 "/" 启用反向代理
location / {
proxy_pass http://127.0.0.1:88;
proxy_redirect off;
proxy_set_header X-Real-IP $remote_addr;

#后端的 Web 服务器可以通过 X-Forwarded-For 获取用户真实 IP
proxy_set_header X-Forwarded-For $proxy_add_x_forwarded_for;
```

```
#以下是一些反向代理的配置，可选

proxy_set_header Host $host;
client_max_body_size 10m; #允许客户端请求的最大单文件字节数
client_body_buffer_size 128k; #缓冲区代理缓冲用户端请求的最大字节数
proxy_connect_timeout 90; #Nginx 跟后端服务器连接超时时间(代理连接超时)
proxy_send_timeout 90; #后端服务器数据回传时间(代理发送超时)
proxy_read_timeout 90; #连接成功后，后端服务器响应时间(代理接收超时)
proxy_buffer_size 4k; #设置代理服务器 (nginx) 保存用户头信息的缓冲区大小
proxy_buffers 4 32k; #proxy_buffers 缓冲区，网页平均在32k 以下的设置
proxy_busy_buffers_size 64k; #高负荷下缓冲大小 (proxy_buffers*2)
proxy_temp_file_write_size 64k;

#设定缓冲文件夹大小，大于这个值，将从 upstream 服务器传
}

#设定查看 Nginx 状态的地址
location /NginxStatus {
stub_status on;
access_log on;
auth_basic "NginxStatus";
auth_basic_user_file confpasswd;
#htpasswd 文件的内容可以用 apache 提供的 htpasswd 工具来产生
}

#本地动静分离反向代理配置
#所有 jsp 的页面均交由 tomcat 或 resin 处理
location ~ .(jsp|jspx|do)?$ {
proxy_set_header Host $host;
proxy_set_header X-Real-IP $remote_addr;
proxy_set_header X-Forwarded-For $proxy_add_x_forwarded_for;
proxy_pass http://127.0.0.1:8080;
}

#所有静态文件由 nginx 直接读取不经过，tomcat 或 resin
location ~ .*.(htm|html|gif|jpg|jpeg|png|bmp|swf|ioc|rar|zip|txt|flv|mid|doc|ppt|pdf|xls|mp3|wma)$
{ expires 15d; }
location ~ .*.(js|css)?$
{ expires 1h; }
}
}
```

更详细的模块参数请参考 http://wiki.nginx.org/Main。

6.6.10　集群案例

这里通过一个 Demo 来演示 Nginx+IIS 服务器搭建服务器集群。

为了演示，创建 2 个 Web 项目，然后部署在同一台 IIS 服务器上面，只是在 IIS 上面创建了 2 个站点，这 2 个站点的内容让其稍微有点不同以做区分。

而在生产环境中，通常是把相同的 Web 项目部署在不同的 IIS 服务器上面。

（1）新建 MVC 站点 MvcAppNginx1。

（2）添加控制器 Home：

```
public ActionResult Index()
{
        return View();
}
```

（3）添加 Index 视图：

```
@{
   ViewBag.Title = "Index";
}

<h2>这是第一个网站</h2>
```

（4）发布站点。这里将其发布到 D:\WorkSpace\publish\NginxSite 目录下面。

（5）创建一个 Web 站点。把 NginxSite 目录下面的内容复制到 NginxSite1 目录下面，然后修改 NginxSite1\Views\Home\Index.cshtml 内容：

```
@{
   ViewBag.Title = "Index";
}

<h2>这是第二个网站</h2>
```

（6）把这两个站点部署到本地 IIS 上面。打开 IIS，分别创建如图 6-15、图 6-16 所示的两个 Web 站点。

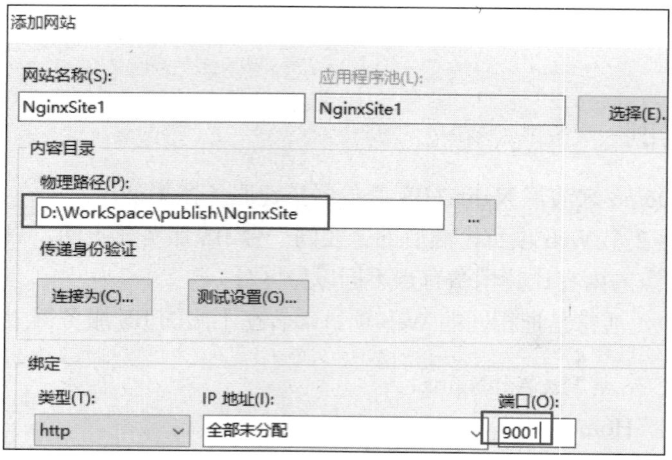

图 6-15

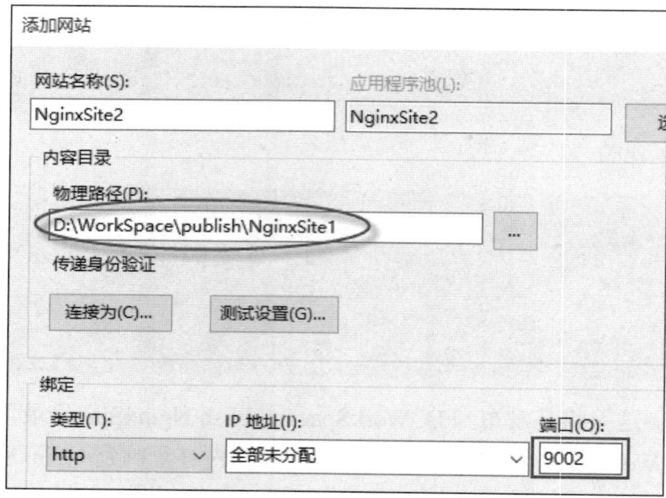

图 6-16

（7）修改配置文件 nginx.conf。

添加如下代码：

```
#服务器的集群
upstream  netittest.com {  #服务器集群名字
        #server   172.16.21.13:8081 weight=1;
        #服务器配置，weight 是权重的意思，权重越大，分配的概率越大
        server    127.0.0.1:9001  weight=1;
        server    127.0.0.1:9002  weight=2;
}
```

修改 location 配置节点：

```
server {
      listen       8088;
```

```
    server_name  127.0.0.1;

    location / {
        #root   html;
        #index  index.html index.htm;
        proxy_pass http://netittest.com; #这个名称要和 upstream 后面的名称一致
        proxy_redirect default;
    }
```

（8）启动 ngnix 服务。以管理员身份运行 CMD 窗口，然后分别执行如下命令。

```
C:\Windows\system32>d:

D:\>cd \WorkSpace\nginx-1.11.1
D:\WorkSpace\nginx-1.11.1>nginx.exe
```

（9）在浏览器中输入“http://127.0.0.1:8088/”（这个地址就是 Nginx 配置文件中配置的服务器监听终结点），尝试刷新浏览器，结果如图 6-17 所示。

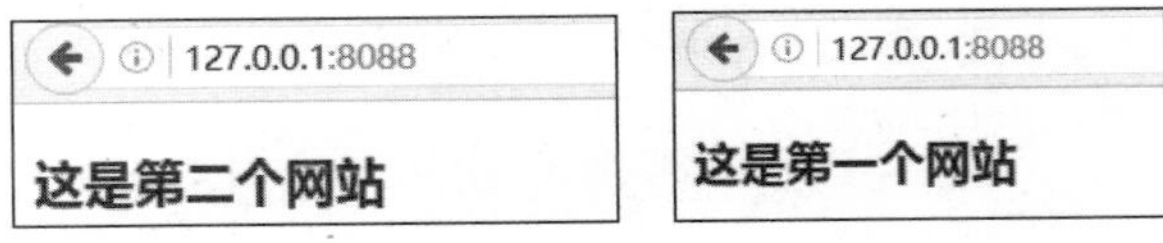

图 6-17

其实现原理如图 6-18 所示。

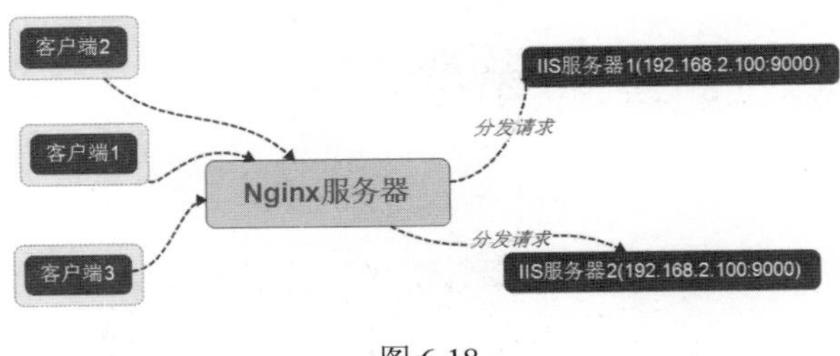

图 6-18

6.7 常用的 Web 安全技术手段

下面列举一些常用的 Web 安全技术手段：

- 提示客户使用比较复杂的密码。
- 使用邮箱或者手机验证码验证。
- 使用复杂的验证码。
- 密码加密。

- 限制每日登录错误次数。
- 限制异常 IP。
- 系统界面出现异常时显示友好界面，避免敏感信息泄漏。
- 及时释放资源，避免内存溢出。
- 记录重要信息的操作日志。
- 避免跨站脚本攻击。
- 过滤查询，防止 SQL 注入。
- 敏感信息不要存在 Cookie 中，防止 Cookie 劫持。
- 前台 js 和后台代码双验证。

第 7 章

◂ NHibernate ▸

7.1 NHibernate 简介

7.1.1 什么是 NHibernate

NHibernate 是一个面向.NET 环境的对象/关系数据库映射工具。对象/关系数据库映射（object/relational mapping，ORM）表示一种技术，用来把对象模型表示的对象映射到基于 SQL 的关系模型数据结构中去。

NHibernate 是一个基于.Net 针对关系型数据库的对象持久化类库。NHibernate 来源于非常优秀的基于 Java 的 Hibernate 关系型持久化工具。

NHibernate 的目标主要用于与数据持久化相关的编程任务，能够使开发人员从原来枯燥的 SQL 语句编写中解放出来，解放出来的精力可以让开发人员投入到业务逻辑的实现上。

7.1.2 NHibernate 的架构

NHibernate 到底是什么样子？下面摘取官方文档中的 3 幅结构图稍做说明。

一个非常简要的 NHibernate 体系结构概要图如图 7-1 所示。

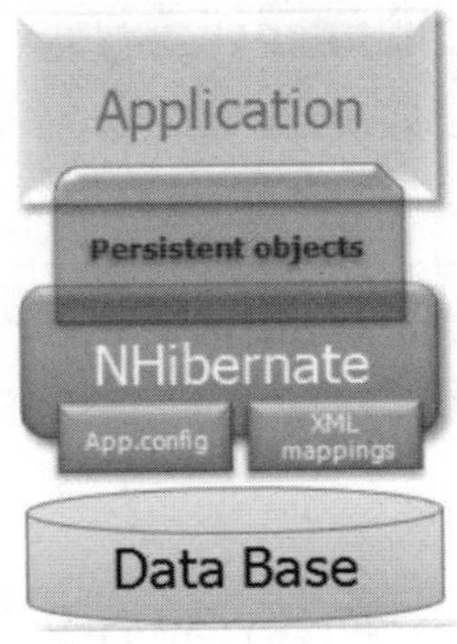

图 7-1

图 7-1 说明了什么？

- NHibernate 为数据库和应用程序提供了一个持久层。

- NHibernate 在应用程序中充当数据访问层。
- 使用 app.config 作为数据库配置文件（数据库连接，日志等）。
- 使用 xml mappings 文件作为数据库映射配置文件（也有支持其他映射方法：属性映射、Fluent Mapping、Auto-Mapping）。

图 7-1 看上去好像非常简单，其实 NHibernate 是比较复杂的。我们了解一下两种极端情况，即轻量级和重量级架构。

轻量级体系如图 7-2 所示，应用程序自己提供 ADO.NET 连接，并且自行管理事务。这种方案使用了 NHibernate API 的最小子集。

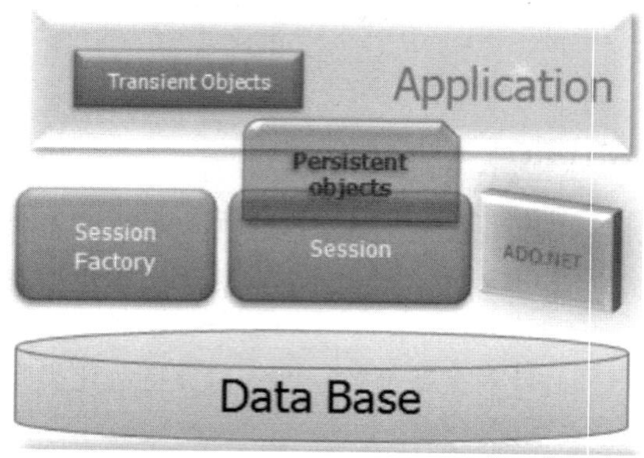

图 7-2

重量级体系如图 7-3 所示，所有的底层 ADO.NET API 都被抽象了，让 NHibernate 来处理这些细节。

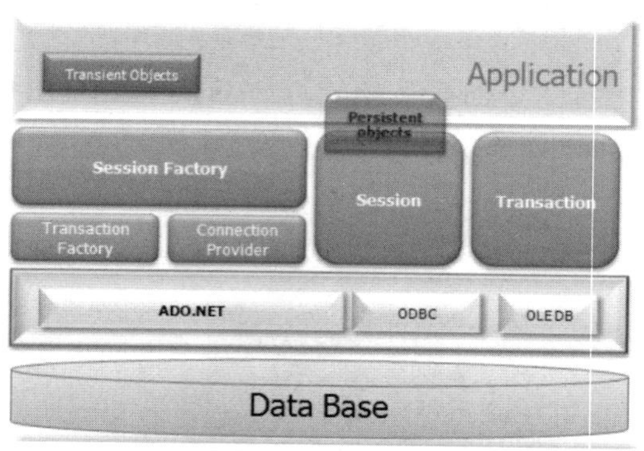

图 7-3

7.1.3 NHibernate 与其 Entity Framework 框架比较

- NHibernate 提供二级缓存。
- NHibernate 支持字典数据类型。
- NHibernate 支持 Batch Update/Insert 优化。

- NHibernate 支持更多种类数据库，支持 SQL Server、Oracle、DB2、Firebird、MySQL、PostgreSQL、SQL Lite，支持 ODBC 和 OLEDB drivers。
- NHibernate 支持更好的并发访问（乐观并发控制）。
- NHibernate 支持多样的查询方式，如 HQL、Critirial、Linq Query、Query Over 和 SQL Query。

7.2 第一个 NHibernate 应用程序

实践出真知，这里将以最直观的形式来表达。下面开始动手构建我们的 NHibernate 应用程序。

开发环境：Windows 10、VS2012、SQL Server 2012。

7.2.1 搭建项目基本框架

（1）创建一个 MVC 项目，并命名为 Shop.Web。

（2）依次分别新建 3 个类库项目：Shop.Domain，Shop.Data，Shop.Business。最终效果如图 7-4 所示。

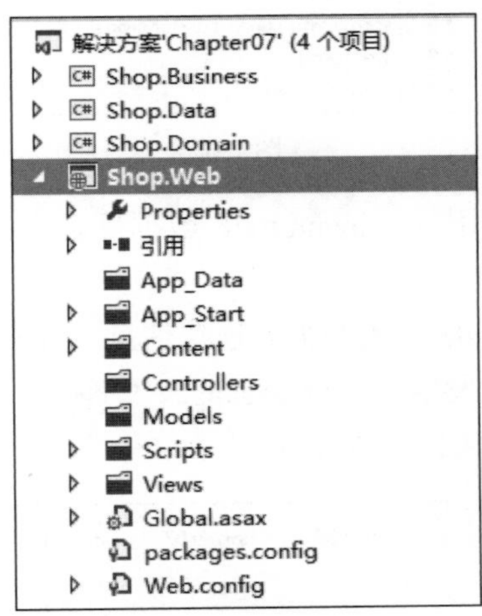

图 7-4

此项目采用传统三层架构：

- Shop.Domain：数据实体和数据库映射文件，也有人称之为领域层。
- Shop.Data：数据层，存放数据库的操作及 NHibernate 辅助类，引用 Iesi.Collections.dll、NHibernate.dll 和类库 Shop.Domain。
- Shop.Business：业务逻辑类，引用类库项目 Shop.Domain 和 Shop.Data。
- Shop.Web：Web 项目，需引用 Iesi.Collections.dll、NHibernate.dll 和类库项目

Shop.Domain、Shop.Business。

（3）下载 NHibernate。右击项目 Shop.Data 中的引用，选择“管理 NuGet 程序包(N)…→联机”，输入“NHibernate”，然后单击“安装”按钮，如图 7-5 所示。

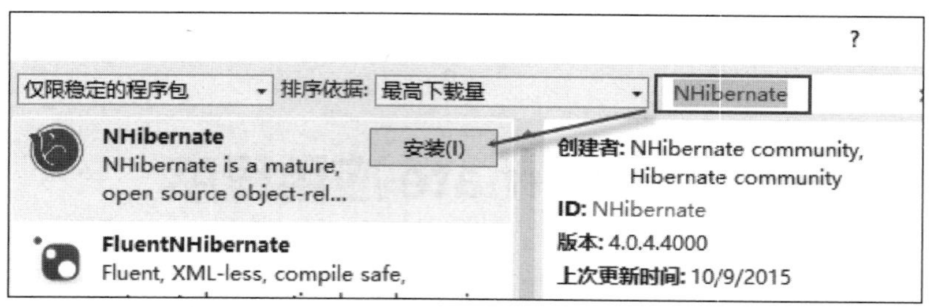

图 7-5

安装完成之后自动添加了如图 7-6 所示的引用。

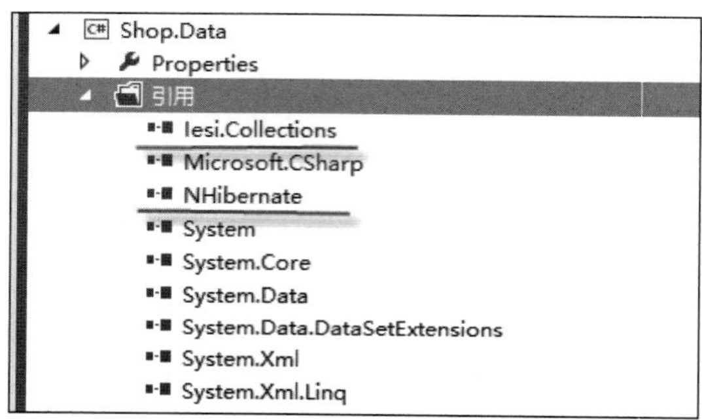

图 7-6

（4）数据库设计。这里使用 northwnd.mdf 数据库，关于 northwnd 的详细介绍请参见 2.3.1。

（5）接下来开始为 NHibernate 配置数据库连接信息。打开本项目解决方案所在文件夹位置 D:\Study\Chapter07\Chapter07，会发现有一个 packages 文件夹，打开此 packages 文件夹会看到一个 NHibernate.4.0.4.4000 文件夹，如图 7-7 所示。

脑 › 本地磁盘 (D:) › Study › Chapter07 › Chapter07 › packages › NHibernate.4.0.4.4000

名称	修改日期	类型	大小
ConfigurationTemplates	2016/6/5 21:35	文件夹	
lib	2016/6/5 21:35	文件夹	
NHibernate.4.0.4.4000.nupkg	2016/6/5 21:35	NUPKG 文件	1,437 KB
NHibernate.license.txt	2016/6/5 21:35	文本文档	25 KB
NHibernate.readme.html	2016/6/5 21:35	Firefox HTML D...	4 KB
NHibernate.releasenotes.txt	2016/6/5 21:35	文本文档	181 KB
nhibernate-configuration.xsd	2016/6/5 21:35	XML Schema File	10 KB
nhibernate-mapping.xsd	2016/6/5 21:35	XML Schema File	66 KB

图 7-7

数据库的配置信息模板可以在图 7-7 中的 Configuration_Templates 目录下找到，这里使用的是 SQL Server 2012，所以我们使用 MSSQL.cfg.xml 文件，将此文件复制到 Shop.WebSite 应用程序根目录，然后重命名为 hibernate.cfg.xml.，并修改 hibernate.cfg.xml。

注意：需根据自己数据库的实例名修改，并添加 mapping 节点，其他的设置可根据需要进行添加。修改后如下（配置文件都加上了注释）：

```
<?xml version="1.0" encoding="utf-8"?>
<!--
This template was written to work with NHibernate.Test.
Copy the template to your NHibernate.Test project folder and rename it in
hibernate.cfg.xml and change it for your own use before compile tests in VisualStudio.
-->
<!-- This is the System.Data.dll provider for SQL Server -->
<hibernate-configuration  xmlns="urn:nhibernate-configuration-2.2" >
  <session-factory>
    <!--定制数据库 IDriver 的类型.-->
    <property
name="connection.driver_class">NHibernate.Driver.SqlClientDriver</property>
    <!--连接字符串-->
    <property name="connection.connection_string">
      Server= .\MSSQLSERVER2012;database=Northwind;uid=sa;pwd=yujie1127
    </property>
    <!--NHibernate 方言（Dialect）的类名 - 可以让 NHibernate 使用某些特定的数据库平台的特性-
->
    <property name="dialect">NHibernate.Dialect.MsSql2012Dialect</property>
    <!--指定映射文档中所在程序集-->
    <mapping  assembly="Shop.Domain"/>
  </session-factory>
</hibernate-configuration>
```

（6）修改 hibernate.cfg.xml 文件的属性，如图 7-8 所示。

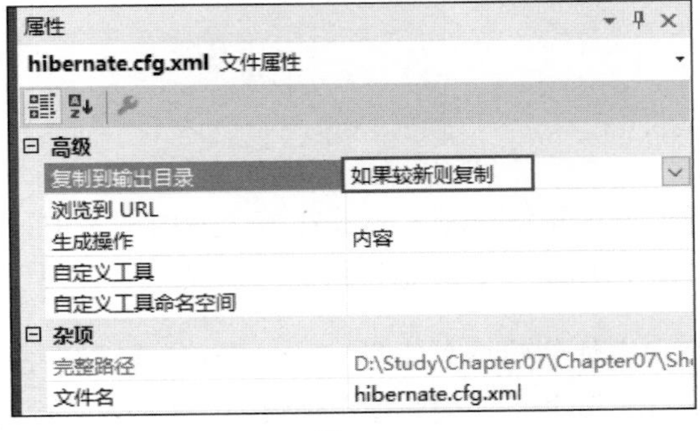

图 7-8

（7）编写 NHibernateHelper 辅助类。

这里编写一个简单的辅助类 NHibernateHelper，用于创建 ISessionFactory 和配置 ISessionFactory，并打开一个新的 ISession 方法。一个 Session 代表一个单线程的单元操作。ISessionFactory 是线程安全的，很多线程可以同时访问它。ISession 不是线程安全的，代表与数据库之间的一次操作。ISession 通过 ISessionFactory 打开，在所有的工作完成后需要将其关闭。ISessionFactory 通常是一个线程安全的全局对象，只需要被实例化一次。我们可以使用 GoF23 中的单例（Singleton）模式在程序中创建 ISessionFactory。在这个实例中我们编写了一个辅助类 NHibernateHelper，用于创建并配置 ISessionFactory、打开一个新的 Session 单线程的方法，之后在每个数据操作类中均可使用这个辅助类创建 ISession 。

在 Shop.Data 项目下新建一个类 NHibernateHelper，代码如下：

```
using NHibernate;
using NHibernate.Cfg;

namespace Shop.Data
{
    public class NHibernateHelper
    {
        private static ISessionFactory _sessionFactory;

        /// <summary>
        /// 创建 ISessionFactory
        /// </summary>
        public static ISessionFactory SessionFactory
        {
            get
            {
                //配置 ISessionFactory
                return _sessionFactory == null ? (new
Configuration()).Configure().BuildSessionFactory() : _sessionFactory;
            }
        }
    }
}
```

SessionFactory 的创建很占用系统资源，一般在整个应用程序中只创建一次。因此，这里通过判断_sessionFactory == null 来实现一个最简单的单例模式。

（8）持久化类。为客户实体创建持久化类 Customers。在项目 Shop.Domain 中新建文件夹 Entities，然后新建类 Customers。这里为了偷懒使用动软工具生成，程序员要学会偷懒，动软自带着有 Nhibernat 模板，如图 7-9 所示。我们可以修改模板或者新建模板，然后通过模板来生成代码。

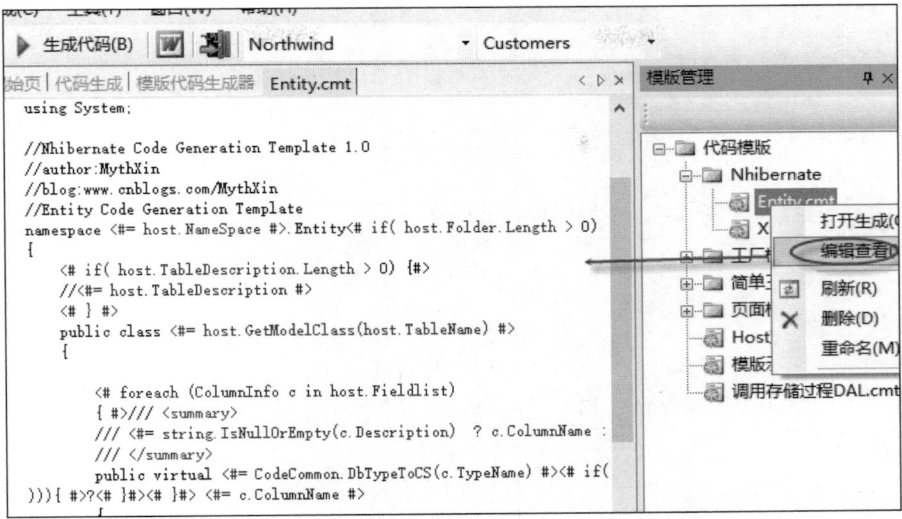
```
using System;

//Nhibernate Code Generation Template 1.0
//author:MythXin
//blog:www.cnblogs.com/MythXin
//Entity Code Generation Template
namespace <#= host.NameSpace #>.Entity<# if( host.Folder.Length > 0)
{
    <# if( host.TableDescription.Length > 0) {#>
    //<#= host.TableDescription #>
    <# } #>
    public class <#= host.GetModelClass(host.TableName) #>
    {

        <# foreach (ColumnInfo c in host.Fieldlist)
        { #>/// <summary>
        /// <#= string.IsNullOrEmpty(c.Description)  ? c.ColumnName :
        /// </summary>
        public virtual <#= CodeCommon.DbTypeToCS(c.TypeName) #><# if(
))){ #>?<# }#><# }#> <#= c.ColumnName #>
```

图 7-9

生成的代码如下：

```
using System;

//Nhibernate Code Generation Template 1.0
//author:MythXin
//blog:www.cnblogs.com/MythXin
//Entity Code Generation Template
namespace Shop.Domain.Entities
{
    //Customers
    public class Customers
    {
        /// <summary>
        /// CustomerID
        /// </summary>
        public virtual string CustomerID { get; set; }
        /// <summary>
        /// CompanyName
        /// </summary>
        public virtual string CompanyName { get; set; }
        /// <summary>
        /// ContactName
        /// </summary>
        public virtual string ContactName { get; set; }
        /// <summary>
        /// ContactTitle
```

```
        /// </summary>
        public virtual string ContactTitle { get; set; }
        /// <summary>
        /// Address
        /// </summary>
        public virtual string Address { get; set; }
        /// <summary>
        /// City
        /// </summary>
        public virtual string City { get; set; }
        /// <summary>
        /// Region
        /// </summary>
        public virtual string Region { get; set; }
        /// <summary>
        /// PostalCode
        /// </summary>
        public virtual string PostalCode { get; set; }
        /// <summary>
        /// Country
        /// </summary>
        public virtual string Country { get; set; }
        /// <summary>
        /// Phone
        /// </summary>
        public virtual string Phone { get; set; }
        /// <summary>
        /// Fax
        /// </summary>
        public virtual string Fax { get; set; }
    }
}
```

注意：NHibernate 默认使用代理功能，要求持久化类不是 sealed 的，而且其公共方法、属性和事件声明为 virtual。在这里，类中的字段要设置为 virtual，否则会出现“‘NHibernate.InvalidProxyTypeException’类型的异常在 Shop.Data.dll 中发生，但未在用户代码中进行处理”的问题。

7.2.2 编写映射文件

习惯了编码时有智能提示的我们如果想要在编写 NHibernate 配置文件时也具备智能提示功

能，只要在下载的 NHibernate 里找到 configuration.xsd 和 nhibernate-mapping.xsd 两个文件（见图 7-10），然后复制到 vs 安装目录下，如 C:\Program Files (x86)\Microsoft Visual Studio 11.0\Xml\Schemas 目录即可。

本地磁盘 (D:) › Study › Chapter07 › Chapter07 › packages › NHibernate.4.0.4.4000

名称	修改日期	类型	大小
ConfigurationTemplates	2016/6/5 21:35	文件夹	
lib	2016/6/5 21:35	文件夹	
NHibernate.4.0.4.4000.nupkg	2016/6/5 21:35	NUPKG 文件	1,437 KB
NHibernate.license.txt	2016/6/5 21:35	文本文档	25 KB
NHibernate.readme.html	2016/6/5 21:35	Firefox HTML D...	4 KB
NHibernate.releasenotes.txt	2016/6/5 21:35	文本文档	181 KB
nhibernate-configuration.xsd	2016/6/5 21:35	XML Schema File	10 KB
nhibernate-mapping.xsd	2016/6/5 21:35	XML Schema File	66 KB

图 7-10

此时，在 NHibernate 的配置文件中就有智能提示功能了，如图 7-11 所示。

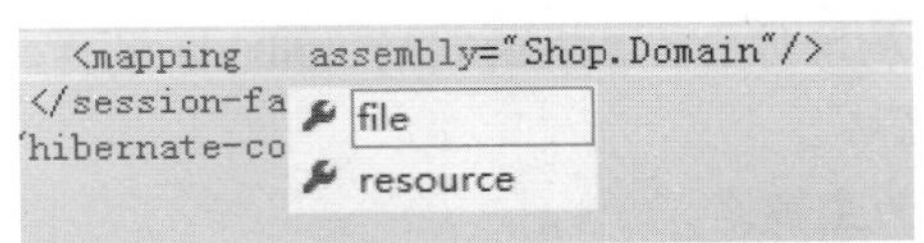

图 7-11

NHibernate 是如何知道持久化类和数据库表的对应关系的呢？这就要通过映射文件来完成这个任务了，映射文件包含了对象/关系映射所需的元数据。元数据包含持久化类的声明和属性到数据库的映射。映射文件告诉 NHibernate 它应该访问数据库里面的哪个表及使用表里面的哪些字段。

编写 Customers 持久化类的映射文件，注意映射文件以.hbm.xml 结尾，如 Customers.hbm.xml。

在项目 Shop.Domain 下新建文件 Customers.hbm.xml。同样使用动软代码生成工具生成映射文件代码，然后复制到 Customers.hbm.xml 中，稍微修改一下，主要是注意程序集名称和类全称，最终代码如下：

```
<?xml version="1.0" encoding="utf-8" ?>
<hibernate-mapping xmlns="urn:nhibernate-mapping-2.2" assembly="Shop.Domain"
namespace="Shop.Domain.Entities">
  <!--类的全称，程序集，数据库表名称-->
  <class name="Shop.Domain.Entities.Customers, Shop.Domain" table="Customers">
    <id name="CustomerID" column="CustomerID" type="string" />
    <property name="CompanyName" column="CompanyName" type="string" />
    <property name="ContactName" column="ContactName" type="string" />
    <property name="ContactTitle" column="ContactTitle" type="string" />
    <property name="Address" column="Address" type="string" />
```

```
    <property name="City" column="City" type="string"/>
    <property name="Region" column="Region" type="string"/>
    <property name="PostalCode" column="PostalCode" type="string"/>
    <property name="Country" column="Country" type="string"  />
    <property name="Phone" column="Phone" type="string"  />
    <property name="Fax" column="Fax" type="string"  />
  </class>
</hibernate-mapping>
```

- hibernate-mapping 的 assembly、namespace 属性要填写正确。
- class 的 name 属性表示类名，table 是映射的表名，如果类名称和表名称相同，可以省略 table 属性。
- property 的 name 属性是类的属性名，如果类属性名和表的列名相同，可以省略 column 属性。
- property 的 type 属性表示.net 类属性映射的 NHibernate 数据类型，如果是 int、bool、double、DateTime、string 这样的.NET 基础数据类型，就可以省略。
- property 的 not-null 属性对应关系表的列的 nullable 属性，默认值是 false。因此，如果允许为空，就可以省略。
- id 表示主键，name 为主键名。
- 文件名必须以.hbm.xml 结尾。

最后不要忘记给此映射文件设置属性，如图 7-12 所示。

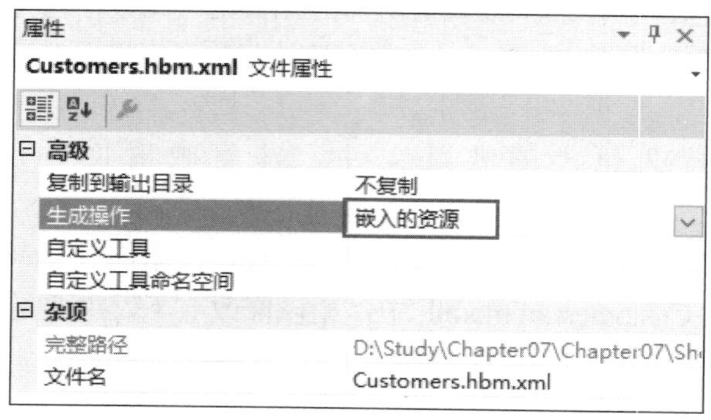

图 7-12

NHibernate 的 Generator 主键生成方式有如下几种。

- assigned：主键由外部程序负责生成，无须 NHibernate 参与。
- hilo：通过 hi/lo 算法实现的主键生成机制，需要额外的数据库表保存主键生成历史状态。
- seqhilo：与 hilo 类似，通过 hi/lo 算法实现的主键生成机制，只是主键历史状态保存在 Sequence 中，适用于支持 Sequence 的数据库，如 Oracle。
- increment：主键按数值顺序递增。此方式的实现机制为在当前应用实例中维持一个变量，以保存当前的最大值，之后每次需要生成主键的时候将此值加 1 作为主键。这种方式可能产生的问题是：如果当前有多个实例访问同一个数据库，那么由于各个实例各自维护主键状态，不同实例可能生成同样的主键，从而造成主键重复异常。因此，如果同一数据库有多个实例访问，那么此方式必须避免使用。

- identity：采用数据库提供的主键生成机制，如 DB2、SQL Server、MySQL 中的主键生成机制。
- sequence：采用数据库提供的 sequence 机制生成主键，如 Oracle 中的 Sequence。
- native：由 NHibernate 根据底层数据库自行判断采用 identity、hilo、sequence 中的一种作为主键生成方式。uuid.hex 由 Hibernate 基于 128 位唯一值产生算法生成十六进制数值（编码后以长度为 32 的字符串表示）作为主键。
- foreign：使用外部表的字段作为主键。

7.2.3　添加数据访问层类

NHibernate 有多种不同的方法可以从数据库中取回对象，最灵活的方式是使用 NHibernate 查询语言（HQL），是完全基于面向对象的 SQL。

在项目 Shop.Data 中，新建类 CustomersData.cs，编写一个方法 GetCustomerList 来获取所有的客户信息列表，代码如下：

```
using System;
using System.Collections.Generic;
using System.Linq;
using NHibernate;
using NHibernate.Linq;
using System.Linq.Expressions;
using Shop.Domain.Entities;

namespace Shop.Data
{
    public class CustomersData
    {
        /// <summary>
        /// 根据条件得到客户信息集合
        /// </summary>
        /// <param name="where">条件</param>
        /// <returns>客户信息集合</returns>
        public IList<Customers> GetCustomerList(Expression<Func<Customers, bool>>
        where)
        {
            try
```

```
            {
                using (ISession session = NHibernateHelper.SessionFactory.OpenSession())
                {
                    return session.Query<Customers>().Select(x => new Customers { CustomerID =
                    x.CustomerID, ContactName = x.ContactName, City = x.City, Address = x.Address,
                        Phone = x.Phone, CompanyName = x.CompanyName, Country =
                        x.Country }).Where(where).ToList();
                }
            }
            catch (Exception ex)
            {
                throw ex;
            }
        }
    }
}
```

使用 using 子句，在 using 代码块完成后将自动调用 ISession 的 Dispose 方法关闭 Session。

7.2.4 添加业务逻辑层类

在项目 Shop.Business 中新建类 CustomersBusiness，代码如下：

```
using Shop.Data;
using Shop.Domain.Entities;
using System;
using System.Collections.Generic;
using System.Linq.Expressions;

namespace Shop.Business
{
    public class CustomersBusiness
    {
        private CustomersData _customersData;
        public CustomersBusiness()
        {
            _customersData = new CustomersData();
        }
        /// <summary>
        /// 根据条件得到客户信息集合
        /// </summary>
        /// <param name="where">条件</param>
        /// <returns>客户信息集合</returns>
        public IList<Customers> GetCustomerList(Expression<Func<Customers, bool>>
        where)
        {
```

```
            return _customersData.GetCustomerList(where);
        }
    }
}
```

7.2.5　添加控制器和视图

在 Shop.Web 项目中的 Controllers 下添加 Customers 控制器:

```
using System.Web.Mvc;
using Shop.Business;

namespace Shop.Controllers
{
    public class CustomersController : Controller
    {
        CustomersBusiness customersBusiness = new CustomersBusiness();

        public ActionResult Index()
        {
            var result = customersBusiness.GetCustomerList(c => 1 == 1);
            return View(result);
        }
    }
}
```

添加 Index 视图，如图 7-13 所示。

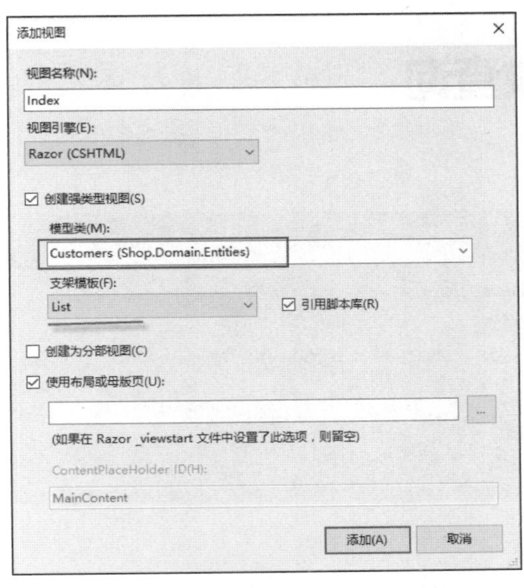

图 7-13

把项目 Shop.Web 设置为启动项，然后运行，修改浏览器地址，在后面添加 Customers 控制器。

最终浏览器地址为“http://localhost:10837/Customers”。当然，这里的端口是启动项目时打开的端口，默认情况下是随机分配的，也可以在项目配置中指定一个固定的端口，如图 7-14 所示。

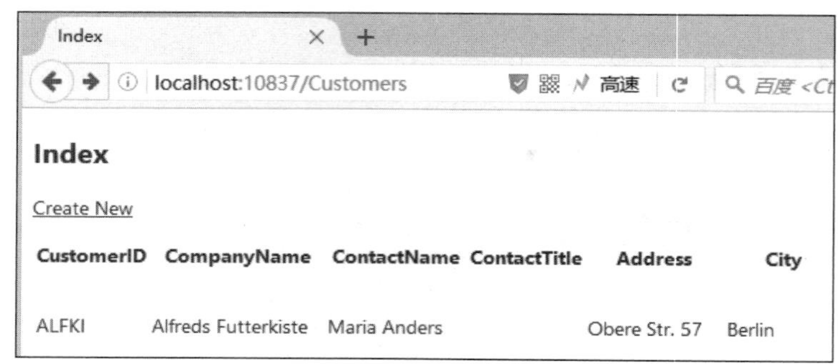

图 7-14

至此，第一个 NHibernate 应用程序就完成了。

如果希望 NHibernate 能够像支持 EF 一样支持可视化操作，就可以下载一个 Mindscape.NHibernateModelDesigner_crack.vsix 插件。下载后，运行 Mindscape.NHibernateModel Designer_crack.vsix 进行安装。安装完成之后，在 VS 中新建项目的时候就会发现多了一个“NHibernate Model”模板。

这里只是使用 NHibernate 构建了一个最基本的项目，没有体现 NHibernate 更多细节，只描述了 NHibernate 的基本面目。当然，NHibernate 有各种各样的程序架构，这里只是按照一般模式构建的。

7.3 增删改查询

获取单条记录代码：

```
public Customers GetById(string id)
{
        using (ISession session = NHibernateHelper.SessionFactory.OpenSession())
        {
            Customers customer = session.Get<Customers>(id);
//session.Load<Customers>(id);
            return customer;
        }
}
```

增删改函数代码：

```
public bool Insert(Customers customer)
{
        using (var session = NHibernateHelper.SessionFactory.OpenSession())
        {
```

```
            var identifier = session.Save(customer);
            session.Flush();
            return string.IsNullOrEmpty(identifier.ToString());
        }
}

public void Update(Customers customer)
{
        using (var session = NHibernateHelper.SessionFactory.OpenSession())
        {
            session.SaveOrUpdate(customer);
            session.Flush();
        }
}

public void Delete(string id)
{
        using (var session = NHibernateHelper.SessionFactory.OpenSession())
        {
            var customer = session.Get<Customers>(id);
            session.Delete(customer);
            session.Flush();
        }
}
```

- session.Save: 插入新记录，返回新记录主键值。
- session.SaveOrUpdate: 如果被调用的 Customers 对象在数据库里不存在（新记录），就插入新记录，否则修改该记录。
- session.Delete: 传入 Customers 对象进行删除。

增删改操作完成之后需要调用 session.Flush()方法将对象持久化写入数据库。如果不调用此方法，那么操作结束后修改记录不能写入数据库中。

7.4 使用代码映射

NHibernate 通过配置文件进行映射的方式在说明文档 NHibernate.Reference.chm 中已经讲解得非常详细了，如果要看中文版，只有 HNibernate 2.2 版的 NHibernate.chm。网上关于 NHibernate 采用配置文件进行映射的文章也很多，就不赘述了。这里要说的是 Nhibernate 采用代码映射的方式，从 NHibernate 3.0 开始就支持代码映射了，当然也可以使用 Fluent NHibernate 来进行代码映射，关于 Fluent NHibernate 的使用可以参考：http://www.cnblogs.com/uncle_danny/p/5700765.html。

7.4.1 NHibernate 入职 Demo

在“各种开发工具组件”文件夹中找到 NHibernateDemo.zip 并解压，打开 Uuch.Demo.sln，如图 7-15 所示。

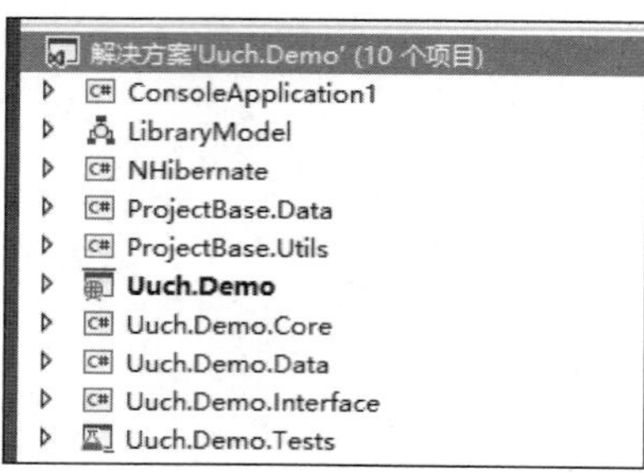

图 7-15

这是一个入职 Demo，是给刚进公司的新开发人员熟悉公司项目框架用的，原本只是搭建好了框架，并实现了一个最基本的功能，其余的功能要让新员工依样画瓢地将其实现。要求是入职的新员工实现一个简单的购物网站，功能包括：商品展示、添加到购物车、生成订单。这里提供的代码是一个新入职的员工写的，代码中演示了一些常用的基本映射方式。

（1）所包含组件

- NHibernate.3.3.3.4001（ORM 组件）
- Unity.2.1.505.2（IOC 组件）
- MVC4.0
- Moq.4.2.1402.2112（单元测试）

（2）项目说明

- LibraryModel——类图例子
- packages——引用 NuGet 包文件夹
- Nhibernate——NHibernate 源码
- ProjectBase.Data——基类数据层
- ProjectBase.Utils——基类扩展层
- Uuch.Demo——Web 网站
- Uuch.Demo.Core——业务逻辑层
- Uuch.Demo.Data——数据实现层
- Uuch.Demo.Interface——调用外部接口实现层
- Uuch.Demo.Tests——单元测试

打开 Navicat for MySql，新建数据库 MyExample，如图 7-16 所示，然后执行 myexample.sql 脚本。

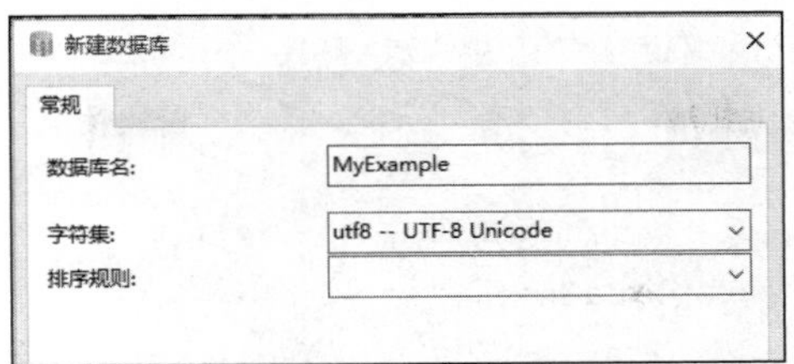

图 7-16

用户名为“Test”，密码为“1”。

IOC

业务逻辑层 Uuch.Demo.Core 需要调用数据实现层 Uuch.Demo.Data。你可能会想到在 Uuch.Demo.Core 中直接新建一个 Uuch.Demo.Data 对象，然后调用；又或者通过工厂或者抽象工厂来创建 Uuch.Demo.Data 对象。其实工厂可以做的事情通过 IOC 都可以实现，而且做得更好。

项目 Uuch.Demo.Core 和 Uuch.Demo.Data 的结构如图 7-17 所示。

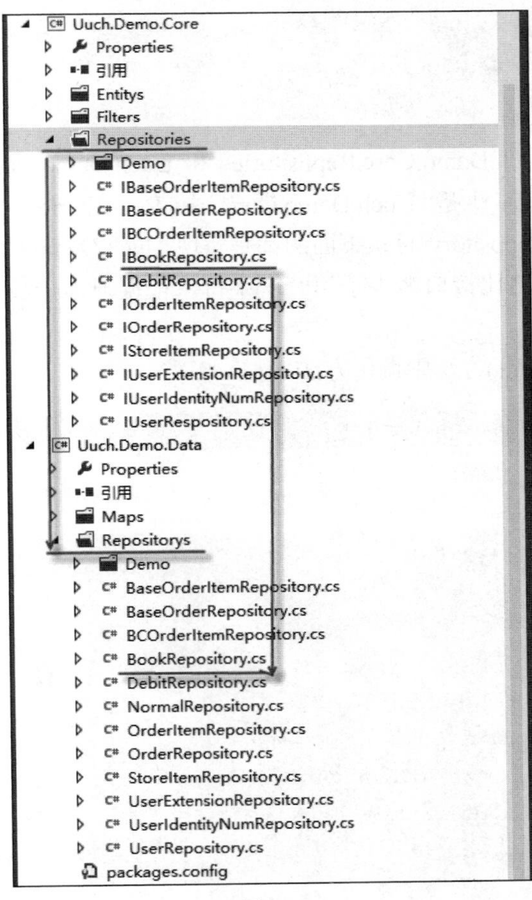

图 7-17

在项目 Uuch.Demo 的 Global.asax 文件中添加如下方法：

```
public static void RegisterContainer(UnityContainer container)
{
    UnityConfigurationSection section = (UnityConfigurationSection)
    ConfigurationManager.GetSection("unity");
    section.Configure(container, "");
    var iRepositories = Assembly.Load("Uuch.Demo.Core").GetTypes()
        .Where(p => p.FullName.Contains("Repositories")).ToList();
    var repositories = Assembly.Load("Uuch.Demo.Data").GetTypes()
        .Where(p => p.FullName.Contains("Repository")).ToList();
    iRepositories.ForEach(
        p => container.RegisterType(p, repositories.Where(r => r.Name ==
        p.Name.Substring(1, p.Name.Length - 1)).FirstOrDefault()));

    container.RegisterType(typeof(ProjectBase.Data.IDao<Uuch.Demo.Core.Enti
    tys.News,                                                            int>),
    typeof(Uuch.Demo.Data.Repositorys.NormalRepository<Uuch.Demo.Core.Entit
    ys.News, int>));
}
```

上面的代码将项目 Uuch.Demo.Core.Repositories 和 Uuch.Demo.Data. Repository 命名空间下面的类进行关联，前提条件是 Uuch.Demo.Core 中 Repositories 目录下面的都是接口，Uuch.Demo.Data 项目中 Repository 目录下面的都是实现 Uuch.Demo.Core 中 Repositories 目录下面接口的具体类，并且类名比接口名少了第一个字母，根据习惯一般把接口名的第一个字母设置为 I。

然后在 Application_Start()方法中调用方法：

```
RegisterContainer(ProjectBase.Data.IocContainer.Instance.Container);
```

代码映射示例：

```
public class UserMap : ClassMapping<User>
{
    public UserMap()
    {
        Table("Demo_User");//对应数据库中的表名
        Lazy(true); //true 表示使用延时加载
        Id(p => p.ID, p => { p.Generator(Generators.Identity); }); //设置自增长主键
        Property(p => p.UserName); //映射普通字段
        Property(p => p.PassWord); //如果表中字段名和 Model 中的属性名不一致，就可以在
        //m.Column 中指定表字段名称
        Property(p => p.CreateTime);
        Property(p => p.Status, p => { p.Type<NHibernate.Type.EnumType<UserStatu
        //s>>(); }); //映射枚举字段
```

```
        Bag(p => p.DebitList, m => { m.Key(k => k.Column("UserID")); m.Cascade(C
        ascade.DeleteOrphans); }, rel => rel.OneToMany()); //一对多映射
        OneToOne(p => p.ExtensionBy, m => { m.Cascade(Cascade.All); });//主键和外
        键是一对一映射,Cascade.All=Cascade.DeleteOrphans 表示外键对象先更新为 null 再删除
        OneToOne(p => p.NumMessage, m => { m.PropertyReference(typeof(UserIdenti
        tyNum).GetProperty("UserBy")); m.Constrained(true); m.Cascade(Cascade.A
        ll); });//非主键方式的一对一映射
        //Bag(p => p.CartList, m => { m.Key(k => k.Column("Buyer")); m.Cascade(C
        ascade.All); }, rel => rel.OneToMany());
    }
}
```

注意所有映射的类都继承了 ClassCustomizer<T>，T 代表实体模型类。

Demo 非常简单，这里就不再说明了，大家有兴趣的话可以自己看源码。

7.4.2　NHibernate 代码映射高级功能

之前的 NHibernate 入职 Demo 代码有点乱，后来饶成龙先生将 Demo 进行了优化，并新增了父子映射。优化后的源码为 NHibernateDemoNew.rar。解压源码后，找到 Mysql 数据库备份文件 161022160410.psc 进行还原。（在前面中是用 SQL 脚本进行的数据库迁移，这里使用备份还原的方式进行数据库迁移。）

打开 Navicat for MySql，新建数据库，并命名为 mydemo，然后选中数据库 mydemo，单击“备份”按钮，在弹出框中选择数据库备份文件 161022160410.psc。操作步骤如图 7-18~图 7-19 所示。

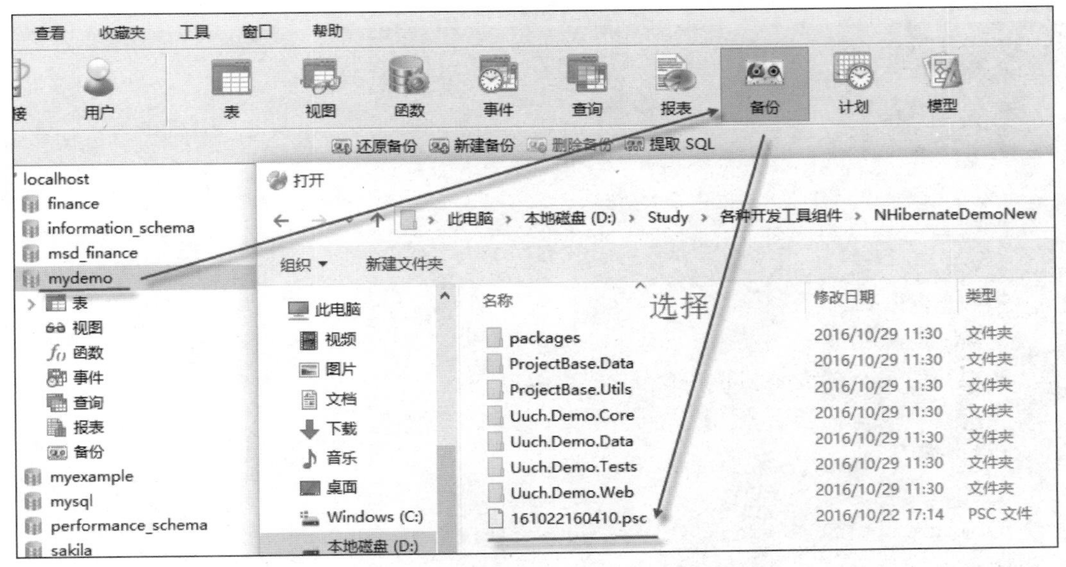

图 7-18

图 7-19

打开 NHibernateDemoNew 目录中的项目 Uuch.Demo.sln，项目中故意只完成了一小部分功能供大家参考，但是框架已经基本搭建好，剩下的需要自己去完善。大家可以发散思维，在现有的代码框架上进行修改，开发出一个属于自己的项目。

1. 视图映射

NHibernate 无须对应表的映射，创建一个实体对象。这个实体对象既可以是对应数据库中的视图，也可以是直接对应一个 SQL 查询结果集，但是 SQL 语句查询对象中至少要有一个唯一性对象。

```
public class PublishView
{
        /// <summary>
        /// 主键
        /// </summary>
        public virtual int ID { get; set; }
        /// <summary>
        /// 用户名
        /// </summary>
        public virtual string UserName { get; set; }
        /// <summary>
        /// 标题
        /// </summary>
        public virtual string Title { get; set; }
        /// <summary>
        /// 创建时间
```

```
        /// </summary>
        public virtual DateTime CreateTime { get; set; }
}
```

代码映射如下：

```
public class PublishViewMap : ClassMapping<PublishView>
{
        public PublishViewMap()
        {
            Lazy(true);
            Id(p => p.ID, p => { p.Generator(Generators.Identity); });
            Property(p => p.Title);
            Property(p => p.UserName);
            Property(p => p.CreateTime);
        }
}
```

2. 父子映射

在 C#类中有继承关系，在 NHibernate 中也有继承关系，反映到数据库中就有多种实现形式了。子类和父类可以映射到同一张表中，子类也可以单独映射成一张表，但是用不同的标签实现，子类表和父类表的关系也不同。在映射文件中，有 3 个标签可以实现继承关系，分别是 subclass、joined-subclass、union-subclass。

先陈述一下这 3 个标签的区别：

- subclass 标签，父类和子类存在同一张表中，它就是为子类嵌入父类的表中而设计的，而且要指定鉴别器，即 subclass 一定要有判断类型的列，用鉴别器指定记录属于子类还是父类。虽然 subclass 也提供了为子类设置单独表的功能，即 join 标签，但是不管是内嵌表还是外部表都要指定鉴别器。注意：这两个标签不能混用，若是只想要内嵌表，或者是既想要内嵌表又想要外部表，又或者是只想要外部表，则只能使用 subclass 标签。需要注意的是，不论使用哪种方式，都要指定鉴别器，即父类对应的表中一定有一个判断类型的列；若是只想要外部表，又不想在父类对应的表中保留判断类型的列，则只能使用 join-subclass。
- joined-subclass 标签，提供的功能是只能为子类设定外部表，而且没有鉴别器，即子类一张表，父类一张表，子类以父类的主键为外键，父类对应的表中没有判断类型的列。
- union-subclass 标签父类表存储父类字段，子类表既要存储父类字段又要存储子类字段。它将父类中的属性添加到子类对应的表中，包括父类中的外键。union-class 和前两个的区别就在于外键的不同，前两个标签中如果子类有单独对应的表，那么这个表的外键就是其父类中的主键，而使用 union-subclass，子类对应表中的外键则和父类的外键是一样的，因为子类把父类继承的属性也加到了自己的表中。这样子类和父类的地位就相当了。不过这不是一种好的理解方式。如果是这样，就可以把父类设为抽象类，并在映射文件中把父类设置为 abstract="true"，这样就不会在数据库中生成父类对应的表了，父类也只起到一个抽象的作用。

（1）subclass 形式

假设有如下两个类，其中 Worker 类继承 Person 类:

```
public class PersonMap
{
        private int id;
        private String name;
        private int age;

}
public class WorkerMap : SubclassMapping<Person>
{
        private String job;
        private String unit;
}
```

数据库中的表结构如图 7-20 所示。

pid	role	name	age	job	unit
1	PERSON	tom	10	(NULL)	(NULL)
2	WORKER	jerry	20	washer	bj

图 7-20

可以看到父类和子类存在同一张表中，其中的 role 字段是辨别者列，用于辨别这条记录是父类还是子类。

subclass 的缺点：

- 子类独有的字段无法设置 not-null 约束，即 job 和 unit 无法非空，毕竟父类中没有这两个字段。
- 如果继承树过于庞大，有 *n* 个实体之间继承，不便于管理时将会很麻烦。

（2）joined-subclass 的继承方式

父类 Order 及其映射代码如下：

```
public class Order : DomainObject<Order, int, IOrderRepository>
{
        public virtual string OrderCode { get; set; }
        public virtual string Receipter { get; set; }
        public virtual DateTime CreateTime { get; set; }
        public virtual OrderStatus Status { get; set; }
```

```
        public virtual OrderType OrderType { get; set; }
        public virtual IList<OrderItem> Items { get; set; }
        public static Order GetByCode(string code)
        {
            return Dao.GetByCode(code);
        }
}
public enum OrderType
{
        BBC=0,
        BC=1
}
public enum OrderStatus
{
        Create=0,
        End=1
}
//父类映射
public class OrderMap : ClassMapping<Order>
{
        public OrderMap()
        {
            Table("Orders");
            Lazy(false);
            Id(p => p.ID, p => { p.Generator(Generators.Identity); });
            Property(p=>p.OrderCode);
            Property(p=>p.Receipter);
            Property(p=>p.CreateTime);
            Property(p=>p.Status,p => { p.Type<NHibernate.Type.EnumType<OrderStatus>>
             (); });
          Property(p => p.OrderType, p => { p.Type<NHibernate.Type.EnumType<OrderType>>
          (); });
          Bag(p => p.Items, m => { m.Key(k => k.Column("OrderID")); m.Cascade
          (Cascade.All); }, rel => rel.OneToMany());
            Discriminator(p => p.Column("OrderType"));
            DiscriminatorValue(OrderType.BBC);
        }
}
```

子类 BCOrder 继承父类 Order，代码如下：

```
public class BCOrder : Order
{
```

```
        public virtual string TaxRateCode { get; set; }

        public static BCOrder GetByCode(string code)
        {
            return Dao.GetChildByCode(code);
        }
}
```

joined-subclass 方式的父子关系是所有数据在一张父表内，其他子类的附加对象在子表内。

```
public class BCOrderMap : JoinedSubclassMapping<BCOrder>
{
        public BCOrderMap()
        {
            Table("BCOrders");
            Lazy(false);
            Key(a => a.Column("ID"));
            Property(p => p.ID);
            Property(p => p.TaxRateCode);
        }
}
```

Orders 表如图 7-21 所示。

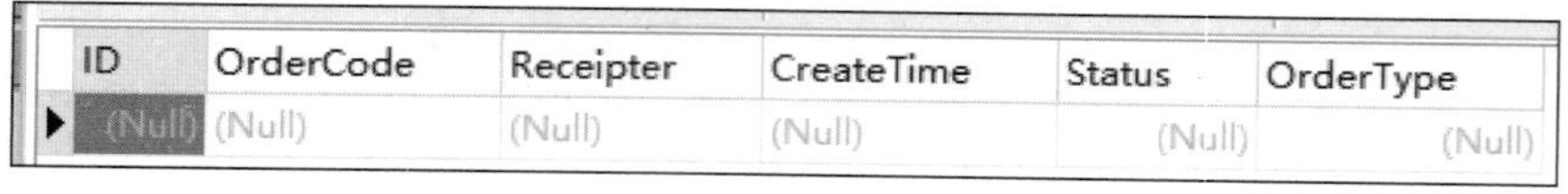

ID	OrderCode	Receipter	CreateTime	Status	OrderType
(Null)	(Null)	(Null)	(Null)	(Null)	(Null)

图 7-21

BCOrders 表如图 7-22 所示。

ID	TaxRateCode
(Null)	(Null)

图 7-22

（3）union-subclass 继承方式

父类 BaseOrder 代码：

```
public abstract class BaseOrder : DomainObject<BaseOrder, int, IBaseOrderRepository>
{
        public virtual string OrderCode { get; set; }
        public virtual string Receipter { get; set; }
        public virtual DateTime CreateTime { get; set; }
        public virtual OrderStatus Status { get; set; }
}
```

父类映射代码：

```
public class BaseOrderMap : ClassMapping<BaseOrder>
{
    public BaseOrderMap()
    {
        Id(p => p.ID, p => { p.Generator(Generators.HighLow); });
        Property(p => p.OrderCode);
        Property(p => p.Receipter);
        Property(p => p.CreateTime);
        Property(p => p.Status, p => { p.Type<NHibernate.Type.EnumType<OrderStatus>>
        (); });
    }
}
```

子类 BBCOrder：

```
public class BBCOrder:BaseOrder
{
    public virtual string SenderName { get; set; }
    public virtual IList<BBCOrderItem> Items { get; set; }
    public static BBCOrder GetByCode(string code)
    {
        return Dao.GetBBCByCode(code);
    }
}
```

子类 BBCOrder 映射的映射代码继承 BaseOrder 的父子关系，直接继承父类的 map，但是由于继承泛型父类的关系，因此需要在子类中重写获取方法。

```
public class BBCOrderMap : UnionSubclassMapping<BBCOrder>
{
    public BBCOrderMap()
    {
        Table("BBCOrders");
        Lazy(false);
        Property(p => p.SenderName);
        Bag(p => p.Items, m => { m.Key(k => k.Column("OrderID"));
         m.Cascade(Cascade.All); }, rel => rel.OneToMany());
    }
}
```

BaseOrders 表如图 7-23 所示。

OrderCode	Receipter	CreateTime	Status
(Null)	(Null)	(Null)	(Null)

图 7-23

BBCOrders 表如图 7-24 所示。

ID	OrderCode	Receipter	CreateTime	Status	SenderName
(Null)	(Null)	(Null)	(Null)	(Null)	(Null)

图 7-24

由此可见，union-subclass 是父类表存储父类字段，而子类表则既要存储父类字段又要存储子类字段。

joined-subclass 与 union-subclass 的缺点：

- 插入与查询效率都低下。
- union-subclass 在更新父类表时效率低下。

7.5 监听 NHibernate 生成的 SQL

有时我们使用 NHibernate 框架需要监听所写代码最终生成的 SQL 语句是什么样的。如果采用的是 SQL Server 数据库，就可以使用 SQL Server Profiler 工具监听生成的 SQL 语句。如果是其他数据库呢？下面介绍两种在 NHibernate 中监听 SQL 语句的方式。

- 使用 show_sql
- 使用 NhibernateProfile

7.5.1 使用 show_sql

在 hibernate.cfg.xml 文件内对应的显示 SQL 执行语句的属性是 show_sql：

```
<property name="show_sql">true</property>
```

在解决方案 Chapter07 中新建控制台项目 ConsoleApp，添加项目 Shop.Domain、Shop.Business 的引用，修改 Program 代码：

```
using Shop.Business;
using System;

namespace ConsoleApp
{
    class Program
    {
```

```
        static void Main(string[] args)
        {
            CustomersBusiness customersBusiness = new CustomersBusiness();
            var result = customersBusiness.GetCustomerList(c => 1 == 1);
            foreach (var v in result)
            {
                Console.WriteLine(v.CompanyName);
            }
            Console.ReadLine();
        }
    }
}
```

设置项目 ConsoleApp 为启动项，运行结果如图 7-25 所示。

```
C:\Windows\system32\cmd.exe
NHibernate: select customers0_.CustomerID as col_0_0_, customers0_.ContactName as co
l_1_0_, customers0_.City as col_2_0_, customers0_.Address as col_3_0_, customers0_.P
hone as col_4_0_, customers0_.CompanyName as col_5_0_, customers0_.Country as col_6_
0_ from Customers customers0_ where @p0=1;@p0 = True [Type: Boolean (0)]
Alfreds Futterkiste
Ana Trujillo Emparedados y helados
Antonio Moreno Taquería
Around the Horn
```

图 7-25

在项目中，如果需要监听的 SQL 语句都采用此种形式显然不怎么方便，其实还可以使用第三方工具——NHibernateProfile。

7.5.2 使用 NHibernateProfile

（1）NHibernateProfile 的下载地址为 http://www.hibernatingrhinos.com/products/NHProf，目前版本是 3.0，缺点是收费，不过可以免费使用 30 天。

下载后解压缩，执行文件 NHProf.exe，按照提示注册一个 trial license，如图 7-26 所示。

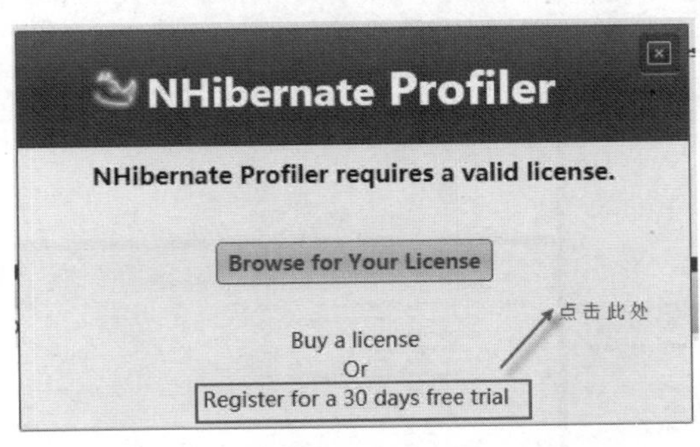

图 7-26

（2）注册后，会发送 license 的 xml 文件到注册邮箱，登录注册邮箱，将 license 的 xml 文件从收件箱下载到本地，导入 license 的 xml 文件。

注册邮箱不要使用 QQ 邮箱，否则可能会收不到，具体原因不详。

导入后，看到的窗口如图 7-27 所示。

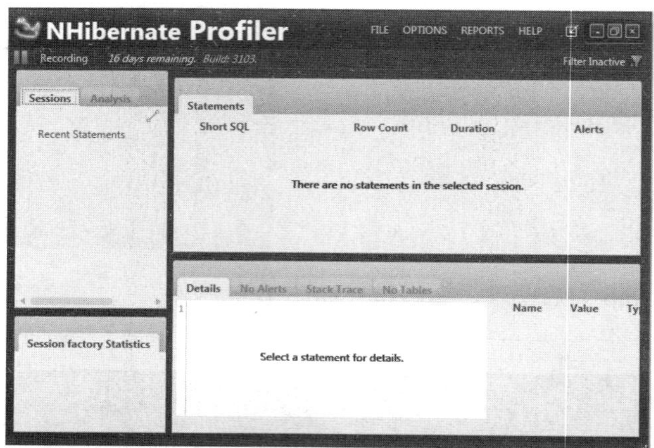

图 7-27

（3）从解压后的文件夹中找到 dll 文件 HibernatingRhinos.Profiler.Appender.dll，将其复制到工程目录的$\packages 文件夹下（这里使用的是 Chapter07\packages），然后在项目 Shop.Web 中添加引用这个文件。

（4）在 Shop.Web 项目中修改 Global.asax 文件，在 Application_Start 方法中添加如下代码：

```
HibernatingRhinos.Profiler.Appender.NHibernate.NHibernateProfiler.Initialize();
```

（5）设置 Shop.Web 项目为启动项并运行，浏览器地址为 http://localhost:10837/Customers。我们可以看到 NHibernateProfile 的监控窗口如图 7-28 所示。

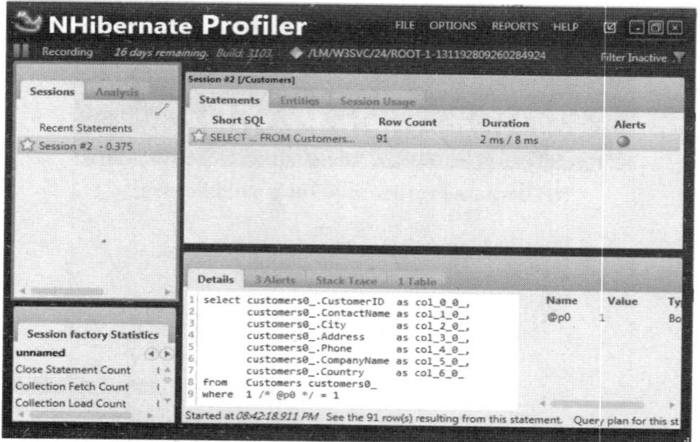

图 7-28

NHibernateProfile 不但可以监控执行的 SQL 语句，而且能够抓取到 NHibernate 执行过程以及显示 NHibernate 异常详细信息，如图 7-29 所示。

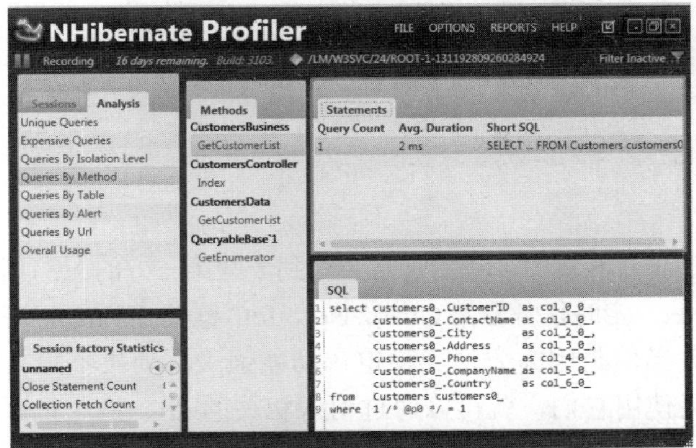

图 7-29

第 8 章

IoC、Log4Net和Quartz.Net

在 ASP.NET MVC 项目开发中，我们常常会用到一些第三方的.NET 框架或者组件，比较常见的有日志框架 Log4Net、定时任务框架 Quartz.Net、IoC 框架 Spring.net 等。

为什么要用它们？当然是因为方便、稳定、功能强大。如果不用它们，许多功能就必须由自己编码实现；如果使用它们，就可以通过引入相关的 dll、调用现成的类库轻松实现相应的功能。

8.1 Unity

在讲解 Unity 之前，我们先来了解一下什么是 IoC。

IoC（Inversion of Control，控制反转）又称为“依赖注入”（Dependence Injection，DI）。

控制反转就是创建对象的权利由开发人员自己控制，转到了由容器来控制。

依赖注入就是通过容器创建对象的时候，在对象的初始化时可以给一些属性、构造方法的参数等注入默认值（可以是复杂的类型）。

IoC 的基本概念是：不创建对象，但是描述创建它们的方式。在代码中不直接与对象和服务连接，但在配置文件中描述哪一个组件需要哪一项服务。容器负责将这些联系在一起。

其原理是基于 OO 设计原则的 The Hollywood Principle：Don't call us, we'll call you（别找我，我会来找你的）。也就是说，所有的组件都是被动的（Passive），所有的组件初始化和调用都由容器负责。组件处在一个容器当中，由容器负责管理。

简单地说，就是应用本身不负责依赖对象的创建和维护，而是将其交给一个外部容器来负责。这样控制权就由应用转移到了外部 IoC 容器，即控制权实现了所谓的反转。比如在类型 A 中需要使用类型 B 的实例，而 B 实例的创建并不由 A 来负责，而是通过外部容器来创建。通过 IoC 的方式实现针对目标 Controller 的激活具有重要的意义。

8.1.1 获取 Unity

Unity 是微软开发的一款 IoC 框架，目前流行的 IoC 框架有 AutoFac、Castle Windsor、Unity、Spring.Net、StructureMap 和 Ninject 等。Unity 在 Codeplex 上的地址为 http://unity.codeplex.com/，我们可以下载相应的安装包和开发文档。当然，如果在 Visual Studio 中安装了 NuGet 包管理器，也可以直接在 NuGet 中获取到最新版本的 Unity，如图 8-1 所示。

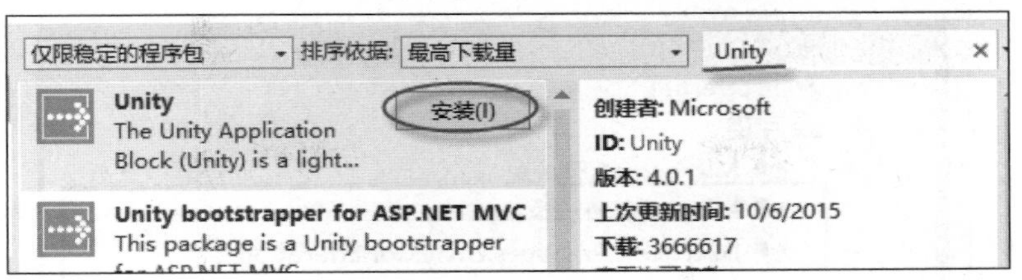

图 8-1

8.1.2　Unity 简介

Unity 是微软 patterns& practices 组用 C#实现的轻量级、可扩展的依赖注入容器，可以通过代码或者 XML 配置文件的形式来配置对象与对象之间的关系，在运行时直接调用 Unity 容器即可获取我们所需的对象，以便建立松散耦合的应用程序。

对于小型项目，用代码的方式实现即可；对于中大型项目，使用配置文件比较好。

Unity 既然是一种 IoC 框架，那么它同样满足 IoC 的共性。依赖注入划分为 3 种形式，即构造器注入、属性（设置）注入和接口注入。

8.1.3　Unity API

- UnityContainer.RegisterType<ITFrom,TTO>();
- UnityContainer.RegisterType< ITFrom, TTO >("keyName");
- IEnumerable<T> databases = UnityContainer.ResolveAll<T>();
- IT instance = UnityContainer.Resolve<IT>();
- T instance = UnityContainer.Resolve<T>("keyName");
- UnityContainer.RegisterInstance<T>("keyName",new T());
- UnityContainer.BuildUp(existingInstance);
- IUnityContainer childContainer1 = parentContainer.CreateChildContainer();

8.1.4　使用 Unity

如果是在项目的引用那里右击选择管理 NuGet 程序包进行安装的，就会自动添加 Microsoft.Practices.Unity.dll、Microsoft.Practices.Unity.Configuration.dll 以及 Microsoft.Practices. Unity.RegistrationByConvention 的引用，如图 8-2 所示。

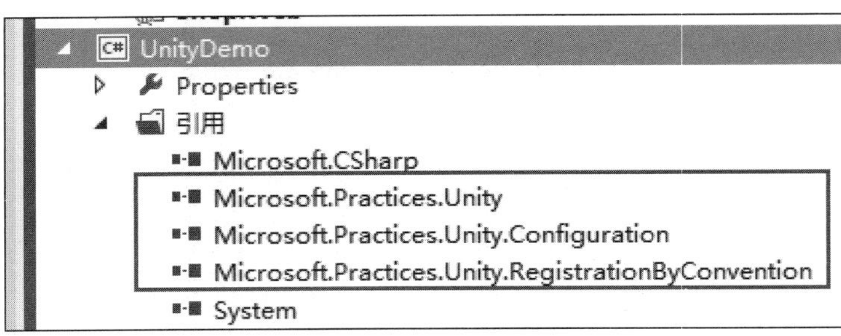

图 8-2

如果是直接在 VS 工具里面使用 NuGet 下载安装，那么相关文件会保存在当前这个解决方案下面的 packages 夹中，必须手动添加引用。

（1）新建控制台项目 UnityDemo。

（2）添加 Unity 引用时使用 NuGet 安装固然方便，但是有时候也会出现安装失败的情况，这时直接从网上下载 dll，然后添加引用反而更快一些。

（3）新建控制台程序 UnityDemo，并依次新建一个接口 IProduct 和两个类 Milk、Sugar。

```
/// <summary>
/// 商品
/// </summary>
public interface IProduct
{
      string ClassName { get; set; }
      void ShowInfo();
}
/// <summary>
/// 牛奶
/// </summary>
public class Milk:IProduct
{
      public string ClassName { get; set; }

      public void ShowInfo()
      {
        Console.WriteLine("牛奶：{0}", ClassName);
      }
}
/// <summary>
/// 糖
/// </summary>
public class Sugar:IProduct
{
```

```
    public string ClassName { get; set; }

    public void ShowInfo()
    {
        Console.WriteLine("糖: {0}", ClassName);
    }
}
```

1. 编程方式

使用 Unity 来管理对象与对象之间的关系可以分为以下几步：

（1）创建一个 UnityContainer 对象。
（2）通过 UnityContainer 对象的 RegisterType 方法来注册对象与对象之间的关系。
（3）通过 UnityContainer 对象的 Resolve 方法来获取指定对象关联的对象。

注入代码如下：

```
/// <summary>
/// 代码注入
/// </summary>
public static void ContainerCode()
{
        IUnityContainer container = new UnityContainer();

        container.RegisterType<IProduct, Milk>();
        //默认注册（无命名），如果后面还有默认注册会覆盖前面的
        container.RegisterType<IProduct, Sugar>("Sugar");  //命名注册

        IProduct _product = container.Resolve<IProduct>();  //解析默认对象
        _product.ClassName = _product.GetType().ToString();
        _product.ShowInfo();

        IProduct _sugar = container.Resolve<IProduct>("Sugar");//指定命名解析对象
        _sugar.ClassName = _sugar.GetType().ToString();
        _sugar.ShowInfo();

        IEnumerable<IProduct> classList = container.ResolveAll<IProduct>();
        //获取容器中所有 IProduct 注册的已命名对象

        foreach (var item in classList)
        {
           item.ClassName = item.GetType().ToString();
           item.ShowInfo();
```

```
        }
}
```

修改 Program.cs 代码：

```
class Program
{
       static void Main(string[] args)
       {
           ContainerCode();
       }
}
```

运行结果如图 8-3 所示。

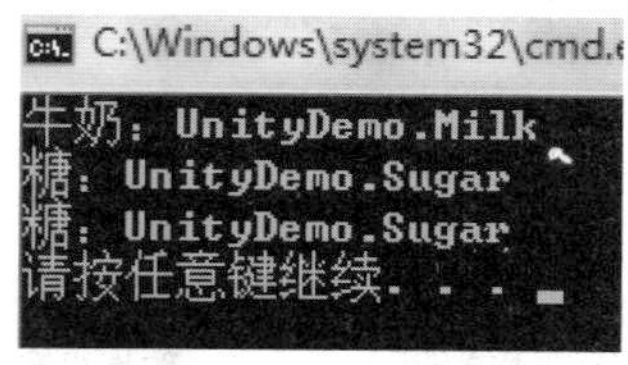

图 8-3

2. 配置文件方式

通过配置文件配置 Unity 信息需要有以下几个步骤：

（1）在配置文件<configSections> 配置节下注册名为 unity 的 section。

（2）在<configuration> 配置节下添加 Unity 配置信息。

（3）在代码中读取配置信息，并将配置载入到 UnityContainer 中。

配置文件内容如下：

```
<?xml version="1.0" encoding="utf-8" ?>
<configuration>
  <configSections>
    <section name="unity" type="Microsoft.Practices.Unity.Configuration.
UnityConfigurationSection,Microsoft.Practices.Unity.Configuration"/>
  </configSections>
  <unity>
    <!--定义类型别名-->
    <aliases>
      <add alias="Iproduct" type="UnityDemo.IProduct,UnityDemo" />
      <add alias="Milk" type="UnityDemo.Milk,UnityDemo" />
      <add alias="Sugar" type="UnityDemo.Sugar,UnityDemo" />
    </aliases>
```

```
    <!--容器-->
    <container name="MyContainer">
      <!--映射关系-->
      <register type="Iproduct"  mapTo="Milk"></register>
      <register type="Iproduct"  mapTo="Sugar" name="Sugar"></register>
    </container>
  </unity>
  <!--<startup>
    <supportedRuntime version="v4.0" sku=".NETFramework,Version=v4.5" />
  </startup>-->
</configuration>
```

先添加 System.Configuration 引用，在配置文件中注入代码：

```
/// <summary>
/// 配置文件注入
/// </summary>
public static void ContainerConfiguration()
{
          IUnityContainer container = new UnityContainer();
          container.LoadConfiguration("MyContainer");
          UnityConfigurationSection section
            =(UnityConfigurationSection)ConfigurationManager.GetSection("unity");
            //获取指定名称的配置节
           section.Configure(container, "MyContainer");
           //获取特定配置节下已命名的配置节<container name='MyContainer'>下的配置信息

          IProduct classInfo = container.Resolve<IProduct>("Sugar");
           classInfo.ClassName = classInfo.GetType().ToString();
          classInfo.ShowInfo();
}
```

如果系统比较庞大，那么对象之间的依赖关系可能会很复杂，最终导致配置文件变得很大，所以我们需要将 Unity 的配置信息从 App.config 或 web.config 中分离出来到某一个单独的配置文件中，比如 Unity.config，实现方式可以参考如下代码：

```
IUnityContainer container = new UnityContainer();
string configFile = "Unity.config";
var fileMap = new ExeConfigurationFileMap { ExeConfigFilename = configFile };
//从 config 文件中读取配置信息
Configuration configuration =ConfigurationManager.
OpenMappedExeConfiguration(fileMap, ConfigurationUserLevel.None);
//获取指定名称的配置节
```

```
UnityConfigurationSection section =
(UnityConfigurationSection)configuration.GetSection("unity");
//载入名称为 FirstClass 的 container 节点
container.LoadConfiguration(section, "MyContainer");
```

8.2 Spring.Net

Spring.Net is an application framework that provides comprehensive infrastructural support for developing enterprise .NET applications。

简单来说，Spring 是一个轻量级的控制反转（IoC）和面向切面（AOP）的容器框架。Spring 通过一种称作控制反转（IoC）的技术促进了松耦合。

Spring 提供了面向切面编程的丰富支持，允许通过分离应用的业务逻辑与系统级服务（例如审计（auditing）和事务（transaction）管理）进行内聚性的开发。应用对象只实现它们应该做的——完成业务逻辑——仅此而已。它们并不负责（甚至是意识）其他系统级关注点，例如日志或事务支持。

比较下 Java 框架和.Net 框架。

- J2EE：SSH=Spring + Struts +Hibernate
- Net：Spring.Net+AspNet MVC+NHibernate

.Net 中的许多框架都是从 Java 框架中演化而来的，看名称就可以知道，如表 8-1 所示。

表 8-1 Java 和.Net 框架比较

Java	.NET
Spring	Spring.Net
Hibernate	NHibernate
Log4	Log4.Net
Quartz	Quartz.Net
Velocity	NVelocity
…	…

8.2.1 Web.config 中的属性注入

（1）新建 MVC 项目 Spring.Net.Demo。

（2）右击项目中的引用，选择“管理 NuGet 程序包”，输入“Spring.Net”，如图 8-4 所示。

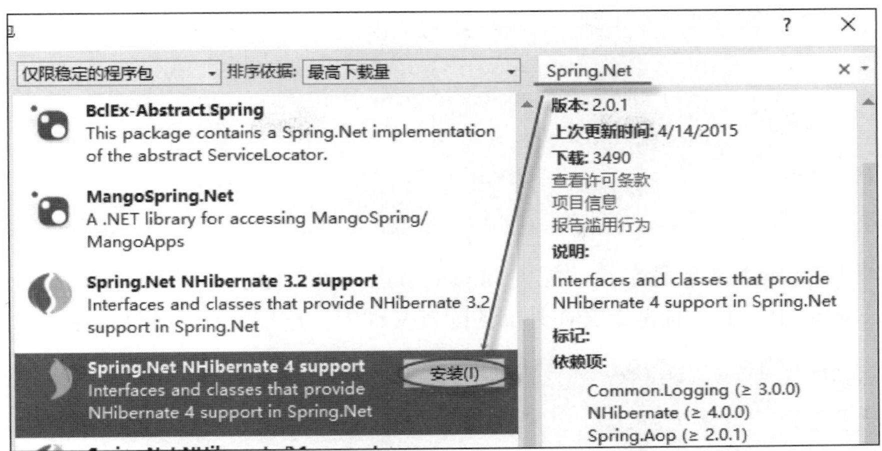

图 8-4

使用 NuGet 程序包进行安装虽然方便，但是会自动下载并添加许多可能用不到的程序集引用，最终都会下载到解决方案根目录下面的 packages 文件夹中。

如果不想通过 NuGet 进行安装，这里在云盘中提供了 Spring.NET-2.0.0-M1.zip，解压后，可以到 Spring.NET-2.0.0-M1\Spring.NET\bin\net\4.0\release 这个目录下面找到所需要的程序集，然后添加程序集 Spring.Core.dll 和 Common.Logging.dll 的引用。

注入分为属性注入和构造函数注入，这里先演示属性注入。

（3）添加 IUserInfo 接口：

```
namespace Spring.Net.Demo
{
   public interface IUserInfo
   {
      string ShowMsg();
   }
}
```

（4）添加 Order 类和 UserInfo 类，继承 IUserInfo 接口：

```
namespace Spring.Net.Demo
{
   public class UserInfo : IUserInfo
   {
      public string UserName { get; set; }
      public Order OrderBy { get; set; }
      public string ShowMsg()
      {
         return "Hello World, " + UserName+"的订单号是: "+OrderBy.OrderNO;
      }
```

```
    }
    public class Order
    {
        public string OrderNO { get; set; }
    }
}
```

（5）添加 Spring.Net 配置节点，配置 object 节点。定义对象主要有两种方式，一种是直接定义在 Web.config 中，另一种是定义在外部的配置文件中，这里直接定义在 Web.config 中。

```
<configSections>
    <!-- For more information on Entity Framework configuration, visit
    http://go.microsoft.com/fwlink/?LinkID=237468 -->
    <section name="entityFramework"
     type="System.Data.Entity.Internal.ConfigFile.EntityFrameworkSection,
    EntityFramework, Version=5.0.0.0, Culture=neutral,
    PublicKeyToken=b77a5c561934e089" requirePermission="false" />
    <!--跟下面 Spring.Net 节点配置是一一对应关系-->
    <sectionGroup name="spring">
    <!--配置解析 Spring 块的对象-->
      <section name="context" type="Spring.Context.Support.ContextHandler,
      Spring.Core"/>
    <!—配置解析 Spring 存放对象的容器集合-->
      <section name="objects" type="Spring.Context.Support.DefaultSectionHandler,
      Spring.Core" />
    </sectionGroup>
</configSections>
<!--Spring.Net 节点配置-->
<spring>
    <context>
      <!--容器配置，配置当前容器对象放置在什么位置：当前是放置在当前配置文件中-->
      <resource uri="config://spring/objects"/>
    </context>
    <objects xmlns="http://www.springframework.net">
      <!--这里放容器里面的所有节点-->
      <description>An  example that demonstrates simple IoC
      features.</description>
      <!--name 必须要唯一的，此处可以随意命名，不过一般直接命名为类名称，方便区分，type=类的全
      名称，所在的程序集，目的是为了让容器能够轻松地通过反射的方式创建对象-->
      <object name="UserInfo" type="Spring.Net.Demo.UserInfo, Spring.Net.Demo">
      <!—这里的 name 值要和属性名称一致-->
        <property name="UserName" value="神刀张三"/>
```

```
        <!--ref 指向下面的属性注入-->
        <property name="OrderBy" ref="Order"/>
      </object>
      <!--复杂属性注入-->
      <object name="Order"  type="Spring.Net.Demo.Order, Spring.Net.Demo">
        <property name="OrderNO" value="201606122226"/>
      </object>
    </objects>
</spring>
```

（6）添加 Home 控制器和名为 Index 的 Action 方法。

```
using Spring.Context;
using Spring.Context.Support;
using System.Web.Mvc;

namespace Spring.Net.Demo.Controllers
{
    public class HomeController : Controller
    {
        public ActionResult Index()
        {
            IApplicationContext ctx = ContextRegistry.GetContext();
            IUserInfo lister = (IUserInfo)ctx.GetObject("UserInfo");
            ViewBag.Msg = lister.ShowMsg();
            return View();
        }
    }
}
```

（7）添加 Index 视图：

```
@{
    ViewBag.Title = "Index";
}

<h2>@ViewBag.Msg</h2>
```

（8）运行结果如图 8-5 所示。

localhost:9970

Hello World，神刀张三的订单号是：201606122226

图 8-5

8.2.2 在单独的配置文件中构造函数注入

（1）在项目 Spring.Net.Demo 中创建一个名为 Config 的文件夹，以保存独立的配置文件。

在 Config 文件夹中创建一个名为 objects.xml 的 xml 配置文件。添加名为 objects 的根元素，添加默认命名空间 xmlns=http://www.springframework.net。

在目录 Spring.Net\Spring.NET-2.0.0-M1\Spring.NET\doc\schema 中找到所有扩展名为 xsd 的架构文件，复制到 vs 安装的架构目录，这里是 C:\Program Files (x86)\Microsoft Visual Studio 11.0\Xml\Schemas。这样，我们在 xml 文件中就具备智能感知功能了。

（2）新建类 NewUserInfo 继承自 IuserInfo 接口。

```
namespace Spring.Net.Demo
{
    public class NewUserInfo : IUserInfo
    {
        public NewUserInfo(string name, Order order)
        {
            this.UserName = name;
            this.OrderBy = order;
        }
        public string UserName { get; set; }
        public Order OrderBy { get; set; }
        public string ShowMsg()
        {
            return "Hello World, " + UserName + "的订单号是：" + OrderBy.OrderNO;
        }
        public class Order
        {
            public string OrderNO { get; set; }
        }
    }
}
```

（3）修改 Web.config 中的 context 配置节点。

```
<context>
    <!--容器配置-->
    <!--<resource uri="config://spring/objects"/>-->
    <!--xml 文件方式，更改属性，复制到输出目录：始终复制-->
    <resource uri="~/Config/objects.xml"/>
</context>
```

（4）为 objects.xml 文件添加如下代码，并设置 objects.xml 的属性“复制到输出目录”值为

"始终复制"。

```
<?xml version="1.0" encoding="utf-8" ?>
<objects xmlns="http://www.springframework.net">
  <!--这里放容器里面的所有节点-->
  <description>An  example that demonstrates simple IoC features.</description>
  <!--name 必须要唯一的，type=类的全名称，所在的程序集-->
  <!--构造函数注入-->
  <object name="NewUserInfo" type="Spring.Net.Demo.NewUserInfo, Spring.Net.Demo">
    <constructor-arg index="0" value="神刀张三"/>
    <!--ref 指向下面的属性注入-->
    <constructor-arg index="1" ref="Order"/>
  </object>
  <!--复杂属性注入-->
  <object name="Order"  type="Spring.Net.Demo.Order, Spring.Net.Demo">
    <property name="OrderNO" value="201606122226"/>
  </object>
</objects>
```

（5）在 Home 控制器中，添加名为 NewIndex 的 Action 方法。

```
public ActionResult NewIndex()
{
        IApplicationContext ctx = ContextRegistry.GetContext();
        IUserInfo lister = (IUserInfo)ctx.GetObject("NewUserInfo");
        ViewBag.Msg = lister.ShowMsg();
        return View();
}
```

（6）添加 NewIndex 视图。

```
@{
    ViewBag.Title = "Index";
}

<h2>@ViewBag.Msg</h2>
```

（7）运行结果如图 8-6 所示。

图 8-6

总结：所有对象的创建都可以通过 IoC 的方式来替代，当我们觉得使用简单工厂、抽象工厂不好的时候，都可以通过 IoC 来替代，而且现在的 IoC 框架一般都支持 AOP。

8.3 Log4Net

Log4Net 是用来记录日志的，可以将程序运行过程中的信息输出到一些地方（文件、数据库、EventLog 等）。日志就是程序的黑匣子，可以通过日志查看系统的运行过程，从而发现系统的问题。

日志的作用就是将运行过程的步骤、成功失败记录下来，将关键性的数据记录下来分析系统问题所在。

对于 Web 应用来说，不能把异常信息显示给用户，异常信息只能记录到日志，出了问题把日志文件发给开发人员就能知道问题所在。

8.3.1 配置 Log4Net 环境

（1）新建项目。

这里沿用 8.2 节中的 Spring.Net.Demo 项目，在项目根目录下新建目录 lib。

（2）添加对 log4net.dll 的引用。

我们同样可以使用 NuGet 来安装，也可以直接去官方网站：http://logging.apache.org/log4net/ 下载。这里在云盘中已经提供了 log4net.dll，可以把它复制到项目中的 lib 文件夹下。

（3）在 Web.Config（或 App.Config）中添加配置。

在<configSections>中添加如下配置节点：

```
<section name="log4net" type="log4net.Config.Log4NetConfigurationSectionHandler,
log4net"/>
```

紧靠 </configSections>后，添加如下配置信息：

```
<log4net>
  <!-- OFF, FATAL, ERROR, WARN, INFO, DEBUG, ALL -->
  <!-- Set root logger level to ERROR and its appenders -->
  <root>
    <level value="ALL"/>
    <appender-ref ref="SysAppender"/>
  </root>

  <!-- Print only messages of level DEBUG or above in the packages -->
  <logger name="WebLogger">
    <!--这里进一步限制了日志级别，只有在大于等于 DEBUG 情况下才会记录日志-->
```

```
    <level value="DEBUG"/>
  </logger>
  <!--指定日志记录的方式：以滚动文件的方式-->
  <appender name="SysAppender"
  type="log4net.Appender.RollingFileAppender,log4net" >
    <!--指定日志存放的路径，这里放置到 App_Data 目录是为了安全-->
    <param name="File" value="App_Data/" />
    <!--日志以追加的形式记录-->
    <param name="AppendToFile" value="true" />
    <param name="RollingStyle" value="Date" />
    <!--设置日志文件名称的生成规则-->
    <param name="DatePattern"value="&quot;Logs_&quot;yyyyMMdd&quot;.txt&quot;" />
    <!--日志名称是否静态：否-->
    <param name="StaticLogFileName" value="false" />
    <!--日志内容格式和布局设置-->
    <layout type="log4net.Layout.PatternLayout,log4net">
      <param name="ConversionPattern" value="%d [%t] %-5p %c - %m%n" />
      <param name="Header" value="-----------------header------------------" />
      <param name="Footer" value="-----------------footer------------------" />
    </layout>
  </appender>
  <appender name="consoleApp" type="log4net.Appender.ConsoleAppender,log4net">
    <layout type="log4net.Layout.PatternLayout,log4net">
      <param name="ConversionPattern" value="%d [%t] %-5p %c - %m%n" />
    </layout>
  </appender>
</log4net>
```

（4）初始化。

在程序最开始加入“log4net.Config.XmlConfigurator.Configure();”，这里添加到 Global.asax 文件中的 Application_Start()方法中。

（5）在要打印日志的地方添加“LogManager.GetLogger(typeof(Program)).Debug("信息");”。通过 LogManager.GetLogger 传递要记录的日志类名获得这个类的 ILog（这样在日志文件中就能看到这条日志是哪个类输出的了），然后调用 Debug 方法输出消息。因为一个类内部不止一个地方要打印日志，所以一般把 ILog 声明为一个 static 字段：

```
Private static ILog logger=LogManager.GetLogger(typeof(Test))
```

输出错误信息用 ILog.Error 方法，第二个参数可以传递 Exception 对象，如 log.Error("***错误"+ex)、log.Error("***错误",ex)。

考虑到记录日志会存在并发的问题，这里通过队列的方式来记录日志。

思路：把所有产生的日志信息存放到一个队列里面，然后通过新建一个线程不断地从这个队列里面读取异常信息，然后往日志里面写。这也就是所谓的生产者-消费者模式。

（6）新建一个类 MyErrorAttribute，继承自全局异常类 HandleErrorAttribute。

```
using System;
using System.Collections.Generic;
using System.Web.Mvc;

namespace Spring.Net.Demo
{
    public class MyErrorAttribute : HandleErrorAttribute
    {
        public static Queue<Exception> ExceptionQueue = new Queue<Exception>();
        public override void OnException(ExceptionContext filterContext)
        {
            ExceptionQueue.Enqueue(filterContext.Exception);
            //filterContext.HttpContext.Response.Redirect("~/Error.html");
            //出现异常时可以考虑让系统跳转到友好界面
            base.OnException(filterContext);
        }
    }
}
```

（7）在 FilterConfig 类中进行如下修改：

```
using System.Web;
using System.Web.Mvc;

namespace Spring.Net.Demo
{
    public class FilterConfig
    {
        public static void RegisterGlobalFilters(GlobalFilterCollection filters)
        {
            //filters.Add(new HandleErrorAttribute()); //注释掉系统默认的
            filters.Add(new MyErrorAttribute()); //添加刚才自定义的
        }
    }
}
```

（8）在 Global.asax 文件中的 Application_Start()方法中添加如下代码：

```
ThreadPool.QueueUserWorkItem(o =>
{
```

```
            while (true)
            {
                if (MyErrorAttribute.ExceptionQueue.Count > 0)
                {
                    Exception ex = MyErrorAttribute.ExceptionQueue.Dequeue();
                    if (ex != null)
                    {
                        ILog logger = LogManager.GetLogger("testError");
                        logger.Error(ex.ToString());//将异常信息写入 Log4Net 中
                    }
                    else
                    {
                        Thread.Sleep(50);
                    }
                }
                else
                {
                    Thread.Sleep(50);
                }
            }
});
```

（9）在控制器 Home 中故意添加一个引发异常的 Action 方法 TestLog。

```
public ActionResult TestLog()
{
        int result = 0;
        int x = 1, y = 0;
        result = x / y;
        return View();
}
```

（10）在浏览器地址栏中输入“http://localhost:9970/home/testlog”，界面会报异常，这时项目根目录下的 App_Data 目录中会多出一个 txt 的日志文件 Logs_20160615.txt，这个日志文件的名称是根据我们在配置文件中配置的规则生成的，以下是日志部分内容：

```
----------------------header------------------------2016-06-15 21:43:37,372 [15]
ERROR testError - System.DivideByZeroException: 尝试除以零。
   在 Spring.Net.Demo.Controllers.HomeController.TestLog()
   位置 d:\Study\Chapter08\Spring.Net.Demo\Controllers\HomeController.cs:行号 27
   在 lambda_method(Closure , ControllerBase , Object[] )
   在 System.Web.Mvc.ActionMethodDispatcher.Execute(ControllerBase controller,
   Object[] parameters)
```

......

8.3.2 Log4Net 相关概念

Log4net 有 3 个主要组件：loggers，appenders 和 layouts。这 3 个组件一起工作使得开发者能够根据信息类型和等级（Level）记录信息，以及在运行时控制信息的格式化和信息的写入位置（如控制台、文件、内存、数据库等）。过滤器帮助这些组件控制追加器（appender）的行为和把对象转换成字符串的对象渲染。

Appender：可以将日志输出到不同的地方，不同的输出目标对应不同的 Appender，如 RollingFileAppender（滚动文件）、AdoNetAppender（数据库）、SmtpAppender （邮件）等。

level（级别）：标识这条日志信息的重要级别。None>Fatal>ERROR>WARN>DEBUG>INFO>ALL，设定一个 Level，那么低于这个 Level 的日志是不会被写到 Appender 中的。

Log4Net 还可以设定多个 Appender，可以实现同时将日志记录到文件、数据、发送邮件等；可以设定不同 Appender 的不同 Level，可以实现普通级别都记录到文件、Error 以上级别都发送邮件；可以实现对不同的类设定不同的 Appender；还可以自定义 Appender，自己实现将 Error 信息发短信等。

日志框架除了 Log4Net 外，还有 Enterprise Library 中的 Logging Application Block、Apache 的 CommonLog 以及 NLog 等，使用起来都差不多。

8.4 Quartz.Net

8.4.1 Quartz.Net 概述

在平时的工作中经常会遇到做轮询调度的任务，比如定时轮询数据库同步、定时邮件通知、定时处理数据等。大家可能通过 Windows 计划任务或 Windows 服务来实现，这里将介绍一个开源的调度框架 Quartz.Net。

Quartz.Net 是一个定时任务框架，可以实现异常灵活的定时任务，开发人员只要编写少量的代码就可以实现“每隔 1 小时执行”、“每天 2 点执行”、“每月 27 日的下午执行 8 次”等各种定时任务。它实现了作业和触发器的多对多关系，还能把多个作业与不同的触发器关联。整合了 Quartz.Net 的应用程序可以重用来自不同事件的作业，还可以为一个事件组合多个作业。

8.4.2 参考资料

官方学习文档：http://www.quartz-scheduler.net/documentation/index.html。

使用实例介绍：http://www.quartz-scheduler.net/documentation/quartz-2.x/quick-start.html。

官方的源代码下载：http://sourceforge.net/projects/quartznet/files/quartznet/。

下载官方的源代码（这里使用 2.4.1 版），解压后如图 8-7 所示。

E:\WorkSpace\Quartz.NET-2.4.1

名称	修改日期	类型	大小
bin	2016/9/23 18:48	文件夹	
database	2016/9/23 18:48	文件夹	
doc	2016/9/23 18:48	文件夹	
lib	2016/9/23 18:48	文件夹	
server	2016/9/23 18:48	文件夹	
src	2016/9/23 18:48	文件夹	
changelog.txt	2016/8/24 5:25	文本文档	29 KB
license.txt	2014/11/2 9:20	文本文档	12 KB
Quartz.2010.sln	2016/8/24 5:16	Microsoft Visual...	5 KB
Quartz.build	2016/8/24 5:40	BUILD 文件	25 KB
Quartz.Server.2010.sln	2016/8/24 5:16	Microsoft Visual...	2 KB
readme.txt	2016/8/24 5:25	文本文档	2 KB

Quartz源码
官方Demo

图 8-7

打开 Quartz.Server.2010.sln，项目结构如图 8-8 所示。

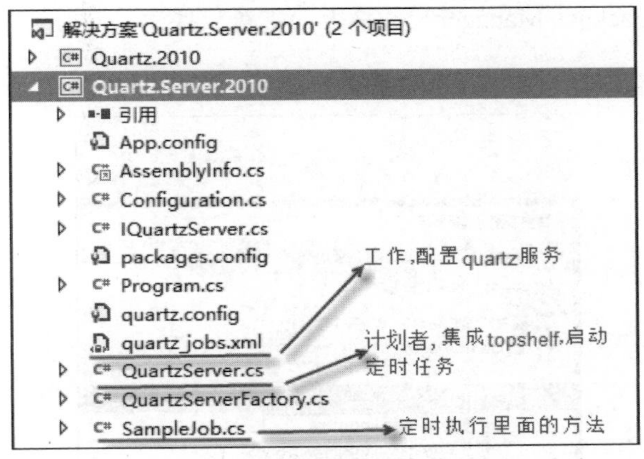

图 8-8

项目无法直接运行，因为少了一些 dll 引用，但是可以直接参考里面的操作代码。

8.4.3 Quartz.Net 使用示例

1. 通过代码的形式

（1）新建控制台项目 QuartzDemo，然后通过 NuGet 安装 Quartz.Net，右击项目 QuartzDemo 中的引用，选择“管理 NuGet 安装包”，如图 8-9 所示。

图 8-9

如果使用的是 VS2012，安装过程中会报错：“Quartz”的架构版本与 NuGet 的版本 2.0.30625.9003 不兼容。解决方案如下：

- 在 Visual Studio 中，选择“工具→扩展与更新...”。
- 选择“更新”选项卡。
- 寻找“NuGet Package Manager”，单击“更新”按钮。

安装完 Quartz.Net 后，项目 QuartzDemo 会自动添加引用，如图 8-10 所示。

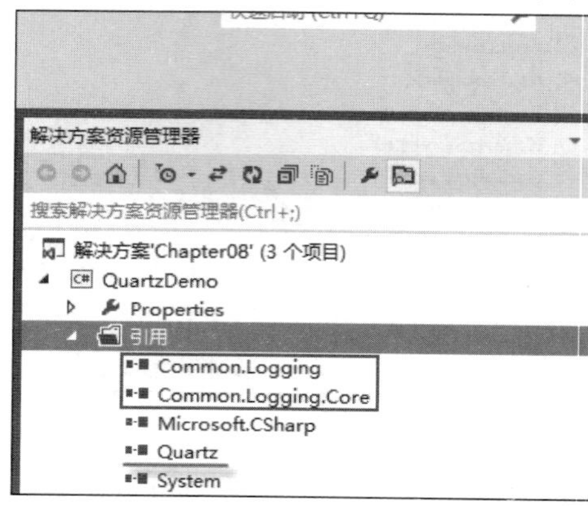

图 8-10

（2）修改 Program 代码：

```
using Quartz;
using Quartz.Impl;
using System;

namespace QuartzDemo
{
    class Program
    {
        static void Main(string[] args)
        {
```

```
    CodeMethod(); //通过代码的方式调用
}

private static void CodeMethod()
{
    Console.WriteLine(DateTime.Now.ToString());
    //1.首先创建一个作业调度池
    ISchedulerFactory scheduler = new StdSchedulerFactory();
    IScheduler sched = scheduler.GetScheduler();
    //2.创建一个具体的作业
    IJobDetail job = JobBuilder.Create<JobDemo>().Build();
    //3.创建并配置一个触发器
    #region 每隔5秒执行一次Execute方法，无休止
    //ISimpleTrigger trigger = (ISimpleTrigger)
    TriggerBuilder.Create().WithSimpleSchedule(x =>
    x.WithIntervalInSeconds(5).WithRepeatCount(int.MaxValue)).Build();
    #endregion

    #region 程序每5秒执行一次，一共执行50次，开始执行时间设定在当前时间，结束时间设定
    并在1小时后，不管50次有没有执行完，1小时后程序都不再继续执行

    DateTimeOffset startTime =
    DateBuilder.NextGivenSecondDate(DateTime.Now.AddSeconds(1), 2);
    DateTimeOffset endTime =
    DateBuilder.NextGivenSecondDate(DateTime.Now.AddHours(1), 5);
    //ISimpleTrigger trigger =
    //(ISimpleTrigger)TriggerBuilder.Create().StartAt(startTime).EndAt(endTime)
    //                          .WithSimpleSchedule(x =>
    //x.WithIntervalInSeconds(5).WithRepeatCount(50))
    //                          .Build();
    #endregion

    #region  实现各种时间纬度的调用（使用cron-like）：在每小时的第10、20、30分钟，
    每分钟的第11、27秒执行一次
    ICronTrigger trigger =
    (ICronTrigger)TriggerBuilder.Create().StartAt(startTime).EndAt(endTime)
                              .WithCronSchedule("11,27 10,20,30 * * * ? ")
                              .Build();
    #endregion

    //4.加入作业调度池中
    sched.ScheduleJob(job, trigger);
```

```
            //5.开始运行
            sched.Start();
            Console.ReadKey();
        }
    }
    public class JobDemo : IJob
    {
        /// <summary>
        /// 这里是作业调度每次定时执行方法
        /// </summary>
        /// <param name="context"></param>
        public void Execute(IJobExecutionContext context)
        {
            Console.WriteLine(DateTime.Now.ToString());
        }
    }
}
```

Quartz.Net 中的概念：计划者（IScheduler）、工作（IJob）、触发器（Trigger）。给计划者一个工作，让他在 Trigger（什么条件下做这件事）触发的条件下执行这个工作。将要定时执行的任务代码写到实现 IJob 接口的 Execute 方法中即可，时间到来的时候 Execute 方法会被调用。

CrondTrigger 是通过 Crond 表达式设置的触发器，还有 SimpleTrigger 等简单的触发器。可以通过 TriggerUtils 的 MakeDailyTrigger、MakeHourlyTrigger 等方法简化调用。

此外，Quartz.Net 支持 cron-like 表达式。

2. 通过配置文件的形式

（1）添加 log4net.dll 的引用。

（2）实现 IJob，新建 Job 类文件 SampleJob。

```
using System;
using log4net;
using Quartz;

namespace QuartzDemo
{
    public sealed class SampleJob : IJob
    {
        private static readonly ILog _logger =
        log4net.LogManager.GetLogger("SampleLog");

        public void Execute(IJobExecutionContext context)
        {
            _logger.Info("SampleJob 测试");
```

```
            Console.WriteLine("执行任务"+DateTime.Now.ToLongTimeString());
        }
    }
}
```

（3）应用程序入口：

```
static void Main(string[] args)
{
        ConfigMethod();//通过配置文件的方式调用
}

private static void ConfigMethod()
{
        var path = string.Format("{0}log4net.xml",
        AppDomain.CurrentDomain.BaseDirectory);
        try
        {
            //工厂
            ISchedulerFactory factory = new StdSchedulerFactory();
            //启动
            IScheduler scheduler = factory.GetScheduler();
            scheduler.GetJobGroupNames();

            log4net.Config.XmlConfigurator.Configure(new  System.IO.FileInfo(path));
            scheduler.Start();
            //Console.ReadKey();
        }
        catch (Exception ex)
        {
            throw;
        }

        //log4net.Config.XmlConfigurator.Configure(new System.IO.FileInfo(path));
        //SampleJob w = new SampleJob();
        //w.Execute(null);
}
```

（4）配置 log4net.xml、quartz_jobs.xml。

从源码项目 Quartz.Server.2010 中复制 quartz_jobs.xml 文件和 quartz_jobs.xml 文件。同时选中这两个文件，右击，选择属性并将复制到输入目录设为“始终复制”。

修改 log4net.xml 可参考 8.3.1 小节：

```
<log4net>
  <!-- Print only messages of level DEBUG or above in the packages -->
  <logger name="SampleLog">
    <level value="INFO"/>
```

```
    <appender-ref ref="SampleInfoFileAppender" />
    <appender-ref ref="SampleErrorFileAppender" />
  </logger>
  <appender name="SampleInfoFileAppender"
  type="log4net.Appender.RollingFileAppender">
    <!--<appender name="ErrorAppender"
    type="log4net.Appender.FileAppender,log4net">-->
    <file value="log/error.txt" />
    <appendToFile value="true" />
    <rollingStyle value="Size" />
    <maxSizeRollBackups value="100" />
    <maximumFileSize value="10240KB" />
    <staticLogFileName value="true" />
    <Encoding value="UTF-8" />
    <filter type="log4net.Filter.LevelRangeFilter">
      <param name="LevelMin" value="WARN" />
      <param name="LevelMax" value="FATAL" />
    </filter>
    <layout type="log4net.Layout.PatternLayout">
      <conversionPattern value="%date %-5level %logger - %message%newline" />
      <param name="Header" value="-------------header---------------&#xD;&#xA;" />
      <param name="Footer" value="-------------footer---------------&#xD;&#xA;" />
    </layout>
  </appender>
  <appender name="SampleErrorFileAppender"
  type="log4net.Appender.RollingFileAppender">
    <!--<appender name="ErrorAppender"
    type="log4net.Appender.FileAppender,log4net">-->
    <file value="log/error.txt" />
    <appendToFile value="true" />
    <rollingStyle value="Size" />
    <maxSizeRollBackups value="100" />
    <maximumFileSize value="10240KB" />
    <staticLogFileName value="true" />
    <Encoding value="UTF-8" />
    <filter type="log4net.Filter.LevelRangeFilter">
      <param name="LevelMin" value="WARN" />
      <param name="LevelMax" value="FATAL" />
    </filter>
    <layout type="log4net.Layout.PatternLayout">
      <conversionPattern value="%date %-5level %logger - %message%newline" />
      <param name="Header" value="------------header----------------&#xD;&#xA;" />
      <param name="Footer" value="------------footer---------------&#xD;&#xA;" />
    </layout>
  </appender>
</log4net>
```

修改 quartz_jobs.xml：

```
<?xml version="1.0" encoding="UTF-8"?>
<!-- This file contains job definitions in schema version 2.0 format -->
<job-scheduling-data   xmlns="http://quartznet.sourceforge.net/JobSchedulingData"
xmlns:xsi="http://www.w3.org/2001/XMLSchema-instance" version="2.0">
  <processing-directives>
    <overwrite-existing-data>true</overwrite-existing-data>
  </processing-directives>
  <schedule>
    <!--SampleJob 测试 任务配置-->
    <job>
      <name>SampleJob</name>
      <group>Sample</group>
      <description>SampleJob 测试</description>
      <job-type>QuartzDemo.SampleJob,QuartzDemo</job-type>
      <durable>true</durable>
      <recover>false</recover>
    </job>
    <!--每隔3秒执行一次-->
    <trigger>
      <cron>
        <name>SampleJobTrigger</name>
        <group>Sample</group>
        <job-name>SampleJob</job-name>
        <job-group>Sample</job-group>
        <start-time>2015-01-22T00:00:00+08:00</start-time>
        <cron-expression>0/3 * * * * ?</cron-expression>
      </cron>
    </trigger>
  </schedule>
</job-scheduling-data>
```

（5）配置 App.config。

```
<?xml version="1.0"?>
<configuration>
  <configSections>
    <section name="quartz" type="System.Configuration.NameValueSectionHandler"/>
    <section name="log4net" type="log4net.Config.Log4NetConfigurationSectionHandler,
    log4net"/>
  </configSections>
  <quartz>
    <add key="quartz.scheduler.instanceName"
    value="ExampleDefaultQuartzScheduler"/>
    <add key="quartz.threadPool.type" value="Quartz.Simpl.SimpleThreadPool,
    Quartz"/>
```

```
    <add key="quartz.threadPool.threadCount" value="20"/>
    <add key="quartz.threadPool.threadPriority" value="4"/>
    <add key="quartz.jobStore.misfireThreshold" value="60000"/>
    <add key="quartz.jobStore.type" value="Quartz.Simpl.RAMJobStore, Quartz"/>
    <add key="quartz.scheduler.exporter.channelName" value="httpQuartz"/>
    <!--*********************Plugin 配置*********************************-->
    <add key="quartz.plugin.xml.type"
    value="Quartz.Plugin.Xml.XMLSchedulingDataProcessorPlugin, Quartz" />
    <add key="quartz.plugin.xml.fileNames" value=" ~/quartz_jobs.xml"/>
  </quartz>
</configuration>
```

运行结果如图 8-11 所示。

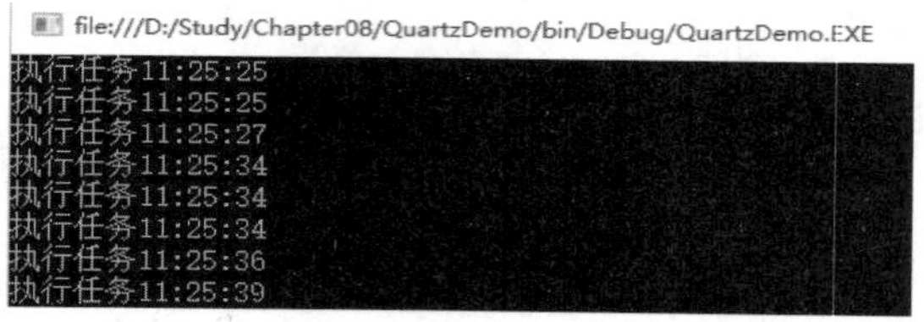

图 8-11

quartz_jobs.xml 的详细配置说明如下。

（1）job 配置节点：其实就是 1.x 版本中的<job-detail>，这个节点是用来定义每个具体任务的，多个任务创建多个 job 节点即可。

- Name（必填）：任务名称，同一个 group 中多个 job 的 name 不能相同，若未设置 group 则所有未设置 group 的 job 为同一个分组，如<name>sampleJob</name>。
- Group（选填）：任务所属分组，用于标识任务所属分组，如<group>sampleGroup</group>。
- Description（选填）：任务描述，用于描述任务具体内容，如<description>Sample job for Quartz Server</description>。
- job-type（必填）：任务类型，任务的具体类型及所属程序集，格式为“实现了 IJob 接口包含完整命名空间的类名，程序集名称”，如<job- type>Quartz.Server.SampleJob, Quartz.Server</job-type>。
- durable（选填）：具体作用不知，官方示例中默认为 true，如<durable>true</durable>。
- recover（选填）：具体作用不知，官方示例中默认为 false，如<recover>false</recover>。

（2）trigger 任务触发器：用于定义使用何种方式触发任务（job），同一个 job 可以定义多个 trigger，多个 trigger 各自独立地执行调度，每个 trigger 中必须且只能定义一种触发器类型（calendar-interval、simple、cron）。

- calendar-interval：一种触发器类型，使用较少。
- simple：简单任务的触发器，可以调度用于重复执行的任务。
- name（必填）：触发器名称，同一个分组中的名称必须不同。

- group（选填）：触发器组。
- description（选填）：触发器描述。
- job-name（必填）：要调度的任务名称，该 job-name 必须和对应 job 节点中的 name 完全相同。
- job-group（选填）：调度任务（job）所属分组，该值必须和 job 中的 group 完全相同。
- start-time（选填）：任务开始执行时间 utc 时间，北京时间需要+08:00，如<start-time>2012-04-01T08:00:00+08:00</start-time>表示北京时间 2012 年 4 月 1 日上午 8:00 开始执行，注意服务启动或重启时都会检测此属性，若没有设置此属性或者 start-time 设置的时间比当前时间早，则服务启动后会立即执行一 次调度，若设置的时间比当前时间晚，服务会等到设置时间相同后才第一次执行任务，一般若无特殊需要请不要设置此属性。
- repeat-count（必填）：任务执行次数，如<repeat-count>-1</repeat-count>表示无限次执行、<repeat- count>10</repeat-count>表示执行 10 次。
- repeat-interval（必填）：任务触发间隔（毫秒），如<repeat-interval>10000</repeat-interval>表示每 10 秒执行一次。

（3）cron 复杂任务触发器——使用 cron 表达式定制任务调度（强烈推荐）。

- name（必填）：触发器名称，同一个分组中的名称必须不同。
- group（选填）：触发器组。
- description（选填）：触发器描述。
- job-name（必填）：要调度的任务名称，该 job-name 必须和对应 job 节点中的 name 完全相同。
- job-group（选填）：调度任务（job）所属分组，该值必须和 job 中的 group 完全相同。
- start-time（选填）：任务开始执行时间 utc 时间，北京时间需要+08:00，如<start-time>2012-04-01T08:00:00+08:00</start-time>表示北京时间 2012 年 4 月 1 日上午 8:00 开始执行，注意服务启动或重启时都会检测此属性，若没有设置此属性，服务会根据 cron-expression 的设置执行任务调度；若 start-time 设置的时间比当前时间早，则服务启动后会忽略掉 cron-expression 设置，立即执行一次调度，之后再根据 cron- expression 执行任务调度；若设置的时间比当前时间晚，则服务会在到达设置时间相同后才应用 cron-expression，根据规则执行任务调度，一般若无特殊需要请不要设置此属性。
- cron-expression（必填）：cron 表达式，如<cron-expression>0/10 * * * * ?</cron-expression>表示每 10 秒执行一次。

cron 表达式由 7 段构成：秒 分 时 日 月 星期 年（可选）。cron 表达式相关内容参看表 8-2、表 8-3、表 8-4 所示。

表 8-2 cron 表达式中的特殊字符说明

特殊字符	说明
-	表示范围，如“MON-WED”表示星期一到星期三、“10-12”表示“10 点、11 点、12 点”
,	表示列举，如 MON,WEB 表示星期一和星期三
*	表示“每”，如每月、每天、每周、每年等
/	表示增量如 0/15（处于分钟段里面）表示每 15 分钟，在 0 分以后开始，3/20 表示每 20 分钟，从 3 分钟以后开始
?	只能出现在日、星期段里面，表示不指定具体的值
L	只能出现在日、星期段里面，是 Last 的缩写，一个月的最后一天，一个星期的最后一天（星期六）
W	表示工作日，距离给定值最近的工作日
#	表示一个月的第几个星期几，例如，"6#3"表示每个月的第三个星期五（1=SUN...6=FRI,7=SAT）
LW	L 和 W 可以在日期域中联合使用，LW 表示这个月最后一周的工作日
C	允许在日期域和星期域出现。这个字符依靠一个指定的 “日历”。也就是说这个表达式的值依赖于相关的“日历”的计算结果，如果没有“日历”关联，则等价于所有包含的“日历”。例如，日期域是“5C”表示关联 “日历”中的第一天，或者这个月开始的第一天的后 5 天；星期域是“1C”表示关联“日历”中的第一天，或者星期的第一天的后 1 天，也就是周日的后一天（周一）

表 8-3 cron 表达式范围说明

字段	允许值	说明
秒	0-59	, - * /
分	0-59	, - * /
小时	0-23	, - * /
月内日期	1-31	, - * ? / L W C
月	1-12 或者 JAN-DEC	, - * /
周内日期	1-7 或者 SUN-SAT	, - * ? / L C #
年（可选）	留空，1970-2099	, - * /

表 8-4 cron 表达式官方实例

表达式	说明
0 0 12 * * ?	每天中午 12 点触发
0 15 10 ? * *	每天上午 10:15 触发
0 15 10 * * ?	每天上午 10:15 触发
0 15 10 * * ? *	每天上午 10:15 触发
0 15 10 * * ? 2005	2005 年的每天上午 10:15 触发
0 * 14 * * ?	在每天下午 2 点到下午 2:59 期间的每 1 分钟触发
0 0/5 14 * * ?	在每天下午 2 点到下午 2:55 期间的每 5 分钟触发
0 0/5 14,18 * * ?	在每天下午 2 点到 2:55 期间和下午 6 点到 6:55 期间的每 5 分钟触发
0 0-5 14 * * ?	在每天下午 2 点到下午 2:05 期间的每 1 分钟触发

（续表）

表达式	说明
0 10,44 14 ? 3 WED	每年三月的星期三的下午 2:10 和 2:44 触发
0 15 10 ? * MON-FRI	周一至周五的上午 10:15 触发
0 15 10 15 * ?	每月 15 日上午 10:15 触发
0 15 10 L * ?	每月最后一日的上午 10:15 触发
0 15 10 L-2 * ?	每个月的最后 2 天的上午 10:15 触发
0 15 10 ? * 6L	每月的最后一个星期五上午 10:15 触发
0 15 10 ? * 6L	每月的最后一个星期五上午 10:15 触发
0 15 10 ? * 6L 2002-2005	2002 年至 2005 年的每月最后一个星期五上午 10:15 触发
0 15 10 ? * 6#3	每月的第三个星期五上午 10:15 触发
0 0 12 1/5 * ?	从每个月的第一天开始，每隔 5 天，中午 12 点时触发
0 11 11 11 11 ?	每年 11 月上午 11:11 触发

扩展：在前两节的介绍中，我们都是将 Quartz.Net 寄宿在 Windows 控制台程序中，而通常在项目中我们是将其寄宿在 Windows 服务中的，要寄宿在 Windows 服务中，我们既可以采用直接用 VS 开发 Windows 服务的形式（参考 http://www.cnblogs.com/jiekzou/p/4393886.html），也可以通过使用第三方组件的形式，如使用 Topshelf 创建 Windows 服务（参考 http://www.cnblogs.com/jys509/p/4614975.html）。

第 9 章 分布式技术

本章将会为大家介绍目前在 ASP.NET 中比较常见的分布式开发技术，并通过 Demo 来演示它们的用法。各项技术都不会做很深入的讲解，因为目标很明确，就是为了让大家能够快速将这些技术运用在工作中，要知道，.NET 开发涉及的技术非常多，每一项技术都可以去网上找到 *N* 本书，而我们在一家公司通常只用到了极少一部分技术，那么我们需要做的就是公司用什么技术就去钻研什么技术，公司没有用到技术可以去了解、开阔视野，这样当遇到各种需求时，你可以想到更多的技术方案。但是没必要花太多的时间去钻研工作中用不到的东西，因为一个人的时间和精力都是有限的，说句现实的，你学的任何东西，只有真正用到了、有了产出才能体现出价值，对企业而言尤其如此。

问题是，我们跳槽换工作，去了新的公司，可能以前公司用的技术用不上了，还要学习新技术。现实就是这样的，所以我们只有练好基本功，以不变应万变，才能满足企业的需求。

9.1 WebService

Web Service 是一个平台独立的、低耦合的、自包含的、基于可编程的 Web 的应用程序，可使用开放的 XML（标准通用标记语言下的一个子集）标准来描述、发布、发现、协调和配置这些应用程序，用于开发分布式的互操作的应用程序。

Web Service 技术能使得运行在不同机器上的不同应用无须借助附加的、专门的第三方软件或硬件就可相互交换数据或集成。依据 WebService 规范实施的应用之间，无论它们所使用的语言、平台或内部协议是什么，都可以相互交换数据。WebService 是自描述、自包含的可用网络模块，可以执行具体的业务功能。WebService 也很容易部署，因为它们基于一些常规的产业标准以及已有的一些技术，诸如标准通用标记语言下的子集 XML、HTTP。WebService 减少了应用接口的花费。WebService 为整个企业甚至多个组织之间的业务流程集成提供了一个通用机制。

9.1.1 创建一个 WebService 并调用

（1）打开 VS2012，新建空项目 WebAppService，注意选择.NET Framework 的版本，这里我选择的是.NET Framework 4.5，.NET 4.6 中找不到 Web 服务模板了。

（2）右击项目 WebAppService，选择“添加新项→Web→Web 服务”，命名为“MyWebService.asmx”。

（3）在 MyWebService 类中默认会添加一个 HelloWorld 方法，继续添加如下方法：

```
/// <summary>
/// 乘法运算
/// </summary>
/// <param name="a"></param>
/// <param name="b"></param>
/// <returns></returns>
[WebMethod]
[WebMethod(Description = "乘法运算（计算两个数相乘的结果）")]
public int Multiplier(int a, int b)
{
   return a * b;
}
```

（4）直接在浏览器中浏览 MyWebService.asmx，如图 9-1 所示。

（5）单击这个方法 Multiplier，如图 9-2 所示。

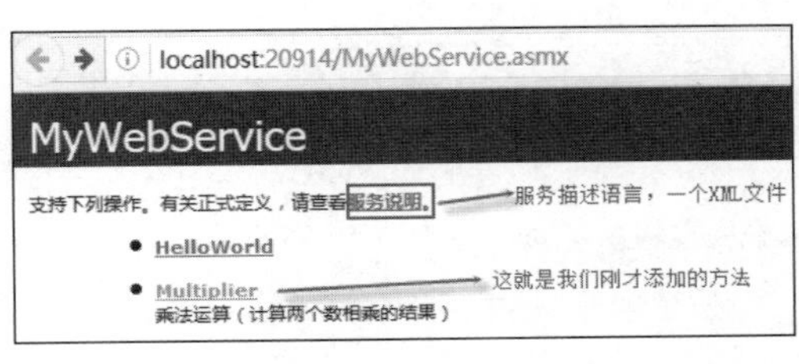

图 9-1

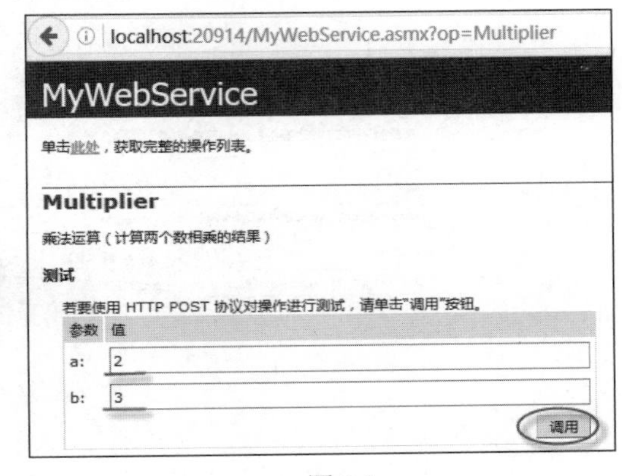

图 9-2

（6）直接调用，效果如图 9-3 所示。

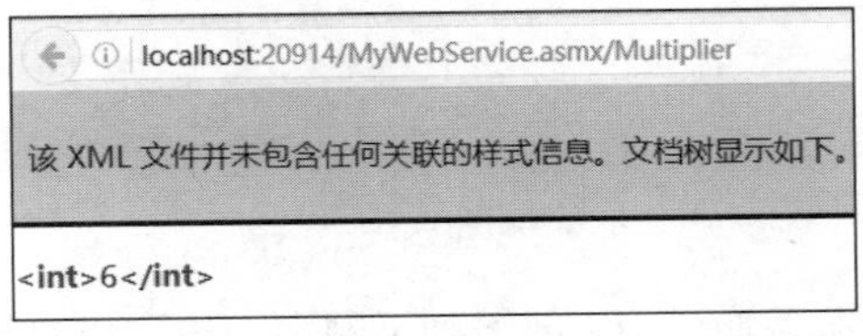

图 9-3

返回的是一个 XML 文件，在项目开发中，我们一般不这样调用，而是通过添加 Web 引用的方式。

在上面我们已经添加了一个 Multiplier 的 WebService 方法，但是这个时候别人还是无法访问的，我们需要把这个 Web 服务发布到 IIS 上面去。在这里，选中 MyWebService.asmx 后右击，选择“在浏览器中查看”，就把 Web 服务寄宿在 IIS Express 里面了。

（7）右击项目 WebAppService，选择“添加新项→Web 窗体”，命名为 WebForm1。WebForm1.aspx 的页面代码如下：

```
<div>
    <asp:TextBox ID="txtA" runat="server"></asp:TextBox>
    *<asp:TextBox ID="txtB" runat="server"></asp:TextBox>
    <asp:Button ID="btnGetResult" runat="server" Text="="
    OnClick="btnGetResult_Click" /><asp:TextBox ID="txtResult"
    runat="server"></asp:TextBox>
</div>
```

（8）在项目中添加服务引用，如图 9-4、图 9-5 所示。

图 9-4

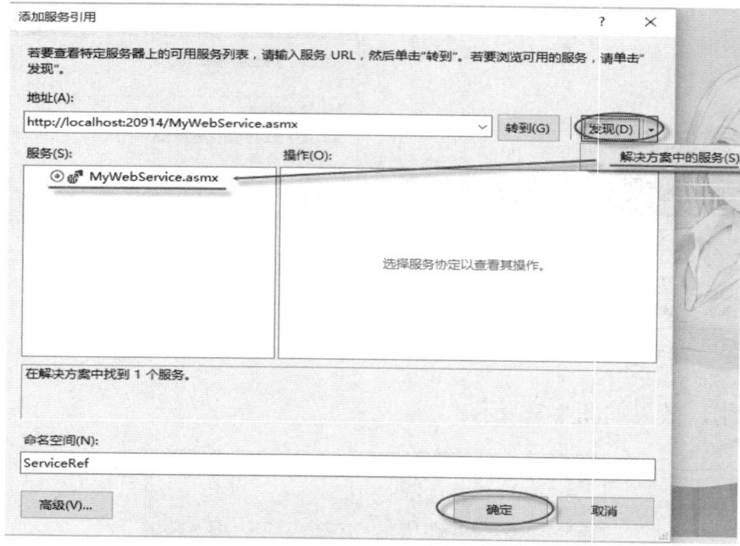

图 9-5

当然，如果我们把 Web 服务通过 IIS 部署到了网上，也可以直接在地址栏中输入 Web 服务地址，然后单击“转到(G)”按钮。

（9）WebForm1 后台代码如下：

```
protected void btnGetResult_Click(object sender, EventArgs e)
```

```
{
        ServiceRef.MyWebServiceSoapClient _client = new
         ServiceRef.MyWebServiceSoapClient();
        txtResult.Text = _client.Multiplier(int.Parse(txtA.Text.Trim()),
        int.Parse(txtB.Text.Trim())).ToString();
}
```

（10）右击 WebForm1.aspx，在浏览器中查看，运行结果如图 9-6 所示。（注意，不要关闭在浏览器中打开的 http://localhost:20914/MyWebService.asmx 界面。）

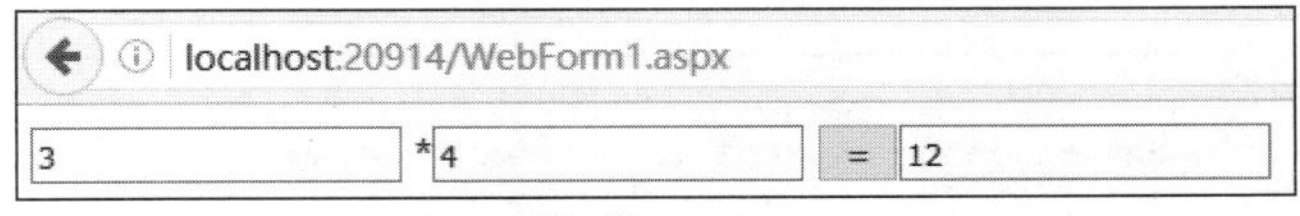

图 9-6

为了方便，这里直接把 Web 服务寄宿在 IIS Express 上了。一般在项目中，我们会把所有的 Web 服务单独放到一个 Web 站点，然后发布到 IIS 上面供用户调用。

9.1.2　调用天气预报服务

在项目开发中，我们除了发布 WebService 供客户调用外，也经常需要调用一些客户或者第三方的 WebService 服务，这里就通过一个 Demo 来演示调用一个第三方的天气预报服务。

（1）寻找天气预报服务。可以百度搜索“天气预报服务”，这里以 http://www.webxml.com.cn/WebServices/WeatherWebService.asmx 这个天气预报服务为例，此服务数据来源于中国气象局 http://www.cma.gov.cn/ ，数据每 2.5 小时左右自动更新一次，准确可靠，包括 340 多个中国主要城市和 60 多个国外主要城市三日内的天气预报数据。

（2）引入 Web 服务。在 VS 中的项目上右击，选择“添加服务引用”，如图 9-7 所示。

图 9-7

（3）在弹出的添加服务引用窗口中单击“高级”按钮。如图 9-8 所示。

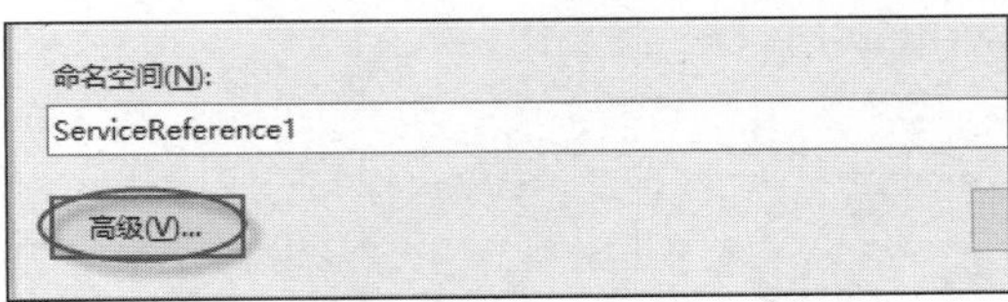

图 9-8

在新弹出的窗体中单击“添加 Web 引用(W)...”按钮，如图 9-9 所示。

图 9-9

（4）输入 Web 服务地址，如图 9-10 所示。

图 9-10

（5）从 http://www.webxml.com.cn/images/weather.zip 中下载天气图标，然后解压文件，将文件夹 weather 复制到根目录文件夹 Content 下。

（6）添加 Home 控制器和两个 GetWeather 方法，实现代码。核心代码如下：

```
public ActionResult GetWeather()
{
        return View();
}
/// <summary>
/// 获得文本框录入的查询城市
/// </summary>
/// <param name="city"></param>
/// <returns></returns>
[HttpPost]
public ActionResult GetWeather(string city)
{
        //把 webservice 当作一个类来操作
        Weather.WeatherWebService client = new Weather.WeatherWebService();
        var s = client.getWeatherbyCityName(city); //string 数组存放返回结果
        //以文本框内容为变量实现 getWeatherbyCityName 方法
```

```
        if (s[8] == "")
        {
            ViewBag.Msg = "暂时不支持您查询的城市";
        }
        else
        {
            ViewBag.ImgUrl = @"/Content/weather/" + s[8];
            //Server.MapPath(@"~\Content\weather\") + s[8];
            ViewBag.General = s[1] + " " + s[6];
            ViewBag.Actually = s[10];
        }
        return View();
}
```

（7）添加 GetWeather 视图，代码如下：

```
@{
    ViewBag.Title = "GetWeather";
}
@using (Html.BeginForm("GetWeather", "Home", FormMethod.Post))
{
<div>请输入查询的城市：@Html.TextBox("city")<input type="submit" value="查询"
/></div>
<div>天气概况：@ViewBag.General @{if(!string.IsNullOrEmpty(ViewBag.ImgUrl as
string)){<img src="@ViewBag.ImgUrl"/>}}</div>
<div>天气实况：@ViewBag.Actually</div>
<div>@ViewBag.Msg</div>
}
```

（8）运行，在浏览器中输入地址“http://localhost:62563/Home/GetWeather”，运行结果如图 9-11 所示。

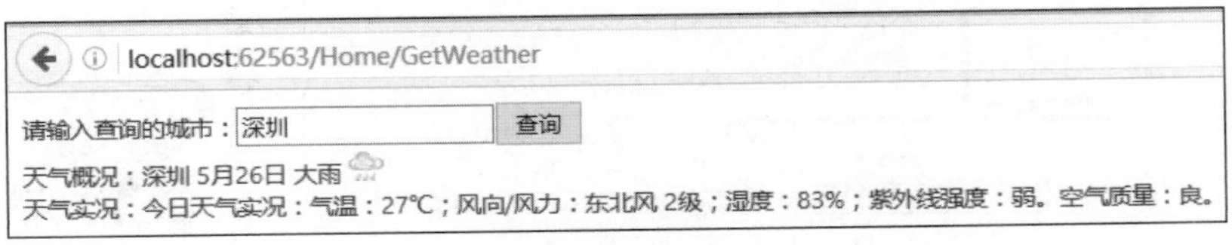

localhost:62563/Home/GetWeather

请输入查询的城市：深圳 查询

天气概况：深圳 5月26日 大雨

天气实况：今日天气实况：气温：27℃；风向/风力：东北风 2级；湿度：83%；紫外线强度：弱。空气质量：良。

图 9-11

9.2 WCF

9.2.1 什么是 WCF

WCF（Windows Communication Foundation）是由微软开发的一系列支持数据通信的应用程序框架，可以翻译为 Windows 通信开发平台。

WCF 整合了原有的 Windows 通信的.net Remoting、WebService、Socket 机制，并融合有 HTTP 和 FTP 的相关技术。WCF 是对这些技术的统一，如图 9-12 所示。

MSDN 上的定义是：WCF 为.NetFramework 提供了一个基础，使其能够编写代码，以在组件、应用程序、系统之间进行通信。WCF 设计遵循的是面向服务原则。服务是指可以通过消息与之进行交互的一段代码。服务是被动的，它们等待传入消息之后才开始工作。客户端是发起者，客户端将消息发送给服务来请求工作。

WCF 是 Windows 平台上开发分布式应用最佳的实践方式。

假定我们要为一家汽车租赁公司开发一个新的应用程序，用于租车预约服务。该租车预约服务会被多个应用程序访问，包括呼叫中心（Call Center）、基于 J2EE 的租车预约服务以及合作伙伴的应用程序（Partner Application）。

使用 WCF，该解决方案的实现就容易多了。如图 9-13 所示，WCF 可用于前述所有情况。因此，租车预定应用程序使用这一种技术就可以实现其所有应用程序间的通信。

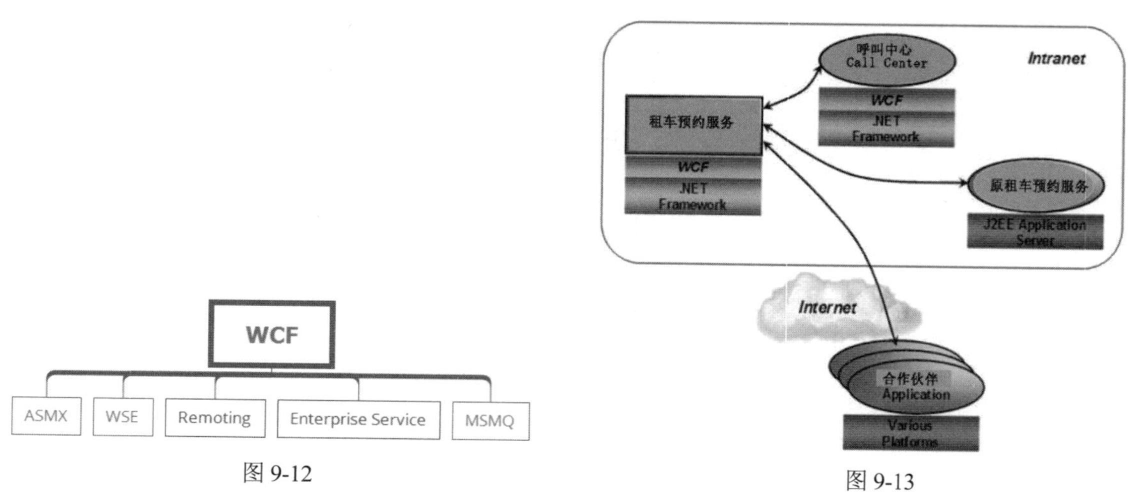

图 9-12

图 9-13

WCF 可使用 Web 服务进行通信，因此与同样支持 SOAP 的其他平台（例如基于 J2EE 的主流应用程序服务器）间的互操作性就变得简单明了了。

我们还可以对 WCF 进行配置和扩展，以便于使用并非基于 SOAP 的消息的 Web 服务进行通信。

性能是大多数业务中至关重要的考虑事项。开发 WCF 的目标就是要使之成为 Microsoft 所开发的速度最快的分布式应用程序平台之一。

WCF 是提供统一的可用于建立安全、可靠的面向服务应用的高效开发平台。

WCF 具有如下优势：

- 统一性
- 互操作性
- 安全与可信赖
- 兼容性

9.2.2　理解面向服务

SOA（Service-Oriented-Architecture，面向服务架构）是指为了解决在 Internet 环境下业务集成的需要，通过连接能完成特定任务的独立功能实体实现的一种软件系统架构。SOA 是一个组件模型，将应用程序的不同功能单元（称为服务）通过这些服务之间定义良好的接口和契约联系起来。

SOA 指出当前系统应该足够灵活，从而允许在不打乱当前成功运行的体系结构和基础结构前提下，改动已有的体系结构。

SOA 有如下原则：

- 边界清晰
- 服务自治
- 兼容性基于策略
- 共享模式（schema）和契约

9.2.3　WCF 体系架构简介

WCF 框架组成如图 9-14 所示。

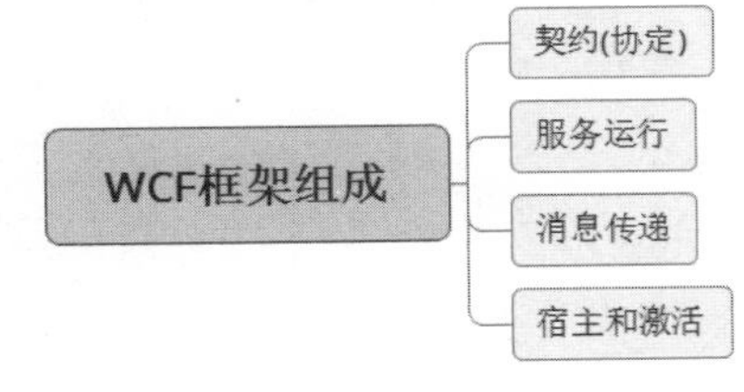

图 9-14

1. 契约（协定）

契约定义消息系统的各个方面。

- 数据契约：服务中的参数。
- 消息契约：使用 SOAP 协议特定的消息部分。
- 服务契约：服务中的方法。
- 策略与绑定：策略设置安全或其他条件，绑定指定传输方式与编码。

2. 服务运行

服务运行期间的行为控制。

- 限制行为：控制处理的消息数。

- 错误行为：出现内部错误时所处理的操作。
- 元数据行为：是否向外提供元数据及元数据的提供方式。
- 实例行为：可运行的服务实例数目。
- 事务行为：处理事务。
- 调度行为：控制 WCF 处理消息的方式。

3. 消息传递

消息传递层说明数据的交换格式和传输模式。消息传递层由通道（信道）组成，通道是对消息进行处理的组件，负责以一致的方式对消息进行整理和传送。通道用于传输层、协议层及消息获取。各层次的通道组成了信道栈。

通道对消息和消息头进行操作，服务运行时对消息正文进行操作。通道包括两种类型：传输通道与协议通道。

- 传输通道：读取和写入来自网络的消息，传输通道通过编码器将消息转换为网络传输使用的字节流以及将字节流转换为消息。传输通道包括 HTTP 通道、命名管道、TCP、MSMQ 等。
- 协议通道：通过读取或写入消息头的方式来实现消息协议，比如 WS-Security、WS-Reliability。

4. 宿主和激活

服务宿主负责 WCF 服务的生命周期和上下文的操作系统进程，负责启动和停止 WCF 服务，并提供控制服务的基本管理功能。

9.2.4 WCF 的基础概念介绍

WCF 的几个重要基础概念如图 9-15 所示。

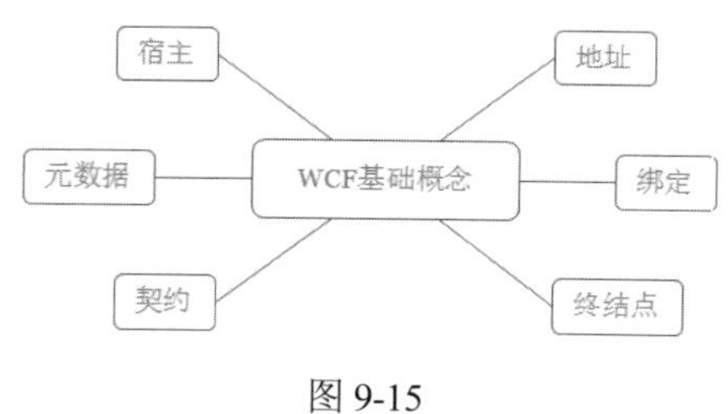

图 9-15

1. 地址

地址指定了接收消息的位置，WCF 中的地址以统一资源标识符（URI）的形式指定。URI 由通信协议和位置路径两部分组成，如 http://localhost:8000/表明通信协议为 http，位置是 localhost（本机）的 8000 端口。

上述提到的消息是指一个独立的数据单元，一般由消息正文和消息头组成，而服务端与客户端的交互都是通过消息来进行的。

WCF 中支持的传输协议包括 HTTP、TCP、Peer network（对等网）、IPC（基于命名管道的内部进程通信）以及 MSMQ（微软消息队列），每个协议对应一个地址类型：

- HTTP 地址：使用 HTTP 协议进行传输（包括 https 安全传输协议），其地址形式为 http://localhost:8000/。如果地址中未指定端口号，则默认端口为 80。
- TCP 地址：使用 TCP 协议进行传输，其地址形式为 net.tcp://localhost:8000/。
- IPC 地址：使用 net.pipe 进行传输，其地址形式为 net.pipe://localhost/。
- MSMQ 地址：使用 Microsoft Message Queue 机制进行传输，其地址形式为 net.msmq://localhost/。
- 对等网地址：使用 net.p2p 进行传输，其地址形式为 net.p2p://localhost/。

2. 绑定（Binding）

由于 WCF 支持 HTTP、TCP、Named Pipe、MSMQ、Peer-To-Peer TCP 等协议，HTTP 又分为基本 HTTP 支持（BasicHttpBinding）以及 WS-HTTP 支持（WsHttpBinding），而 TCP 亦支持 NetTcpBinding、NetPeerTcpBinding 等通信方式，因此双方必须要统一通信的协议，并且要在编码以及格式上保持一致。

一个设置通信协议绑定的示例如下：

```
<?xml version="1.0" encoding="utf-8" ?>
<configuration>
  <system.serviceModel>
    <!-- 包含应用中驻留的所有 service 的配置要求 -->
    <services>
      <service name=" CalculatorService" >
        <endpoint address="" binding="wsHttpBinding"bindingConfiguration="Binding1"
        contract="ICalculator"/>
      </service>
    </services>
    <!-- 指定一个或多个系统内置的 binding，当然也可以指定自定义的 customBinding -->
    <bindings>
      <wsHttpBinding>
        <binding name="Binding1">
        </binding>
      </wsHttpBinding>
    </bindings>
  </system.serviceModel>
</configuration>
```

虽然 WCF 也可以使用 SOAP 做通信格式，但它和以往的 ASP.NET XML WebService 不同，因此有部分技术文章中会将 ASP.NET 的 XML WebService 称为 ASMX Service。

WCF 的服务可以挂载于 Console Application、WindowsApplication、IIS（ASP.NET）Application、Windows Service 以及 Windows Activation Services 中，但大多都会挂在 Windows

Service。

主要的系统内置绑定如表 9-1 所示。

表 9-1 系统内置绑定

绑定	配置元素	说明	传输协议	编码格式
BasicHttpBinding	<basicHttpBinding>	一个绑定，适用于与符合 WS-Basic Profile 的 Web 服务（例如基于 ASP.NET Web 服务（ASMX）的服务）进行的通信。此绑定使用 HTTP 作为传输协议，并使用文本/XML 作为默认的消息编码	HTTP/HTTPS	Text，MTOM
WSHttpBinding	<wsHttpBinding>	一个安全且可互操作的绑定，适合于非双工服务约定	HTTP/HTTPS	Text，MTOM
WS2007HttpBinding	<ws2007HttpBinding>	一个安全且可互操作的绑定，可为 Security、ReliableSession 的正确版本和 TransactionFlow 绑定元素提供支持	HTTP/HTTPS	Text，MTOM
WSDualHttpBinding	<wsDualHttpBinding>	一个安全且可互操作的绑定，适用于双工服务协定或通过 SOAP 媒介进行的通信	HTTP	Text，MTOM
WSFederationHttpBinding	<wsFederationHttpBinding>	一个安全且可互操作的绑定，支持 WS 联合协议并使联合中的组织可以高效地对用户进行身份验证和授权	HTTP/HTTPS	Text，MTOM
WS2007FederationHttpBinding	<ws2007FederationHttpBinding>	一个安全且可互操作的绑定，派生自 WS2007HttpBinding 并支持联合安全性	HTTP/HTTPS	Text，MTOM
NetTcpBinding	<netTcpBinding>	一个安全且经过优化的绑定，适用于 WCF 应用程序之间跨计算机的通信	TCP	Binary
NetNamedPipeBinding	<netNamedPipeBinding>	一个安全、可靠且经过优化的绑定，适用于 WCF 应用程序之间计算机上的通信	IPC	Binary
NetMsmqBinding	<netMsmqBinding>	一个排队绑定，适用于 WCF 应用程序之间跨计算机的通信	MSMQ	Binary
NetPeerTcpBinding	<netPeerTcpBinding>	一个支持多计算机安全通信的绑定	P2P	Binary
MsmqIntegrationBinding	<msmqIntegrationBinding>	一个绑定，适用于 WCF 应用程序和现有消息队列（也称为 MSMQ）应用程序之间跨计算机的通信	MSMQ	Binary

系统绑定支持的功能如图 9-16 所示。

绑定名称	传输性安全	消息级安全	WS*兼容性	WS*事务支持	持久可靠消息传送	可靠会话	性能	通信方式		
								请求/响应	单向	双工
basicHttpBinding	√	√	√				良好	√	√	
wsHttpBinding	√	√	√	√		√	良好	√	√	
wsDualHttpBinding	√	√	√	√		√	良好	√	√	√
netTcpBinding	√	√		√		√	更佳	√	√	√
netNamedPipeBinding	√			√			最佳		√	√
netMsmqBinding	√	√			√		更佳		√	
netTcpPeerBinding	√						更佳	√	√	√
msmqIntegrationBinding	√				√		更佳		√	
wsFederationHttpBinding	√	√	√			√	良好	√	√	

图 9-16

各种绑定方式的性能比较如图 9-17 所示。

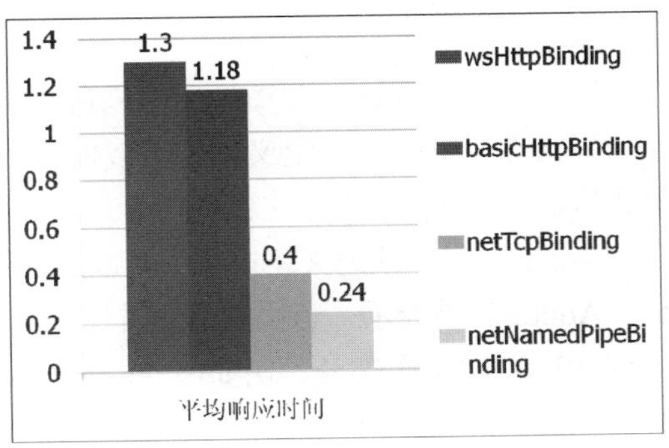

图 9-17

3. 契约（Contract）

WCF 的基本概念是以契约（Contract）来定义双方沟通的协议，契约必须以接口的方式来体现，而实际的服务代码必须要由这些契约接口派生并实现。契约可分为以下 4 种。

- 数据契约（Data Contract）：指定双方沟通时的数据格式。
- 服务契约（Service Contract）：指定服务的定义。
- 操作契约（Operation Contract）：指定服务提供的方法。
- 消息契约（MessageContract）：指定在通信期间改写消息内容的规范。

一个 WCF 中的契约如同下列代码所示：

```
using System.ServiceModel;

namespace HelloService
{
    [ServiceContract(Namespace = "http://Microsoft.ServiceModel.Samples")]// 服务契约
    public interface ICalculator
    {
        [OperationContract] // 操作契约
        double Add(double n1, double n2);

        [OperationContract] // 操作契约
        double Subtract(double n1, double n2);

        [OperationContract] // 操作契约
        double Multiply(double n1, double n2);

        [OperationContract] // 操作契约
        double Divide(double n1, double n2);
    }
}
```

4. 终结点

终结点是用来发送或接收消息（或执行这两种操作）的构造。终结点包括一个定义消息可以发送到的目的地位置（地址）结点，包括一个定义消息可以发送到的目的地位置（地址）、一个描述消息应如何发送的通信机制规范（绑定）以及对于可以在该位置发送或接收（或两者皆可）的一组消息的定义（服务协定）——该定义还描述了可以发送何种消息。

终结点的地址由 EndpointAddress 类表示，该类包含一个表示服务地址的统一资源定位符（URI），大多数传输的地址 URI 包含 4 个部分。例如，“http://www.cnblogs.com:80/jiekzou”这个 URI 具有以下 4 个部分：

- 方案：http。
- 计算机：www.cnblogs.com。
- 端口（可选）：80。
- 路径：/jiekzou。

5. 元数据

所谓的“元数据”就是描述数据的数据，即描述当前服务有哪些服务契约、方法契约和数据契约以及终结点的信息。而“元数据终结点”就是向外界暴露元数据的终结点。当客户端添加 WCF 服务引用的时候会首先通过元数据取得服务器端的契约信息、终结点信息，然后根据这些信息在客户端创建代理类。我们在客户端调用 WCF 服务的过程实际上就是通过代理类调用 WCF 服务的过程。

6. 宿主（Host）

一种应用程序域和进程，服务将在该环境中运行。

9.2.5　创建第一个 WCF 程序

1. 什么是接口

接口如图 9-18 所示。

图 9-18

（1）必须知道的接口特性：

- 接口不可以被实例化（常作为类型使用）。
- 实现类必须实现接口的所有方法（抽象类除外）。
- 实现类可以实现多个接口（Java、C#中的多继承）。
- 接口中的变量都是静态常量。

（2）理解接口：定义一个接口是为了遵循同一种规范，便于程序的扩展。接口是一种能力，是一种约定。

（3）关键字：Interface、public 、abstract。

2. 理解契约式编程

契约合同能保障双方的利益，对客户来说，合同规定了供应者要做的工作；对供应者来说，合同说明了如果约定的条件不满足，供应者没有义务一定要完成规定的任务。该道理同样也适用于软件. 所以契约式编程是编程的一种方法。

引入契约观念之后，图 9-19 所示的 Client 与 Server 关系将被打破，大家都是平等的，你需要我正确提供服务，那么你必须满足我提出的条件，否则我没有义务“排除万难”地保证完成任务。

图 9-19

WCF 服务契约

服务契约描述了暴露给外部的类型（接口或类）、服务所支持的操作、使用的消息交换模式和消息的格式。每个 WCF 服务必须实现至少一个服务契约。使用服务契约必须要引用命名空间 System.ServiceModel 。

ServiceContractAttribute：该特性可被用来作用于子类或者接口之上，并允许重复声明。

OperationContractAttribute：只有定义了该特性的方法才会被放入服务之中。

3. 创建 WCF 程序

（1）新建服务程序。选择“新建项目→类库”，这里不直接新建 WCF 服务，而是新建一个类库，并命名为 HelloService。

（2）添加 System.ServiceModel 程序集引用，如图 9-20 所示。

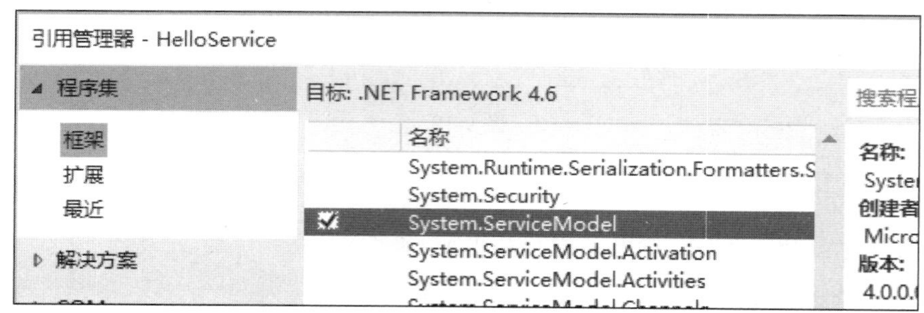

图 9-20

（3）新建一个接口 IHelloService.cs，引入命名空间（using System.ServiceModel;）。

```
using System.ServiceModel;

namespace HelloService
{
    [ServiceContract]
    public interface IHelloService
    {
        [OperationContract]
        string SayHello(string name);
    }
}
```

（4）添加 HelloService 类实现 IHelloService 接口：

```
public class HelloService : IHelloService
{
        public string SayHello(string name)
        {
            return "你好，我是：" + name;
        }
}
```

ServiceHost 类型：当 IIS 和 WAS 作为宿主程序时，IIS 和 WAS 会自动创建 ServiceHost 类型。

手动创建的基本语法是：

```
public ServiceHost(Type serviceType,params Uri[] baseAddresses);
```

（5）新建宿主，选择“新建项目→控制台应用程序”，并命名为 HelloServiceHost。

（6）添加 System.ServiceModel 引用和项目引用 HelloService，引用之前的类库项目。

（7）在 HelloServiceHost 项目中修改 Program.cs 代码：

```
using System;
using System.ServiceModel;
using System.ServiceModel.Channels;

namespace HelloServiceHost
{
    class Program
    {
        static void Main(string[] args)
        {
            using (MyHelloHost host = new MyHelloHost())
            {
                host.Open();
                Console.ReadLine();
            }
        }
    }

    public class MyHelloHost : IDisposable
    {
        /// <summary>
        /// 定义一个服务对象
        /// </summary>
        private ServiceHost _myHelloHost;
        public const string BaseAddress = "net.pipe://localhost"; //基地址
        public const string HelloServiceAddress = "Hello"; //可选地址
        public static readonly Type ServiceType =
        typeof(HelloService.HelloService); //服务契约实现类型
        public static readonly Type ContractType =
        typeof(HelloService.IHelloService); //服务契约接口
        public static readonly Binding HelloBinding = new NetNamedPipeBinding();
//服务定义一个绑定
```

```
        /// <summary>
        /// 构造方法
        /// </summary>
        public MyHelloHost()
        {
            CreateHelloServiceHost();
        }

        /// <summary>
        /// 构造服务对象
        /// </summary>
        protected void CreateHelloServiceHost()
        {
            _myHelloHost = new ServiceHost(ServiceType, new Uri[] { new
            Uri(BaseAddress) });//创建服务对象
            _myHelloHost.AddServiceEndpoint(ContractType, HelloBinding,
            HelloServiceAddress); //添加终结点
        }

        /// <summary>
        /// 打开服务方法
        /// </summary>
        public void Open()
        {
            Console.WriteLine("开始启动服务...");
            _myHelloHost.Open();
            Console.WriteLine("服务已启动");
        }

        /// <summary>
        /// 销毁服务宿主对象实例
        /// </summary>
        public void Dispose()
        {
            if (_myHelloHost != null)
                (_myHelloHost as IDisposable).Dispose();
        }
    }
}
```

（8）新建客户端调用程序，选择“新建项目→控制台应用程序”，并命名为 HelloClient。

（9）添加项目引用，引用项目 HelloService、HelloServiceHost 和程序集 System.ServiceModel，如图 9-21 所示。

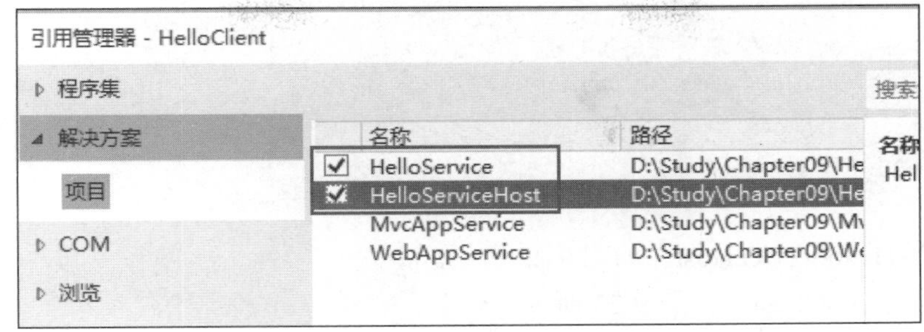

图 9-21

（10）修改 HelloClient 项目中的 Program.cs 代码：

```
using System;
using System.ServiceModel;
using System.ServiceModel.Channels;
using HelloService;

namespace HelloClient
{
    class Program
    {
        static void Main(string[] args)
        {
            using (HelloProxy proxy = new HelloProxy())
            {
                //利用代理调用方法
                Console.WriteLine(proxy.Say("郑少秋"));
                Console.ReadLine();
            }
        }
    }

    [ServiceContract]
    interface IService
    {
        [OperationContract]
        string Say(string name);
    }

    class HelloProxy : ClientBase<IHelloService>, IService
    {
        public static readonly Binding HelloBinding = new NetNamedPipeBinding();
        //硬编码定义绑定
        //硬编码定义地址，注意这里要和之前服务定义的地址保持一致
        public static readonly EndpointAddress HelloAddress = new
        EndpointAddress(new Uri("net.pipe://localhost/Hello"));
```

```
        public HelloProxy() : base(HelloBinding, HelloAddress) { } //构造方法

        public string Say(string name)
        {
            //使用 Channel 属性对服务进行调用
            return Channel.SayHello(name);
        }
    }
}
```

（11）运行 HelloServiceHost，如图 9-22 所示。

（12）运行 HelloClient，如图 9-23 所示。

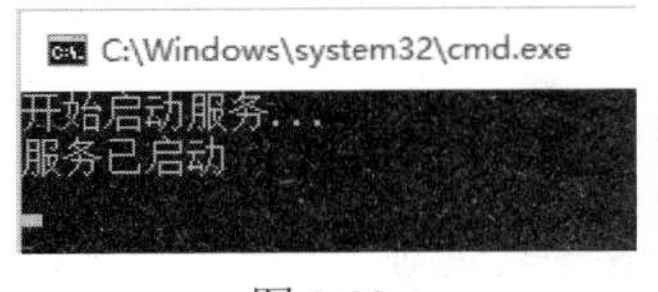

图 9-22

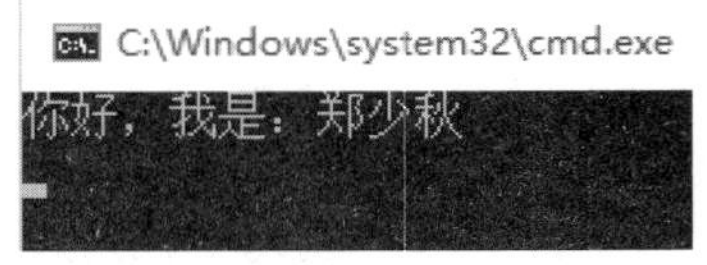

图 9-23

这里，我为了让大家对 WCF 有一个更加深入的理解，并没有直接新建一个 WCF 项目，而在真正的项目开发过程中，我们一般都是直接通过新建一个 WCF 项目，然后把 WCF 项目发布到 IIS 上面，IIS 站点就成为 WCF 的宿主，当我们需要调用的时候，就只需要添加服务引用，然后直接输入这个站点的 http://IP:端口:服务名称。这里的“IP:端口”也可以通过域名直接替换。

（1）新建 WCF 服务，命名为 MyWcfService，如图 9-24 所示。

图 9-24

VS 默认为我们创建了一个 Service1.svc 服务和两个默认的方法 GetData、GetDataUsingDataContract。

（2）在 MyWcfService 项目中，右击 Service1.svc，选择“在浏览器中查看”，如图 9-25 所示。

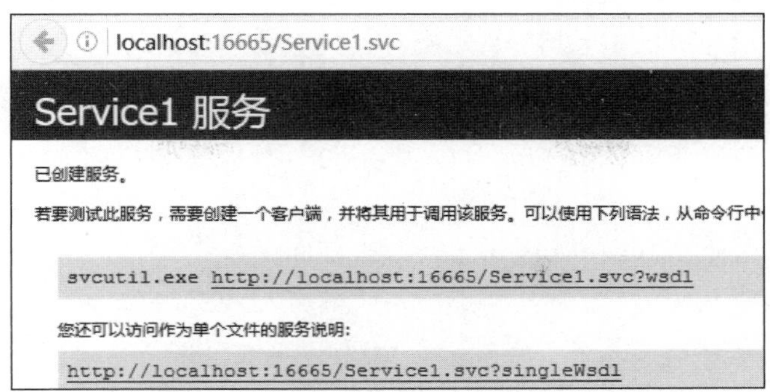

图 9-25

（3）新建控制台程序 WcfClient，并添加服务引用，如图 9-26 所示。在地址栏中输入刚才运行的 WCF 服务地址，然后单击“转到”按钮。此外，因为 WCF 服务和控制台程序 WcfClient 在同一个解决方案下，所以还可以直接单击“转到”按钮后面的“发现”按钮，自动查找解决方案中的所有服务，达到同样的效果。

图 9-26

（4）修改 Program 类中的代码：

```
using System;

namespace WcfClient
{
```

```
    class Program
    {
        static void Main(string[] args)
        {
            ServiceRef.Service1Client _client = new ServiceRef.Service1Client();
            Console.WriteLine(_client.GetData(11));
            Console.ReadLine();
        }
    }
}
```

和之前 Webservice 的调用很像。

（5）运行 WcfClient 项目，结果如图 9-27 所示。

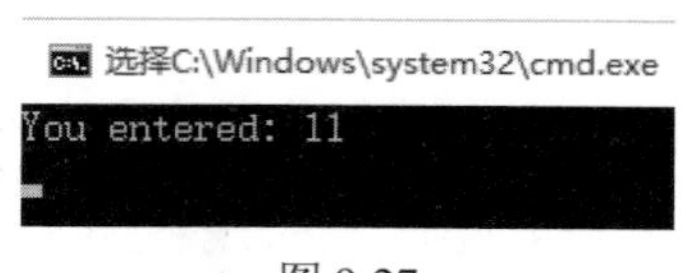

图 9-27

9.2.6 WCF 和 WebService 的区别

（1）WebService：严格来说是行业标准，不是技术，使用 XML 扩展标记语言来表示数据（这个是跨语言和平台的关键）。微软的 Web 服务实现称为 ASP.NETWebService，使用 Soap 简单对象访问协议来实现分布式环境里应用程序之间的数据交互。

WSDL 实现服务接口相关的描述。此外，WebServices 可以注册到 UDDI 中心，供其客户查找使用。后来微软做了 ASP.NET WebService 的安全、性能、数据加密、解密、托管宿主等多方面的扩展，称为 WSE 系列。这个是过渡产品，最高到 WSE 3.0。后来就是 WCF 时代。

（2）WCF：其实一定程度上就是 ASP.NET WebService，因为它支持 WebService 的行业标准和核心协议，因此 ASP.NET WebService 和 WSE 能做的事情它几乎都能胜任，跨平台和语言更不是问题（数据也支持 XML 格式化，而且提供了自己的格式化器）。

9.3 Web API

WebAPI 可以返回 json、xml 类型的数据，对于数据的增、删、改、查，提供对应的资源操作，按照请求的类型进行相应的处理，主要包括 Get（查）、Post（增）、Put（改）、Delete（删），这些都是 HTTP 协议支持的请求方式。

WebAPI 的请求方式：根据路由规则请求。

WebService 和 WebAPI 两种 Web 服务的比较：

- WebService：基于 SOAP 风格的网络服务，使用方法进行请求。
- WebAPI：基于 REST 风格的网络服务，使用资源进行请求。

WebAPI 中 5 个方法分别是查单个、查所有、增加、修改、删除。

微软有了 WebService 和 WCF，为什么还要有 WebAPI？

用过 WCF 的人应该都清楚，面对一大堆复杂的配置文件，万一出问题，真的会叫人抓狂，而且供不同的客户端调用也不是很方便。不得不承认 WCF 的功能确实非常强大，可是有时候我们通常不需要那么复杂的功能，只需要简单的仅通过使用 HTTP 或 HTTPS 来调用的增、删、改、查功能，这时 WebAPI 应运而生。那么什么时候考虑使用 WebAPI 呢？

当你遇到以下这些情况的时候就可以考虑使用 WebAPI 了。

- 需要 WebService 但是不需要 SOAP。
- 需要在已有的 WCF 服务基础上建立 non-soap-based http 服务。
- 只想发布一些简单的 HTTP 服务，不想使用相对复杂的 WCF 配置。
- 发布的服务可能会被带宽受限的设备访问。
- 希望使用开源框架，关键时候可以自己调试或者自定义一下框架。

熟悉 MVC 的朋友可能会觉得 Web API 与 MVC 很类似。遗憾的是 MVC 4 中 WebAPI 和 MVC 并没有共用同一套框架。而在 MVC 6 中，微软把 WebAPI 和 MVC 的框架进行了合并，去掉了一些重复的功能。

9.3.1 创建 WebAPI

（1）新建项目→Web→ASP.NET MVC 4 Web 应用程序→命名为 WebApiApp，在项目模板中选中“Web API”。

（2）在 Models 目录中新建类 Product。

```
public class Product
{
        public int Id { get; set; }
        public string Name { get; set; }
        public string Category { get; set; }
        public decimal Price { get; set; }
}
```

（3）新建 API 控制器 Products，右击目录 Controllers，选择“添加→控制器”。为了演示，这里不连接数据库，而是直接在代码中构造假数据。和普通控制器不一样的地方是这里要继承 ApiController，而不是 Controller，如图 9-28 所示。

添加控制器

控制器名称(C):

ProductsController

基架选项

模板(T):

空 API 控制器

模型类(M):

图 9-28

```
using System.Collections.Generic;
using System.Linq;
using System.Web.Http;
using WebApiApp.Models;

namespace WebApiApp.Controllers
{
    public class ProductsController : ApiController
    {
        Product[] products = new Product[]
        {
            new Product { Id = 1, Name ="Tomato Soup",Category="Groceries",Price = 1 },
          new Product { Id = 2, Name = "Yo-yo", Category = "Toys", Price = 3.75M },
          new Product { Id = 3, Name = "Hammer", Category = "Hardware", Price = 16.99M }
        };

        public IEnumerable<Product> GetAllProducts()
        {
            return products;
        }

        public Product GetProduct(int id)
        {
            var product = products.FirstOrDefault((p) => p.Id == id);
            return product;
        }
    }
}
```

9.3.2 调用 WebAPI

WebAPI 有两种调用方式。

1. 调用方式 1：jQuery 的 Ajax

指定请求的数据类型（contentType）：“application/json;charset=utf-8”。

主要属性如下：

- type：请求方式，包括 Get、Post、Put、Delete。
- url：请求资源，根据路由规则编写。
- data：请求数据，为 json 格式。
- contentType：请求数据的类型及编码。
- dataType：返回的数据类型，可以是 text、json。
- success：成功处理的回调函数。

使用 js 的异步操作不支持跨域访问，也就是说 js 异步调用操作必须和 WebAPI 在同一个站点上。

（1）新建 Index.html 来测试 WebAPI 的调用，代码如下：

```
<!DOCTYPE html>
<html xmlns="http://www.w3.org/1999/xhtml">
<head>
   <title>Product App</title>
</head>
<body>
   <div>
      <h2>All Products</h2>
      <ul id="products" />
   </div>
   <div>
      <h2>Search by ID</h2>
      <input type="text" id="prodId" size="5" />
      <input type="button" value="Search" onclick="find();" />
      <p id="product" />
   </div>
   <script src="Scripts/jquery-1.7.1.min.js"></script>
   <script>
      var uri = 'api/products';

      $(document).ready(function () {
         $.getJSON(uri)
            .done(function (data) {
               $.each(data, function (key, item) {
                  $('<li>', { text:
```

```
                formatItem(item) }).appendTo($('#products'));
            });
        });
    });

    function formatItem(item) {
        return item.Name + ': $' + item.Price;
    }

    function find() {
        var id = $('#prodId').val();
        $.getJSON(uri + '/' + id)
            .done(function (data) {
                $('#product').text(formatItem(data));
            })
            .fail(function (jqXHR, textStatus, err) {
                $('#product').text('Error: ' + err);
            });
    }
  </script>
</body>
</html>
```

（2）先运行 WebApiApp 项目，然后浏览 index.html 页面，运行结果如图 9-29 所示。

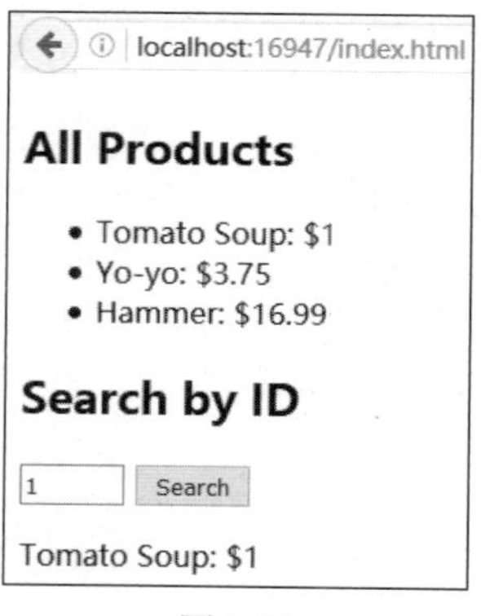

图 9-29

2. 调用方式 2：HttpClient

（1）新建 MVC 项目 WebApiClient。

（2）在 Models 目录下新建 Product 类，其实我们也可以把这个 Product 类放到一个单独的程序集中，然后通过添加程序集引用使 WebApiApp 和 WebApiClient 共用 Product 类。

```
public class Product
```

```
{
    public int Id { get; set; }
    public string Name { get; set; }
    public string Category { get; set; }
    public decimal Price { get; set; }
}
```

（3）添加 Home 控制器和 Index 的 Action 方法：

```
using System.Collections.Generic;
using System.Net.Http;
using System.Net.Http.Headers;
using System.Web.Mvc;
using WebApiClient.Models;

namespace WebApiClient.Controllers
{
    public class HomeController : Controller
    {
        public ActionResult Index()
        {
            //客户端对象的创建与初始化
            HttpClient client = new HttpClient();
            client.DefaultRequestHeaders.Accept.Add(new
MediaTypeWithQualityHeaderValue("application/json"));

            //执行 Get 操作
            HttpResponseMessage response =
            client.GetAsync("http://localhost:16947/api/Products").Result;
            var list = response.Content.ReadAsAsync<List<Product>>().Result;
            ViewData.Model = list;

            return View();
        }
    }
}
```

（4）添加 Index 视图：

```
@{
    ViewBag.Title = "Index";
}
@model List<WebApiClient.Models.Product>
```

```
<ul>
@foreach (var v in Model)
{
    <li>@v.Id", "@v.Name, @v.Price</li>
}
</ul>
```

先运行项目 WebApiApp，然后运行 WebApiClient，结果如图 9-30 所示。

图 9-30

9.3.3 WebAPI 授权

有时候我们对外提供的一些 WebAPI 服务只想对指定的客户开放，不想让所有人都能够调用，所以我们就要进行 WebAPI 授权。项目开发中用得比较多的，一般是通过让调用方传递一个类似于授权码的东西，然后 WebAPI 服务端从数据库中查询此授权码是否有效。如果有加请求 IP 限制，我们还要判断请求的 IP 地址是否是我们允许的 IP。而类似于授权验证的信息，我们可以添加到请求头上面。

（1）新建授权过滤器类 APIAuthorizeAttribute.cs。在 WebApiApp 项目中，新建文件夹 Filter，用于存放过滤器类。

```
using System;
using System.Net;
using System.Net.Http;
using System.Text;
using System.Threading;
using System.Web.Http.Filters;

namespace WebApiApp.Filter
{
    public class APIAuthorizeAttribute : AuthorizationFilterAttribute
    {
        public override void
        OnAuthorization(System.Web.Http.Controllers.HttpActionContext actionContext)
        {
            //如果用户使用了 forms authentication，就不必再做 basic authentication 了
```

```
    if (Thread.CurrentPrincipal.Identity.IsAuthenticated)
    {
        return;
    }

    var authHeader = actionContext.Request.Headers.Authorization;

    if (authHeader != null)
    {
        if (authHeader.Scheme.Equals("basic",
        StringComparison.OrdinalIgnoreCase) &&
            !String.IsNullOrWhiteSpace(authHeader.Parameter))
        {
            var credArray = GetCredentials(authHeader);
            var userName = credArray[0];
            var key = credArray[1];
            string ip =
            System.Web.HttpContext.Current.Request.UserHostAddress;
            //if (IsResourceOwner(userName, actionContext))
            //{
            //You can use Websecurity or asp.net memebrship provider to login, for
        // he sake of keeping example simple, we used out own login functionality
            //if (APIAuthorizeInfoValidate.ValidateApi(userName, key, ip))
            // 这里我们可以自定义权限验证方法
            //{
            //    var currentPrincipal = new GenericPrincipal(new
            // GenericIdentity(userName), null);
            //    Thread.CurrentPrincipal = currentPrincipal;
            //    return;
            //}
            //}
        }
    }

    HandleUnauthorizedRequest(actionContext);
}

private string[]
GetCredentials(System.Net.Http.Headers.AuthenticationHeaderValue authHeader)
{

    //Base 64 encoded string
```

```
            var rawCred = authHeader.Parameter;
            var encoding = Encoding.GetEncoding("iso-8859-1");
            var cred = encoding.GetString(Convert.FromBase64String(rawCred));

            var credArray = cred.Split(':');

            return credArray;
        }

        private bool IsResourceOwner(string userName,
        System.Web.Http.Controllers.HttpActionContext actionContext)
        {
            var routeData = actionContext.Request.GetRouteData();
            var resourceUserName = routeData.Values["userName"] as string;

            if (resourceUserName == userName)
            {
                return true;
            }
            return false;
        }

        private void HandleUnauthorizedRequest
        (System.Web.Http.Controllers.HttpActionContext actionContext)
        {
            actionContext.Response = actionContext.Request.CreateResponse
            (HttpStatusCode.Unauthorized);

            actionContext.Response.Headers.Add("WWW-Authenticate",
                                        "Basic Scheme='eLearning'
                                        location='http://localhost:8323/APITest'");

        }
    }
}
```

（2）把此 WebAPI 的授权验证过滤器添加到全局过滤器中，需要注意的是不要添加到 FilterConfig.cs，而是要添加到 WebApiConfig.cs，因为 FilterConfig 是 MVC 用的，我们这里是 WebAPI。

```
public static void Register(HttpConfiguration config)
{
```

```
        config.Routes.MapHttpRoute(
            name: "DefaultApi",
            routeTemplate: "api/{controller}/{id}",
            defaults: new { id = RouteParameter.Optional }
        );
        config.Filters.Add(new APIAuthorizeAttribute()); //添加授权验证过滤器
}
```

9.3.4　WebAPI 的调试

1. 方式一：VS 插件

当 WebAPI 站点开发完成之后，可以使用 NuGet 安装一个插件自动生成 API 文档、WebAPI 在线测试。

2. 方式二：浏览器插件

浏览器插件（见图 9-31）有火狐的 RESTClient（安装方式请参考 http://jingyan.baidu.com/article/1876c8529b07e3890b137623.html）、谷歌的 PostMan（使用方式请参考 http://jingyan.baidu.com/article/eb9f7b6d861da9869364e83a.html）等。

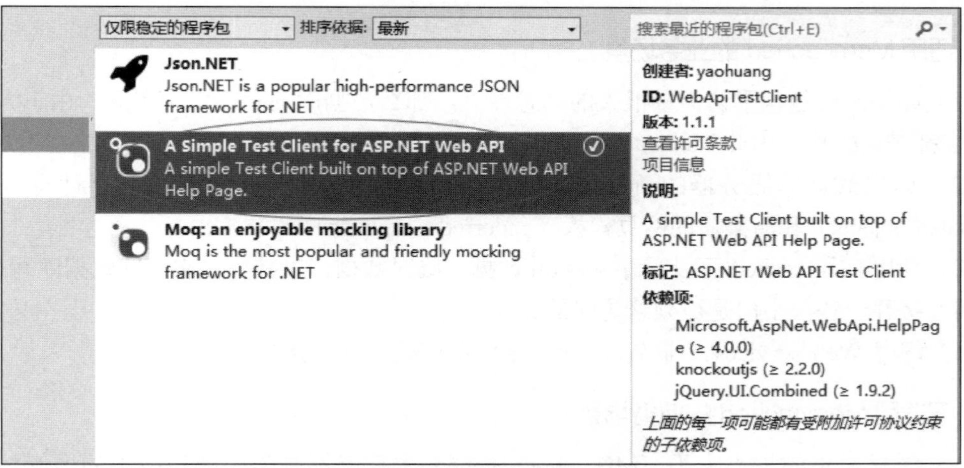

图 9-31

9.4　Memcached

9.4.1　Memcached 简介

Memcached 是一个高性能的分布式内存对象缓存系统，用于动态 Web 应用，以减轻数据库负载。它通过在内存中缓存数据和对象来减少读取数据库的次数，从而提高动态、数据库驱动网站的速度。Memcached 基于一个存储键/值对的 hashmap，其守护进程（daemon ）是用 C 写的，但是客户端可以用任何语言来编写，并通过 memcached 协议与守护进程通信。

Memcached 的缓存是一种分布式的，可以让不同主机上的多个用户同时访问， 因此解决了共享内存只能单机应用的局限，更不会出现使用数据库做类似事情的时候导致磁盘开销和阻塞的发生。

1. 为什么需要 Memcached

- 高并发访问数据库的痛楚：死锁！
- 磁盘 IO 之痛：多客户端共享缓存，Net + Memory >> IO。
- 基于客户端分布式，客户端共享缓存。
- 读写性能完美：1s 可以读取 1 万次，写 10 万次。
- 超简单集群搭建 Cluster。
- 开源 Open Source。
- 学习成本非常低，入门非常容易。
- 丰富的成功案例。

不足的是既没有提供主从复制功能，也没有提供容灾等功能，所有的代码基本都只是考虑性能最佳。

2. 适用 Memcached 的业务场景

（1）如果网站包含了访问量很大的动态网页，那么数据库的负载将会很高。由于大部分数据库请求都是读操作，因此 Memcached 可以显著地减小数据库负载。

（2）如果数据库服务器的负载比较低但 CPU 使用率很高，就可以缓存计算好的结果（computed objects）和渲染后的网页模板（enderred templates）。

（3）利用 Memcached 可以缓存 session 数据、临时数据，以减少对它们的数据库写操作。

（4）缓存一些很小但是被频繁访问的文件。

（5）缓存 Web 'services'（非 WebServices）或 RSS feeds 的结果。

3. 不适用 Memcached 的业务场景

（1）缓存对象的大小大于 1MB。Memcached 本身就不是为了处理庞大的多媒体（large media）和巨大的二进制块（streaming huge blobs）而设计的。

（2）key 的长度大于 250 字符。

（3）虚拟主机不让运行 Memcached 服务。

如果应用本身托管在低端的虚拟私有服务器上，像 vmware、xen 这类虚拟化技术并不适合运行 Memcached。Memcached 需要接管和控制大块的内存，如果 Memcached 管理的内存被 OS 或 hypervisor 交换出去，那么 Memcached 的性能将大打折扣。

（4）应用运行在不安全的环境中。Memcached 未提供任何安全策略，仅仅通过 Telnet 就可以访问到 Memcached。如果应用运行在共享的系统上，需要着重考虑安全问题。

（5）业务本身需要的是持久化数据或者说需要的是 database。

9.4.2 Memcached 基本原理

Memcached 的本质就是一个大的哈希表，通过键值对存储数据，key 的最大长度是 255 个字符，item 最大 1MB，当然 key/item 最好都别太大，最长过期时间是 30 天。

1.Memcached 相关概念

内存模型：Memcache 预先将可支配的内存空间进行分区（Slab），每个分区里再分成多个块（Chunk），但同一个分区里：块的长度（bytes）是固定的。

插入数据：查找适合自己长度的块，然后插入，会有内存浪费。

LRU（Least Recently Used 近期最少使用算法）：闲置>过期>最少访问。

惰性删除：它并没有提供监控数据过期的机制，而是惰性的，当查询到某个 key 数据时，如果过期就直接抛弃。

集群搭建原理：Memcached 服务器端并没有提供集群功能，但是通过客户端的驱动程序实现了集群配置。

客户端实现集群的原理：首先客户端配置多台集群机器的 IP 和端口的列表。然后客户端驱动程序在写入之前先对 key 做哈希处理，得到哈希值后再对总的机器个数进行取余，选择余数对应的机器。将记录从 Memcached 删除后，已经分配的内存（即 Chunk），也不会被释放，而是会重复利用，这样就彻底解决了内存碎片的问题。

Memcached 采用“惰性”方式来应对记录的超期问题。

解决多线程问题：CAS（Compare and Set）。

2. CAS 的基本原理

Memcached 于 1.2.4 版本新增 CAS 协议，类似于 Java 并发包中的 CAS（Compare and Set）原子操作，用来处理同一 item 被多个线程更改过程的并发问题。

基本原理非常简单，简而言之就是“版本号”。每个存储的数据对象都有一个版本号，在 Memcached 中每个 key 关联有一个 64bit 长度的 long 型唯一数值，表示该 key 对应 value 的版本号。

这个数值由 Memcached 产生，从 1 开始，且同一 Memcached 不会重复，在两种情况下这个版本数值会加，即新增与更新，而删除 item 版本值不会减小。

我们可以从下面的例子来理解。

（1）如果不采用 CAS，则有如下的情景：

第一步，A 取出数据对象 X。

第二步，B 取出数据对象 X。

第三步，B 修改数据对象 X，并将其放入缓存。

第四步，A 修改数据对象 X，并将其放入缓存。

我们可以发现，第四步会产生数据写入冲突。

（2）如果采用 CAS 协议，则有如下情景：

第一步，A 取出数据对象 X，并获取到 CAS-ID1。

第二步，B 取出数据对象 X，并获取到 CAS-ID2。

第三步，B 修改数据对象 X，在写入缓存前检查 CAS-ID 与缓存空间中该数据的 CAS-ID 是否一致。如果结果是“一致”，就将修改后带有 CAS-ID2 的 X 写入缓存。

第四步，A 修改数据对象 Y，在写入缓存前检查 CAS-ID 与缓存空间中该数据的 CAS-ID 是否一致。如果结果是“不一致”，就拒绝写入，返回存储失败。

9.4.3 Memcached 服务端的安装

（1）下载 Memcached，本书中已提供。

（2）下载后，将服务程序复制到一个磁盘上的目录然后解压，如 D:\WorkSpace\memcache。

（3）安装服务：cmd→Memcached.exe -d install。打开服务监控窗口，可以查看服务是否启动。

详细说明：以管理员身份运行“cmd”，然后定位到 Memcached 服务安装包所在的目录 D:\WorkSpace\memcache，再运行命令“Memcached.exe -d install”。

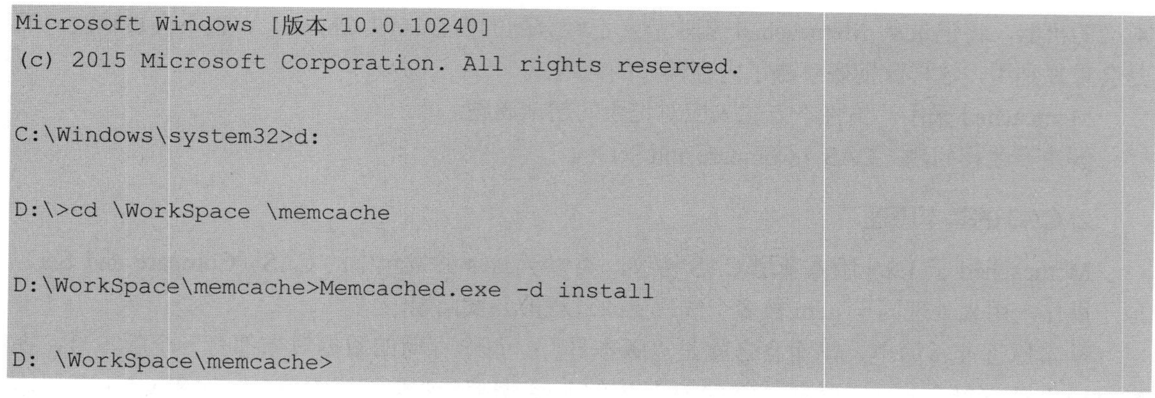

```
Microsoft Windows [版本 10.0.10240]
(c) 2015 Microsoft Corporation. All rights reserved.

C:\Windows\system32>d:

D:\>cd \WorkSpace \memcache

D:\WorkSpace\memcache>Memcached.exe -d install

D: \WorkSpace\memcache>
```

查看服务是否已经成功安装：选择“开始→运行→services.msc”。打开 Windows 服务控制台，如果能看到如图 9-32 所示的服务，就说明安装成功。

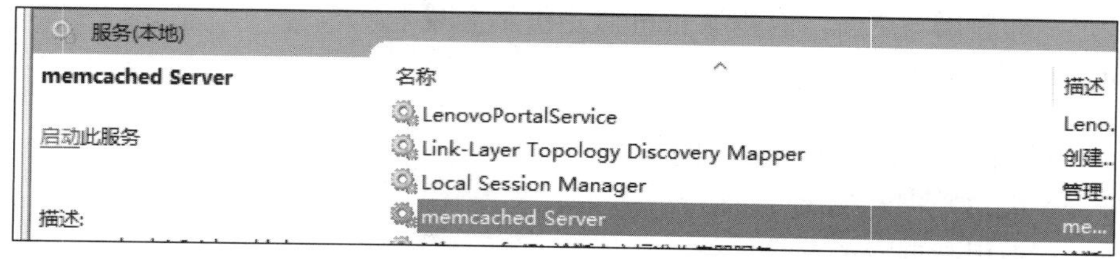

图 9-32

（4）启动服务：cmd→Memcached.exe -d start（restart 重启，stop 关闭服务），如果出现如

图 9-33 所示的界面就说明服务启动成功。

名称	描述	状态	启动类
LenovoPortalService	Leno...		手动
Link-Layer Topology Discovery Mapper	创建...		手动
Local Session Manager	管理...	正在运行	自动
memcached Server	me...	正在运行	自动

图 9-33

（5）检查服务状态。在使用 Telnet 命令之前，我们应当先在电脑上面安装 Telnet 客户端，因为在默认情况下 Windows 是没有安装 Telnet 的。

安装方法：选择“控制面板→程序→程序和功能”，单击“启用或关闭 Windows 功能”，然后勾选 Telnet 客户端，最后单击“确定”按钮，如图 9-34 所示。

连接到 Memcached 控制台“telnet 127.0.0.1 11211”，然后输入命令“stats”，按回车键，检查当前服务状态，如图 9-35 所示。

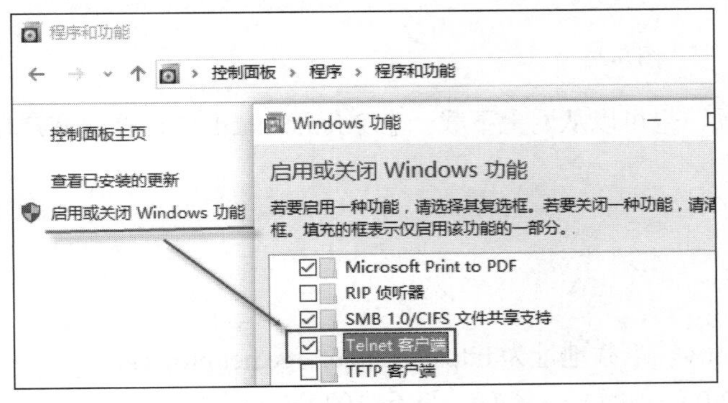

图 9-34

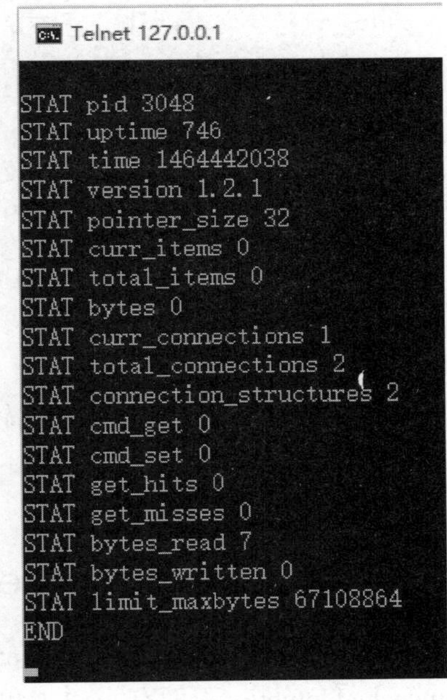

图 9-35

Memcached 卸载服务命令是：Memcached.exe -d uninstall（卸载服务）。

Memcached 的状态属性说明如图 9-36 所示。

stats统计项

名称	描述	
pid	Memcached进程ID	
uptime	Memcached运行时间，单位：秒	
time	Memcached当前的UNIX时间	
version	Memcached的版本号	
rusage_user	该进程累计的用户时间，单位：秒	分析CPU占用是否高
rusage_system	该进程累计的系统时间，单位：秒	
curr_connections	当前连接数量	分析连接数是否太多
total_connections	Memcached运行以来接受的连接总数	
connection_structures	Memcached分配的连接结构的数量	
cmd_get	查询请求总数	分析命中率是否太低
get_hits	查询成功获取数据的总次数	
get_misses	查询成功未获取到数据的总次数	
cmd_set	存储（添加/更新）请求总数	
bytes	Memcached当前存储内容所占用字节数	分析字节数流量
bytes_read	Memcached从网络读取到的总字节数	
bytes_written	Memcached向网络发送的总字节数	
limit_maxbytes	Memcached在存储时被允许使用的字节总数	
curr_items	Memcached当前存储的内容数量	分析对象数LRU频率
total_items	Memcached启动以来存储过的内容总数	
evictions	LRU释放对象数，用来释放内存	

图 9-36

如果觉得这样使用命令操作比较麻烦，也可以从网上下载一个 Memcached 的可视化客户端管理工具。

9.4.4 C#操作 Memcached

（1）下载.NET memcached client library。下载地址为 https://sourceforge.net/projects/memcacheddotnet/，里面有.net 1.1 和.net 2.0 两种版本，还有一个不错的 Demo。

（2）下载 Demo 解压到 D:\WorkSpace。打开 D:\WorkSpace\memcacheddotnet_clientlib-1.1.5\memcacheddotnet\trunk\clientlib\src 里面的 memcachedDOTnet_2.0.sln 解决方案，运行结果如图 9-37 所示。

```
解决方案'memcachedDOTnet_2.0' (4 个项目)
▷ clientlib_2.0
◢ MemCachedBench_2.0
    Properties
  ▷ 引用
    App.config
  ▷ AssemblyInfo.cs
  ▷ MemCachedBench.cs
▷ MemCachedTest_2.0
◢ UnitTests_2.0
    Properties
  ▷ 引用
  ▷ AssemblyInfo.cs
  ▷ UnitTests.cs
```

图 9-37

（3）修改类 MemCachedBench.cs 中的代码：

```
string[] serverlist = { "127.0.0.1:11211" };//{ "140.192.34.72:11211",
"140.192.34.73:11211" };
```

（4）设置项目 MemCachedBench_2.0 为启动项，然后运行，结果如图 9-38 所示。

```
C:\Windows\system32\cmd.exe
100 sets: 00:00:00.0340883ms
100 gets: 00:00:00.0190479ms
Cache hits: 100
Cache misses: 0
127.0.0.1:11211
get_misses:0
curr_connections:3
get_hits:100
bytes:29900
limit_maxbytes:67108864
connection_structures:4
cmd_get:100
total_connections:5
time:1464448341
total_items:100
pointer_size:32
curr_items:100
pid:3048
uptime:7049
bytes_read:28814
version:1.2.1
cmd_set:100
bytes_written:28874

请按任意键继续. . .
```

图 9-38

如果需要在其他项目中使用，可以去目录 D:\WorkSpace\memcacheddotnet_clientlib-1.1.5\memcacheddotnet\trunk\clientlib\src\clientlib\bin\2.0\Debug 中将 Commons.dll，ICSharpCode.SharpZipLib.dll、log4net.dll、Memcached.ClientLibrary.dll 等放到指定的 DLL 目录，然后添加这

4 个程序集的引用，再从项目 memcachedDOTnet_2.0.sln 中复制需要用到的代码。

9.5 Redis

9.5.1 Redis 简介

Redis 是一个开源的、使用 C 语言编写、面向“键/值”对类型数据的分布式 NoSQL 数据库系统，特点是高性能、持久存储，适应高并发的应用场景。Redis 纯粹为应用而产生，是一个高性能的 key-value 数据库，并且提供了多种语言的 API。

性能测试结果表示 SET 操作每秒钟可达 110000 次，GET 操作每秒 81000 次（当然不同的服务器配置性能不同）。

Redis 目前提供 5 种数据类型：string（字符串）、list（链表）、Hash（哈希）、set（集合）及 zset（sorted set，有序集合）。

Redis 开发维护很活跃，虽然只是一个 Key-Value 数据库存储系统，但本身支持 MQ 功能，所以完全可以当作一个轻量级的队列服务来使用。对于 RabbitMQ 和 Redis 的入队和出队操作，各执行 100 万次，每 10 万次记录一次执行时间。测试数据分为 128Bytes、512Bytes、1KB 和 10KB 四个不同大小的数据。实验表明：入队时，当数据比较小时 Redis 的性能要高于 RabbitMQ，而如果数据大小超过了 10KB，Redis 则慢得无法忍受；出队时，无论数据大小，Redis 都表现出非常好的性能，而 RabbitMQ 的出队性能则远低于 Redis。

9.5.2 Redis 与 Memcached 的比较

（1）Memcached 是多线程，而 Redis 使用单线程。

（2）Memcached 使用预分配的内存池的方式，Redis 使用现场申请内存的方式来存储数据，并且可以配置虚拟内存。

（3）Redis 可以实现持久化、主从复制，实现故障恢复。

（4）Memcached 只是简单的 key 与 value，但是 Redis 支持数据类型比较多。

Redis 的存储分为内存存储、磁盘存储，从这一点上也说明了 Redis 与 Memcached 是有区别的。Redis 与 Memcached 一样，为了保证效率，数据都是缓存在内存中，区别是 Redis 会周期性地把更新的数据写入磁盘或者把修改操作写入追加的记录文件，并且在此基础上实现了 master-slave（主从）同步。

Redis 有两种存储方式，默认是 snapshot 方式，实现方法是定时将内存的快照（snapshot）持久化到硬盘，这种方法的缺点是持久化之后如果出现 crash 就会丢失一段数据。因此在完美主义者的推动下作者增加了 aof 方式。aof 即 append only file，在写入内存数据的同时将操作命令保存到日志文件，在一个并发更改上万的系统中，命令日志是一个非常庞大的数据，管理维护成本非常高，恢复重建时间会非常长，这样会失去 aof 的高可用性本意。另外，更重要的是

Redis 是一个内存数据结构模型，所有的优势都是建立在对内存复杂数据结构高效的原子操作上，这样就可以看出 aof 是一个非常不协调的部分。其实 aof 的主要目的是数据可靠性及高可用性。

9.5.3 Redis 环境部署

1. 下载 Redis 服务

云盘里提供了安装包，当然你也可以去网上下载最新版的 Redis，下载地址为 https://github.com/dmajkic/redis/downloads。将服务程序复制到一个磁盘的目录里，这里放到 D:\WorkSpace\Redis\Redis 服务\MasterRedis 目录下，如图 9-39 所示。

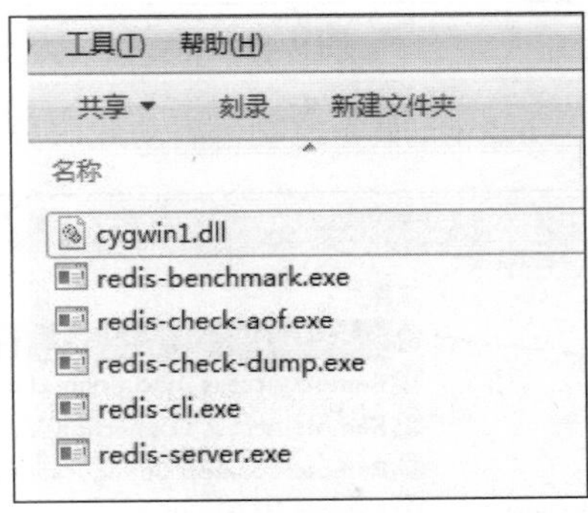

图 9-39

文件说明：

- redis-server.exe：服务程序。
- redis-check-dump.exe：本地数据库检查。
- redis-check-aof.exe：更新日志检查。
- redis-benchmark.exe：性能测试，用于模拟同时由 *N* 个客户端发送 *M* 个 SETs/GETs 查询。
- redis-cli.exe：服务端开启后，在客户端就可以输入各种命令测试了。

2. 安装 Redis 服务

（1）打开一个 CMD 窗口，使用 cd 命令切换到 Redis 服务所在目录（D:\WorkSpace\Redis\Redis 服务\MasterRedis），再运行命令“redis-server.exe redis.conf”。

（2）重新打开一个 CMD 窗口，使用 cd 命令切换到 Redis 服务所在目录（D:\WorkSpace\Redis\Redis 服务\MasterRedis），再运行命令“redis-cli.exe -h 127.0.0.1 -p 6379”。其中，127.0.0.1 是本地 IP，6379 是 Redis 服务端的默认端口（这样可以开启一个客户端程序进行特殊指令的测试）。

每次使用 Redis 都打开一个 CMD 命令窗口显然不是一种好的方式。我们可以将此服务设置为 Windows 系统服务，只需下载 Redis 服务的可视化安装软件，一键安装即可。

安装包放置在 D:\WorkSpace\Redis\服务安装目录下，当然也可以去网站（https://github.com/rgl/redis/downloads）上自行下载。

这里提供了 32bit 和 64bit 两个安装包，根据系统环境选择安装即可。

这里选择安装 redis-2.4.6-setup-64-bit.exe（电脑系统是 win10 64bit），如图 9-40 所示。

图 9-40

安装完成 Redis 服务后，去 Windows 服务控制台启动此服务，如图 9-41 所示。

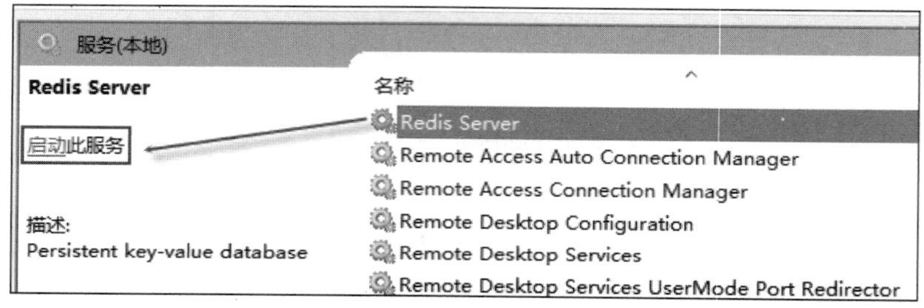

图 9-41

3. 下载 C#驱动

在使用 Redis 时还需要下载 C#驱动（也就是 C#开发库），在 D:\WorkSpace\Redis\RedisClient 目录下提供了两个版本的 dll 库，这里选择 4.0 版本，主要有如图 9-42 所示的 4 个 dll 库。

图 9-42

ServiceStack.Redis 的源码在云盘中已提供，同时也可以去 https://github.com/ServiceStack/ServiceStack.Redis 网址下载。下载方法如图 9-43 所示。

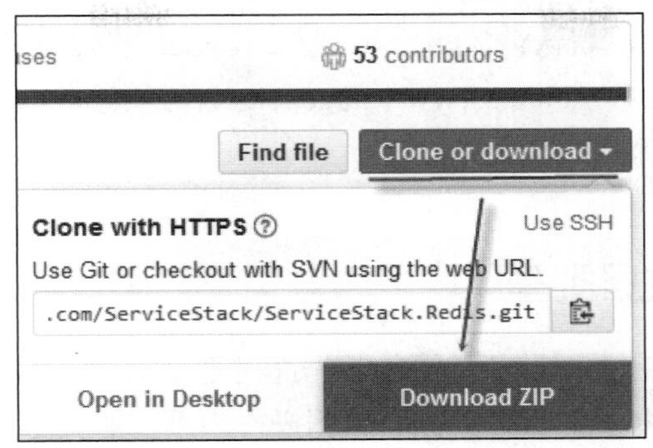

图 9-43

4. 部署开发环境

（1）新建控制台应用程序 RedisConsoleApp。

（2）添加 ServiceStack.Text.dlll、ServiceStack.Common.dll、ServiceStack.Interfaces.dll、ServiceStack.Redis.dll 的程序集引用。

当然，我们还可以通过 NuGet 直接安装。可以这么说，基本上.NET 开发平台需要的各种插件组件都可以通过 NuGet 来进行安装，如图 9-44 所示。

图 9-44

在 NuGet 左下角有一个“设置”按钮，我们可以通过它设置安装包的来源，如图 9-45 所示。

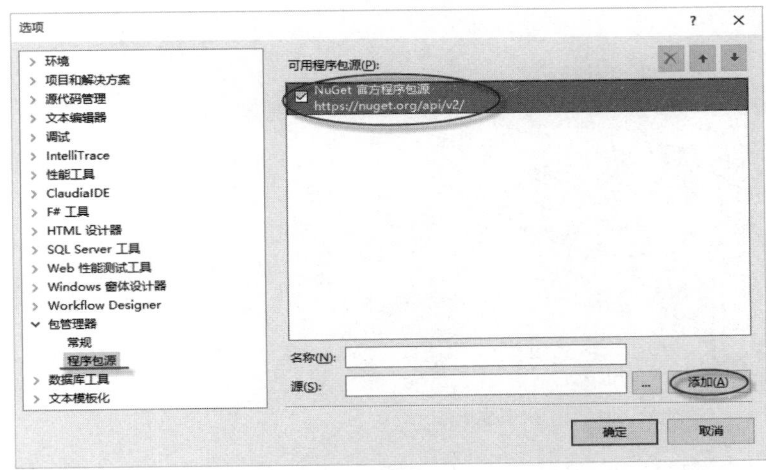

图 9-45

从图 9-45 中可以看出，默认情况下我们通过 NuGet 安装各种组件插件，其实都是通过 https//nuget.org/api/v2/这个服务从 https://www.nuget.org/官方网站获取的各种安装包。当然，我们也可以根据需要手动添加其他的程序包源。

9.5.4 Redis 常用数据类型

使用 Redis，我们不用在面对功能单调的数据库时把精力放在如何把大象放进冰箱这样的问题上，而是利用 Redis 灵活多变的数据结构和数据操作为不同的大象构建不同的冰箱。

Redis 最为常用的数据类型主要有 5 种：String、Hash、List、Set 和 sorted set。

1. String 类型

string 是最常用的一种数据类型，普通的 key/value 存储都可以归为此类 。一个 key 对应一个 value，string 类型是二进制安全的。Redis 的 string 可以包含任何数据，比如 jpg 图片（生成二进制）或者序列化的对象。基本操作如下：

```
using System;
using System.Collections.Generic;
//添加如下引用
using ServiceStack.Redis;

namespace RedisConsoleApp
{
    class Program
    {
        static RedisClient client = new RedisClient("127.0.0.1", 6379);

        static void Main(string[] args)
        {
            StringTest();
        }
        /// <summary>
        /// 1. 字符串测试
        /// </summary>
        private static void StringTest()
        {
            Console.WriteLine("******************字符串类型*********************");
            client.Set<string>("name", "zouqj");
            string userName = client.Get<string>("name");
            Console.WriteLine(userName);

            UserInfo userInfo = new UserInfo() { UserName = "张三", UserPwd =
```

```
        "123" };//(底层使用 json 序列化 )
        client.Set<UserInfo>("userInfo", userInfo);
        UserInfo user = client.Get<UserInfo>("userInfo");
        Console.WriteLine(user.UserName);

        List<UserInfo> list = new List<UserInfo>() { new UserInfo() { UserName
        = "李四", UserPwd = "1234" }, new UserInfo() { UserName = "王五",
        UserPwd = "12345" } };
        client.Set<List<UserInfo>>("list", list);
        List<UserInfo> userInfoList = client.Get<List<UserInfo>>("list");

        foreach (UserInfo u in userInfoList)
        {
            Console.WriteLine(u.UserName);
        }
    }
  }
  public class UserInfo
  {
     public string UserName { get; set; }
     public string UserPwd { get; set; }
  }
}
```

运行结果如图 9-46 所示。

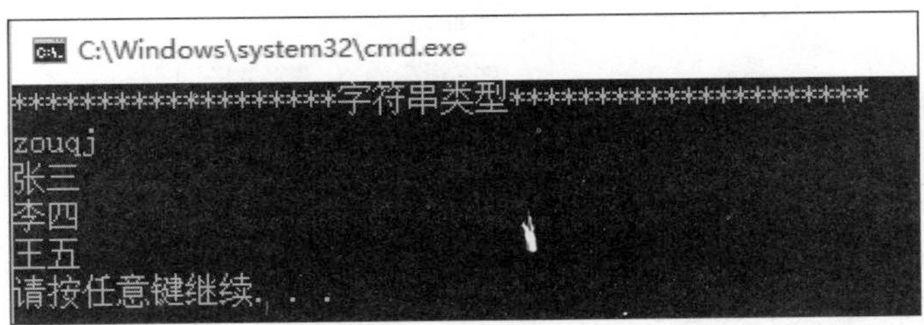

图 9-46

2. hash 类型

hash 是一个 string 类型的 field 和 value 的映射表。hash 特别适合存储对象。相对于将对象的每个字段存成单个 string 类型，一个对象存储在 hash 类型中会占用更少的内存，并且可以更方便地存取整个对象。作为一个 key value 存在，很多开发者会自然地使用 set/get 方式来使用 Redis，如图 9-47 所示。实际上这并不是最优化的使用方法。尤其在未启用 VM 情况下，Redis 全部数据需要放入内存，节约内存尤其重要。此方式增加了序列化/反序列化的开销，并且在修改其中一项信息时需要把整个对象取回 。

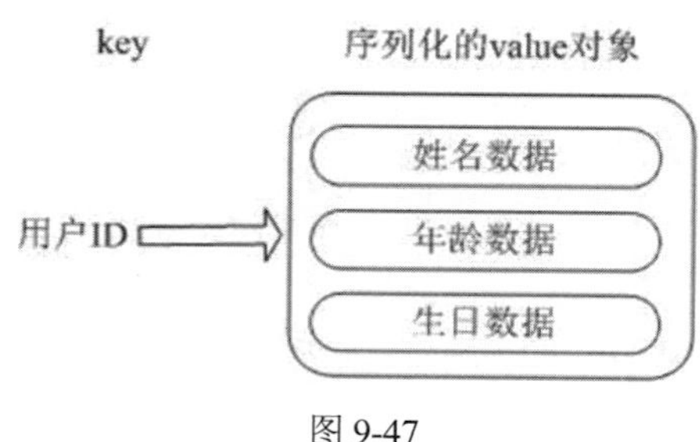

图 9-47

Redis 为单进程单线程模式，采用队列模式将并发访问变为串行访问。Redis 本身没有锁的概念，Redis 对于多个客户端连接并不存在竞争。Redis 是单线程的程序，为什么还会这么快呢?

- 大量线程导致的线程切换开销。
- 锁。
- 非必要的内存备份。
- Redis 多样的数据结构，每种结构只做自己爱做的事。

hash 对应的 value 内部实际就是一个 HashMap，会有两种不同的实现。HashMap 的成员比较少时，Redis 为了节省内存会采用类似一维数组的方式来紧凑存储，而不会采用真正的 HashMap 结构，当成员量增大时会自动转成真正的 HashMap，如图 9-48 所示。

key 仍然是用户 ID，value 是一个 Map，这个 Map 的 key 是成员的属性名，value 是属性值，这样对数据的修改和存取都可以直接通过其内部 Map 的 key（Redis 里称内部 Map 的 key 为 field），也就是通过 key（用户 ID）+field（属性标签）就可以操作对应的属性数据了，既不需要重复存储数据，也不会带来序列化和反序列化。

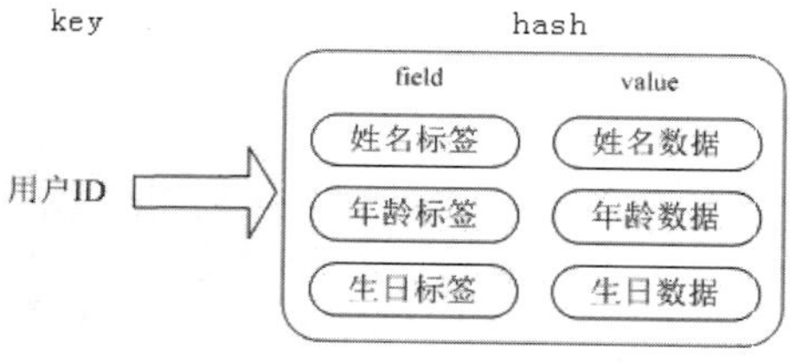

图 9-48

```
/// <summary>
/// 2. hash 测试
/// </summary>
private static void HashTest()
{
        Console.WriteLine("*******************Hash********************");
        client.SetEntryInHash("userInfoId", "name", "zhangsan");
        var lstKeys = client.GetHashKeys("userInfoId");
        lstKeys.ForEach(k => Console.WriteLine(k));
        var lstValues = client.GetHashValues("userInfoId");
```

```
        lstValues.ForEach(v => Console.WriteLine(v));
        client.Remove("userInfoId");
        Console.ReadKey();
}
```

Redis 分别为不同数据类型提供了一组参数来控制内存使用，我们在前面提到过 Redis hash 的 value 内部是一个 HashMap，如果该 Map 的成员比较少，则会采用一维数组的方式来紧凑存储该 MAP，省去了大量指针的内存开销，这个参数在 Redis conf 配置文件中有 Hash-max-zipmap-entries 64 和 Hash-max-zipmap-value 512 两项。

- Hash-max-zipmap-entries 的含义是当 value 这个 Map 内部不超过多少个成员时会采用线性紧凑格式存储，默认是 64，即 value 内部有 64 个以下的成员就使用线性紧凑存储，超过该值自动转成真正的 HashMap。
- Hash-max-zipmap-value 的含义是当 value 这个 MAP 内部的每个成员值长度不超过多少字节时会采用线性紧凑存储来节省空间。任意一个条件超过设置值都会转成真正的 HashMap，也就不会再节省内存了。这个值设置为多少需要权衡利弊。HashMap 的优势就是查找和操作时间短。

一个 key 可对应多个 field，一个 field 对应一个 value。

这里同时需要注意，Redis 提供了接口（hgetall）可以直接取到全部的属性数据，但是如果内部 Map 的成员很多，就会涉及遍历整个内部 Map 的操作。由于 Redis 单线程模型的缘故，这个遍历操作可能会比较耗时，而其他客户端的请求则完全不响应。

建议使用对象类别和 ID 构成键名，使用字段表示对象属性、字段值存储属性值，例如 car:2 price 500。

3. list 类型

list 是一个链表结构，主要功能是 push、pop 以及获取一个范围内所有的值等，操作中的 key 可以理解为链表名字。

Redis 的 list 类型其实就是一个每个子元素都是 string 类型的双向链表。我们可以通过 push、pop 操作从链表的头部或者尾部添加删除元素，这样 list 既可以作为栈，又可以作为队列。

Redis list 的实现为一个双向链表，既可以支持反向查找和遍历，也便于操作，不过带来了部分额外的内存开销。Redis 内部的很多实现（包括发送缓冲队列等）也都是用的这个数据结构。

```
/// <summary>
/// 3. 队列和栈测试
/// </summary>
private static void QueueTest()
{
        Console.WriteLine("*******************队列 先进先出*******************");
```

```
        client.EnqueueItemOnList("test", "饶成龙");//入队
        client.EnqueueItemOnList("test", "周文杰");
        long length = client.GetListCount("test");
        for (int i = 0; i < length; i++)
        {
            Console.WriteLine(client.DequeueItemFromList("test"));//出队
        }
        Console.WriteLine("********************栈 先进后出*****************");
        client.PushItemToList("name1", "邹琼俊");//入栈
        client.PushItemToList("name1", "周文杰");
        long length1 = client.GetListCount("name1");
        for (int i = 0; i < length1; i++)
        {
            Console.WriteLine(client.PopItemFromList("name1"));//出栈
        }
        Console.ReadKey();
}
```

4. set 类型

set 是 string 类型的无序集合。set 是通过 hash table 实现的添加、删除和查找，对于集合可以取并集、交集、差集。

```
/// <summary>
/// 4. set 类型操作测试
/// </summary>
private static void SetTest()
{
        //对 set 类型进行操作
        client.AddItemToSet("HighSchool", "卢沛");
        client.AddItemToSet("HighSchool", "邹琼俊");
        client.AddItemToSet("HighSchool", "周泱");
        client.AddItemToSet("HighSchool", "钟哲颖");
        client.AddItemToSet("HighSchool", "李薇");
        client.AddItemToSet("HighSchool", "刘自珍");
        client.AddItemToSet("HighSchool", "王鹏");
        client.AddItemToSet("HighSchool", "刘娇龙");
        client.AddItemToSet("HighSchool", "姜昆鹏");
        client.AddItemToSet("HighSchool", "吴燕妮");
        System.Collections.Generic.HashSet<string> hashset1 =
        client.GetAllItemsFromSet("HighSchool");
        Console.WriteLine("************邹琼俊和以下人员是高中同学************");
```

```
        ConsoleHashSetInfo(hashset1);
        //求并集
        client.AddItemToSet("college", "邹琼俊");
        client.AddItemToSet("college", "卢沛");
        client.AddItemToSet("college", "熊平");
        client.AddItemToSet("college", "陈望");
        client.AddItemToSet("college", "王小敏");
        System.Collections.Generic.HashSet<string> hashset2 =
        client.GetUnionFromSets(new string[] { "HighSchool", "college" });
        Console.WriteLine("*******邹琼俊和以下人员是高中同学或者大学同学********");

        ConsoleHashSetInfo(hashset2);
        Console.WriteLine("******邹琼俊和以下人员既是高中同学又大学同学*******");
        //求交集
        System.Collections.Generic.HashSet<string> hashset3 =
        client.GetIntersectFromSets(new string[] { "HighSchool", "college" });
        ConsoleHashSetInfo(hashset3);

        Console.WriteLine("*********邹琼俊和以下人员只是高中同学************");
        //求差集.
        System.Collections.Generic.HashSet<string> hashset4 =
        client.GetDifferencesFromSet("HighSchool", new string[] { "college" });
        ConsoleHashSetInfo(hashset4);
}

      private static void ConsoleHashSetInfo
     (System.Collections.Generic.HashSet<string> hs)
{
        foreach (string str in hs)
        {
            if (str == "邹琼俊")
                continue;
            Console.WriteLine(str);
        }
}
```

5. sorted set 类型

sorted set 是 set 的一个升级版本，在 set 的基础上增加了一个顺序的属性，这一属性可以在添加修改元素的时候指定，每次指定后，zset（表示有序集合）会自动重新按新值调整顺序。可以将其理解为有列的表，一列存 value，一列存顺序。操作中 key 可以理解为 zset 的名字。

Redis sorted set 的使用场景与 set 类似，区别是 set 不是自动有序的，而 sorted set 可以通过用户额外提供一个优先级（score）的参数来为成员排序，并且是插入有序的，即自动排序。需要一个有序的并且不重复的集合列表时就可以选择 sorted set 数据结构。

```
/// <summary>
/// 5. sorted set 测试
/// </summary>
private static void SortedSetTest()
{
        client.AddItemToSortedSet("friend", "熊平",1);
        client.AddItemToSortedSet("friend", "陈望",3);
        client.AddItemToSortedSet("friend", "王小敏",5);
        client.AddItemToSortedSet("friend", "刘继豪",2);
        client.AddItemToSortedSet("friend", "侯亮",4);
        System.Collections.Generic.List<string> list =
        client.GetAllItemsFromSortedSet("friend");
        foreach (string str in list)
        {
            Console.WriteLine(str);
        }
}
```

9.5.5 给 Redis 设置密码

安装 Redis 服务器后，不能让其在外网环境下面裸奔，我们可以为其设置一个复杂的访问密码，最好是 20 位以上，这样可以有效防止别人暴力破解。

找到 Redis 的配置文件，这里 Redis 服务的安装目录是 D:\Program Files\Redis，所以配置文件在 D:\Program Files\Redis\conf\redis.conf 目录下。查找 requirepass 选项配置，把这个选项前面的#注释去掉，然后在后面添加一个复杂的密码。

```
# use a very strong password otherwise it will be very easy to break.
#
requirepass 2016@Test.88210_yujie

# Command renaming.
```

带密码的访问，需要修改为如下代码：

```
var client = new RedisClient("127.0.0.1", 6379); //IP 和端口
client.Password = "2016@Test.88210_yujie"; //Redis 访问密码
```

9.5.6　Redis 主从复制

Redis 的主从复制功能非常强大，一个 master 可以拥有多个 slave，而一个 slave 又可以拥有多个 slave，如此下去，形成强大的多级服务器集群架构。

在 master 上进行写操作，在 slave 上面进行读操作，因为我们大多数场景查得多一些。

如果在 9.5.3 小节中已经在 Windows 服务中安装了 Redis，就先把这个服务停一下。因为 Redis 服务默认使用了 6379 端口。当然，如果不想停就直接把 Redis 服务当成 Master 服务，也就不需要再配置下面的 Master 服务了。

实现步骤如下：

（1）在 Windows 某个磁盘上创建两个目录，例如 MasterRedis（存储 Master 服务）和 SlaveRedis（存储 Slave 服务）。把 Reidis 文件各复制一份到这两个目录中。

（2）在 Master 服务中的配置文件 redis.conf 中修改：

```
bind 127.0.0.1
```

（3）在 Slave 服务中的配置文件 redis.conf 修改：

```
port 6381（服务端口号要分开）
bind 127.0.0.1
slaveof 127.0.0.1 6379  （设置Master的Host以及Port）
```

（4）分别启动 Master 服务与 Slave 服务。

启动 Master 服务，选择“开始→运行→cmd”，如图 9-49 所示。

```
D:\Redis服务\MasterRedis>redis-server.exe redis.conf
[6352] 17 Apr 08:36:59 * Server started, Redis version 2.4.2
[6352] 17 Apr 08:36:59 * DB loaded from disk: 0 seconds
[6352] 17 Apr 08:36:59 * The server is now ready to accept connections on port 6379
[6352] 17 Apr 08:37:00 - 0 clients connected (0 slaves), 672784 bytes in use
```

图 9-49

启动 Slave 服务，选择“开始→运行→cmd”，如图 9-50 所示。

```
D:\Redis服务\SlaveRedis>redis-server.exe redis.conf
[5444] 17 Apr 08:39:56 * Server started, Redis version 2.4.2
[5444] 17 Apr 08:39:56 * DB loaded from disk: 0 seconds
[5444] 17 Apr 08:39:56 * The server is now ready to accept connections on port 6381
[5444] 17 Apr 08:39:57 - 0 clients connected (0 slaves), 672832 bytes in use
[5444] 17 Apr 08:39:57 * Connecting to MASTER...
[5444] 17 Apr 08:39:57 * MASTER <-> SLAVE sync started
[5444] 17 Apr 08:39:57 * Non blocking connect for SYNC fired the event.
[5444] 17 Apr 08:39:58 * MASTER <-> SLAVE sync: receiving 10 bytes from master
[5444] 17 Apr 08:39:58 * MASTER <-> SLAVE sync: Loading DB in memory
[5444] 17 Apr 08:39:58 * MASTER <-> SLAVE sync: Finished with success
```

图 9-50

这时会发现主从服务器配置都正确了。

接下来尝试在 Master 服务器里面写一条数据，然后看一下在 Slave 服务器中是否能查询到。再打开一个 CMD 窗口：

```
D:\Redis 服务\MasterRedis>redis-cli.exe -h 127.0.0.1 -p 6379
redis 127.0.0.1:6379> set name "zouqj"
OK
redis 127.0.0.1:6379>
```

由于默认的 Redis 配置为 Save 900 1，表示有 1 个 key 改变，900 秒以后执行快照，因此 15 分钟后才会同步到 Slave 服务器。15 分钟后，打开 Slave 服务器目录下面的 dump.rdb 进行查看，显示“REDIS0002? name-zouqj”，表示已经同步到 Slave 服务器。

注意，先启动 Master，再启动 Slave 的时候，可以发现 Slave 上显示：

```
[5444] 17 Apr 08:39:57 * MASTER <-> SLAVE sync started
[5444] 17 Apr 08:39:57 * Non blocking connect for SYNC fired the event.
[5444] 17 Apr 08:39:58 * MASTER <-> SLAVE sync: receiving 10 bytes from master
[5444] 17 Apr 08:39:58 * MASTER <-> SLAVE sync: Loading DB in memory
[5444] 17 Apr 08:39:58 * MASTER <-> SLAVE sync: Finished with success
```

这会发送一个 SYNC 请求，从 Master 上面进行响应，而且它支持自动重连，即当 Master 掉线时它会处于等待请求的状态。而 Master 上的显示则如图 9-51 所示。

```
[6352] 17 Apr 08:39:57 - Accepted 127.0.0.1:4781
[6352] 17 Apr 08:39:57 * Slave ask for synchronization
[6352] 17 Apr 08:39:57 * Starting BGSAVE for SYNC
[6352] 17 Apr 08:39:57 * Foregroud saving started by pid 6352
[6352] 17 Apr 08:39:58 * DB saved on disk
[6352] 17 Apr 08:39:58 * Background saving terminated with success
[6352] 17 Apr 08:39:58 * Synchronization with slave succeeded
```

图 9-51

第一次 Slave 向 Master 同步的实现是：Slave 向 Master 发出同步请求，Master 先 dump 出 rdb 文件，然后将 rdb 文件全量传输给 Slave，然后 Master 把缓存的命令转发给 Slave，初次同步完成。第二次以及以后的同步实现是：Master 将变量的快照直接实时依次发送给各个 Slave。不管什么原因导致 Slave 和 Master 断开重连都会重复以上过程。Redis 的主从复制是建立在内存快照的持久化基础上的，只要有 Slave 就一定会有内存快照发生。虽然 Redis 宣称主从复制无阻塞，但由于 Redis 使用单线程服务，如果 Master 快照文件比较大，那么第一次全量传输就会耗费比较长的时间，且文件传输过程中 Master 可能无法提供服务，也就是说服务会中断。

（1）Redis 数据快照。数据快照的原理是将整个 Redis 内存中的所有数据遍历一遍，并通过 save 命令存储到一个扩展名为 rdb 的数据文件中，可以执行这个存储操作。数据快照配置如下：

```
Save 900 1
Save 300 10
Save 60 10000
```

以上在 redis.conf 中的配置指出在多长时间内有多少次更新操作，就将数据同步到数据文件中，这个可以多个条件进行配合。上面的含义是 900 秒后有一个 key 发生改变就执行 save，300 秒后有 10 个 key 发生改变就执行 save，60 秒有 10000 个 key 发生改变就执行 save。

数据快照的缺点是持久化之后如果出现系统宕机就会丢失一段数据，因此增加了另外一种追加式的操作日志记录，叫 append only file，其日志文件以 aof 结尾，称之为 aof 文件。要开启 aof 日志的记录，需要在配置文件中进行如下配置：

```
appendonly yes
```

Appendonly 配置不开启，可能在断电时导致一段时间的数据丢失，因为 Redis 本身同步数据文件时是按 Save 条件来同步的，所以有的数据会在一段时间内只存在于内存中。

```
Appendfsync no/always/everysec
```

- no：表示等操作系统进行数据缓存同步到磁盘。性能最好，持久化没有保障。
- always：表示每次更新操作后手动调用 fsync()将数据写到磁盘，每次收到写命令就立即强制写入磁盘，最慢的，但是保障完全的持久化。
- Everysec：表示每秒同步一次，每秒钟强制写入磁盘一次，在性能和持久化方面做了很好的折中。

为了定时减小 AOF 文件的大小，Redis 2.4 以后增加了自动的 bgrewriteaof 的功能，Redis 会选择一个自认为负载低的情况下执行 bgrewriteaof，这个重写 AOF 文件的过程是很影响性能的。解决方案是：Master 关闭 Save 功能，关闭 AOF 日志功能，以求达到性能最佳。Slave 开启 Save 并开启 AOF 日志功能，再开启 bgrewriteaof 功能，不对外提供服务，这样 Slave 的负载总体上会高于 Master 负载，但是 Master 性能达到最好。

bgrewriterof 内部实现过程如下：

① Redis 通过 fork 一个子进程遍历数据写入新临时文件。
② 父进程继续处理 client 请求，子进程继续写临时文件。
③ 父进程把新写入的 AOF 写在缓冲区。
④ 子进程写完退出，父进程接收退出消息，将缓冲区 AOF 写入临时文件。
⑤ 临时文件重命名成 appendonly.aof，原来的文件被覆盖，整个过程完成。

（2）Redis 数据恢复。当 Redis 服务器挂掉以后，重启时将按以下优先级恢复数据到内存：

① 如果只配置了 AOF，重启时加载 AOF 文件恢复数据。
② 如果同时配置了 RBD 和 AOF，启动时只加载 AOF 文件恢复数据。
③ 如果只配置了 RDB，启动时将加载 dump 文件恢复数据。

ServiceStack.Redis 从 4.0 开始商用，有每小时 6000 次访问的限制，如果要在生产环境使用，建议使用低于 4.0 的版本，这里在云盘中提供了 ServiceStack.Redis.3.9.29.0 的版本。

9.6 MongoDB

9.6.1 MongoDB 简介

MongoDB 是一个高性能、开源、无模式的文档型数据库，是当前 NoSQL 数据库中比较热门的一种。它在许多场景下可用于替代传统的关系型数据库或键/值存储方式。

传统的关系数据库一般由数据库（database）、表（table）、记录（record）三个层次概念组成，MongoDB 则是由数据库 （database）、集合（collection）、文档对象（document）三个层次组成。MongoDB 对应关系型数据库里的表，但是集合中没有列、行和关系的概念，这体现了模式自由的特点。

（1）特点

高性能、易部署、易使用，存储数据非常方便。

（2）功能

- 面向集合的存储：适合存储对象及 JSON 形式的数据。
- 模式自由。
- 动态查询：Mongo 支持丰富的查询表达式。查询指令使用 JSON 形式的标记，可轻易查询文档中内嵌的对象及数组。
- 完整的索引支持：包括文档内嵌对象及数组。Mongo 的查询优化器会分析查询表达式，并生成一个高效的查询计划。
- 查询监视：Mongo 包含一个监视工具，用于分析数据库操作的性能。
- 复制及自动故障转移：Mongo 数据库支持服务器之间的数据复制，支持主-从模式及服务器之间的相互复制。复制的主要目标是提供冗余及自动故障转移。
- 高效的传统存储方式：支持二进制数据及大型对象（如照片或图片）。
- 自动分片以支持云级别的伸缩性：自动分片功能支持水平的数据库集群，可动态添加额外的机器。
- 自动处理碎片，以支持云计算层次的扩展性。
- 支持 Python、PHP、Ruby、Java、C、C#、Javascript、Perl 及 C++语言的驱动程序，社区中也提供了对 Erlang 及.NET 等平台的驱动程序。
- 文件存储格式为 BSON（一种 JSON 的扩展）。
- 可通过网络访问。
- 支持完全索引，包含内部对象。

（3）适用场合

- 网站数据：Mongo 非常适合实时插入、更新与查询，并具备网站实时数据存储所需的复制及高度伸缩性。

- 缓存：由于性能很高，因此 Mongo 也适合作为信息基础设施的缓存层。在系统重启之后，由 Mongo 搭建的持久化缓存层可以避免下层的数据源过载。
- 大尺寸、低价值的数据：使用传统的关系型数据库存储一些数据时可能会比较昂贵。在此之前，很多时候程序员往往会选择传统的文件进行存储。
- 高伸缩性的场景：Mongo 非常适合由数十或数百台服务器组成的数据库。Mongo 的路线图中已经包含对 MapReduce 引擎的内置支持。
- 对象及 JSON 数据的存储：Mongo 的 BSON 数据格式非常适合文档化格式的存储及查询。

9.6.2 下载安装和配置

1. 下载 MongoDB 安装包

3.2.6 版下载地址为 https://www.mongodb.com/download-center。其他版本下载地址为 https://www.mongodb.org/dl/win32/x86_64-2008plus-ssl。

由于多款 MongoDB 可视化工具连接 3.2.6 版本时都会出现问题，所以在这里使用比较低的 3.0.7 版本。这也告诉我们一个经验：在进行技术选型的时候，不要追求最新的，而应当选择既能满足需求又运行非常稳定的。因为最新的技术不但相关资料少，而且也没有经过项目的实际验证，学习成本很高，风险较大。当然，如果不需要使用可视化客户端工具，在这里使用 MongoDB 3.2.6 版本进行操作也是没有任何问题的。

这里下载的版本如图 9-52 所示。

https://www.mongodb.org/dl/win32/x86_64-2008plus-ssl

win32/mongodb-win32-x86_64-2008plus-ssl-3.2.0-rc0.zip	2015-10-15 02:26:17
win32/mongodb-win32-x86_64-2008plus-ssl-3.0.7-signed.msi	2015-10-12 23:51:23
win32/mongodb-win32-x86_64-2008plus-ssl-3.0.7.zip	2015-10-12 23:51:18
win32/mongodb-win32-x86_64-2008plus-ssl-3.1.9-signed.msi	2015-10-07 02:21:31

图 9-52

2. 在 Windows 下安装 MongoDB

下载安装包后进行安装，默认会把 MongoDB 安装在目录 C:\Program Files\MongoDB 中。查看 C:\Program Files\MongoDB\Server\3.0\bin 目录，如图 9-53 所示。

脑 › Windows (C:) › Program Files › MongoDB › Server › 3.0 › bin

名称	修改日期	类型	大小
bsondump.exe	2015/10/12 21:27	应用程序	9,553 KB
libeay32.dll	2015/7/14 20:16	应用程序扩展	1,937 KB
mongo.exe	2015/10/12 21:30	应用程序	6,343 KB
mongod.exe	2015/10/12 21:34	应用程序	14,113 KB
mongod.pdb	2015/10/12 21:34	Program Debug...	108,052 KB
mongodump.exe	2015/10/12 21:28	应用程序	10,066 KB
mongoexport.exe	2015/10/12 21:28	应用程序	9,885 KB
mongofiles.exe	2015/10/12 21:27	应用程序	9,824 KB
mongoimport.exe	2015/10/12 21:28	应用程序	10,057 KB
mongooplog.exe	2015/10/12 21:29	应用程序	9,565 KB
mongoperf.exe	2015/10/12 21:35	应用程序	12,303 KB
mongorestore.exe	2015/10/12 21:28	应用程序	10,176 KB
mongos.exe	2015/10/12 21:34	应用程序	6,209 KB
mongos.pdb	2015/10/12 21:34	Program Debug...	55,588 KB
mongostat.exe	2015/10/12 21:27	应用程序	9,776 KB
mongotop.exe	2015/10/12 21:28	应用程序	9,649 KB
ssleay32.dll	2015/7/14 20:16	应用程序扩展	343 KB

图 9-53

mongod.exe 是用来连接到 mongo 数据库服务器的，即服务器端。mongo.exe 是用来启动 MongoDB shell 的，即客户端。其他文件的功能如下：

- mongodump.exe：逻辑备份工具。
- mongorestore.exe：逻辑恢复工具。
- mongoexport.exe：数据导出工具。
- mongoimport.exe：数据导入工具。

（1）配置 MongoDB 环境：在目录 D:\WorkSpace\mongodb 下新建 data 文件夹，data 文件夹将会作为数据存放的根文件夹。

（2）以管理员身份运行“cmd”，分别执行如下两条 cmd 命令：

```
cd \Program Files\MongoDB\Server\3.0\bin
mongod.exe --dbpath D:\WorkSpace\mongodb\data
```

最后一行命令中的-dbpath 参数值就是我们第一步新建的文件夹。这个文件夹一定要在开启服务之前事先建立好，否则会报错，因为 MongoDB 不会自己创建此文件夹。

命令执行成功后会看到如下信息：

```
C:\Windows\system32>cd \Program Files\MongoDB\Server\3.0\bin

C:\Program Files\MongoDB\Server\3.0\bin>mongod.exe -dbpath
D:\WorkSpace\mongodb\data
2016-06-04T00:00:00.153+0800 I JOURNAL  [initandlisten] journal
```

```
dir=D:\WorkSpace\mongodb\data\journal
2016-06-04T00:00:00.155+0800 I JOURNAL  [initandlisten] recover : no journal
files present, no recovery needed
2016-06-04T00:00:00.224+0800 I JOURNAL  [durability] Durability thread started
2016-06-04T00:00:00.224+0800 I JOURNAL  [journal writer] Journal writer thread
started
2016-06-04T00:00:00.323+0800 I CONTROL  [initandlisten] MongoDB starting :
pid=4392 port=27017 dbpath=D:\WorkSpace\mongodb\data 64-bit host=DESKTOP-V7CFIC3
2016-06-04T00:00:00.324+0800 I CONTROL  [initandlisten] targetMinOS: Windows
7/Windows Server 2008 R2
2016-06-04T00:00:00.324+0800 I CONTROL  [initandlisten] db version v3.0.7
2016-06-04T00:00:00.324+0800 I CONTROL  [initandlisten] git version:
6ce7cbe8c6b899552dadd907604559806aa2e9bd
2016-06-04T00:00:00.324+0800 I CONTROL  [initandlisten] build info: windows
sys.getwindowsversion(major=6, minor=1, build=7601, platform=2,
service_pack='Service Pack 1') BOOST_LIB_VERSION=1_49
2016-06-04T00:00:00.324+0800 I CONTROL  [initandlisten] allocator: tcmalloc
2016-06-04T00:00:00.324+0800 I CONTROL  [initandlisten] options: { storage:
{ dbPath: "D:\WorkSpace\mongodb\data" } }
2016-06-04T00:00:00.326+0800 I INDEX    [initandlisten] allocating new ns file
D:\WorkSpace\mongodb\data\local.ns, filling with zeroes...
2016-06-04T00:00:00.586+0800 I STORAGE  [FileAllocator] allocating new datafile
D:\WorkSpace\mongodb\data\local.0, filling with zeroes...
2016-06-04T00:00:00.589+0800 I STORAGE  [FileAllocator] creating directory
D:\WorkSpace\mongodb\data\_tmp
2016-06-04T00:00:00.597+0800 I STORAGE  [FileAllocator] done allocating datafile
D:\WorkSpace\mongodb\data\local.0, size: 64MB,  took 0.004 secs
2016-06-04T00:00:00.623+0800 I NETWORK  [initandlisten] waiting for connections
on port 27017
```

（3）在浏览器中输入“http://localhost:27017/”，可以看到如下提示：

```
It looks like you are trying to access MongoDB over HTTP on the native driver
port.
```

说明 MongoDB 数据库服务已经成功启动了。

（4）再次查看 D:\WorkSpace\mongodb\data 文件夹，就会发现 data 目录下面多了许多文件，如图 9-54 所示。

电脑 › 本地磁盘 (D:) › WorkSpace › mongodb › data

名称	修改日期	类型	大小
_tmp	2016/6/4 0:00	文件夹	
journal	2016/6/4 0:00	文件夹	
local.0	2016/6/4 0:00	0 文件	65,536 KB
local.ns	2016/6/4 0:00	NS 文件	16,384 KB
mongod.lock	2016/6/4 0:00	LOCK 文件	1 KB
storage.bson	2016/6/4 0:00	BSON 文件	1 KB

图 9-54

每次都使用 cmd 命令来开启 MongoDB 服务端和客户端十分麻烦，我们完全可以把服务端做成服务，设置开机自启动，然后让客户端从网上下载可视化客户端管理工具。这里将分别演示如何将其都做成批处理命令和服务。

（1）MongoDB 服务端

① 做成批处理。新建文本文件 mongodb_server.txt，输入如下文本：

```
@echo off
start cmd /k "cd/d C:\Program Files\MongoDB\Server\3.0\bin&&mongod --dbpath
D:\WorkSpace\mongodb\data"
```

然后修改文件后缀名称为 bat，最终文件全名为 mongodb_server.bat。

② 做成 Windows 服务。

安装服务：新建文本文件 mongodb_server_ Installer.txt，输入如下文本，然后另存为.bat 文件。

```
@echo off
echo 正在安装服务 MongoDB...
start cmd /k "cd/d C:\Program Files\MongoDB\Server\3.0\bin&&mongod --install --
serviceName MongoDB -serviceDisplayName MongoDB -logpath
D:\WorkSpace\mongodb\log\MongoDB.Log --dbpath D:\WorkSpace\mongodb\data
echo 服务 MongoDB 安装成功...
echo 正在停止服务...
net start MongoDB
echo 按任意键退出...
pause 启动服务：net start MongoDB
```

以管理员身份运行 mongodb_server_ Installer.bat。

卸载服务：新建文本文件 mongodb_server_ UnInstaller.txt，输入如下文本，然后另存为.bat 文件。

```
echo 正在停止服务...
net stop MongoDB
echo 正在删除服务...
```

```
sc delete MongoDB
echo 按任意键退出...
pause
```

以管理员身份运行 mongodb_server_ UnInstaller.bat。

（2）MongoDB 客户端

新建文本文件 mongodb_client.txt，输入如下文本：

```
@echo off
start cmd /k "cd/d C:\Program Files\MongoDB\Server\3.0\bin&&mongo"
```

然后修改文件后缀名称为 bat，最终文件全名为 mongodb_client.bat。以后需要启动的时候，直接双击批处理命令就可以运行了。

9.6.3　使用 mongo.exe 执行数据库增删改查操作

（1）在 MongoDB 服务已经成功启动的情况下，以管理员身份打开一个 CMD 窗口，执行如下命令：

```
Microsoft Windows [版本 10.0.10240]
(c) 2015 Microsoft Corporation. All rights reserved.

C:\Windows\system32>cd \Program Files\MongoDB\Server\3.0\bin

C:\Program Files\MongoDB\Server\3.0\bin>mongo
MongoDB shell version: 3.0.7
connecting to: test
>
```

可以看到 MongoDB 版本号为 3.0.7，默认连接的数据库为 test。test 数据库是系统默认将要创建的，因为此时并不存在此数据库，或者说它现在还只在内存中，并没有创建在物理磁盘上。每次都使用 cd 命令来跳转到 MongoDB 的安装目录很麻烦，我们也可以把安装目录添加到环境变量中，以后就可以直接在 CMD 窗口中输入 MongodDB 命令了。

（2）创建数据库，输入命令“use demo”，然后按回车键，结果如下：

```
> use demo
switched to db demo
>
```

这里的 use 命令用来切换当前数据库，如果该数据库不存在，就会先新建一个。

（3）创建 collection 并插入数据。在传统关系型数据库中创建完了库后会创建表，但是在

MongoDB 中没有“表”的概念，与其对应的一个概念是集合，即 collection。

在 CMD 窗口输入如下命令：

```
> db.users.insert({'name':'zouqj','age':'27'})
```

按回车键，结果如下：

```
WriteResult({ "nInserted" : 1 })
```

继续输入命令：

```
> db.users.insert({'name':'邹玉杰','age':'27'})
```

按回车键，结果如下：

```
WriteResult({ "nInserted" : 1 })
```

说明：向 users 集合中插入 2 条数据。如果集合 users 不存在，就先新建一个再插入数据，参数以 JSON 格式传入。

（4）查询数据。

```
// 显示所有数据库
> show dbs
demo   0.078GB
local  0.078GB
>
// 显示当前数据库下的所有集合
> show collections
system.indexes
users
>
// 显示 users 集合下的所有数据文档
> db.users.find()
{ "_id" : ObjectId("57520c5cf5755d5a81cf32ad"), "name" : "zouqj", "age" : "27" }
{ "_id" : ObjectId("57520c69f5755d5a81cf32ae"), "name" : "邹玉杰", "age" : "27" }
>
```

我们可以看到系统给每条记录创建了一个唯一主键 _id。这个主键 ID 不是 Guid 类型，而是特有算法生成的唯一标识。

（5）更新数据。假设现在要把第 1 条数据的 name 改成 “邹琼俊”，可在 CMD 窗口输入如下命令：

```
> db.users.update({'name':'zouqj'},{'$set':{ 'name':'邹琼俊
'}},upsert=true,multi=false)
```

```
WriteResult({ "nMatched" : 1, "nUpserted" : 0, "nModified" : 1 })
>
```

这里我们用到了一个 update 方法，各参数所代表的含义如下：

参数 1：查询的条件。
参数 2：更新的字段。
参数 3：如果不存在则插入。
参数 4：是否允许修改多条记录。

（6）删除记录。假设我们现在要把第一条记录即 name 为“邹玉杰”的数据删除，可在 CMD 窗口输入如下命令：

```
> db.users.remove({'name':'邹玉杰'})
WriteResult({ "nRemoved" : 1 })
>
```

再使用 db.users.find()命令查看结果：

```
> db.users.find()
{ "_id" : ObjectId("57520c5cf5755d5a81cf32ad"), "name" : "邹琼俊", "age" : "27" }
>
```

其他命令：

```
db.users.remove() //删除所有记录
db.users.drop()  //删除 collection，如果删除成功会返回“true”，否则返回“false”
db.dropDatabase() //删除当前数据库
```

9.6.4 更多命令

```
db.AddUser(username,password)  //添加用户
db.auth(usrename,password)     //设置数据库连接验证
db.cloneDataBase(fromhost)     //从目标服务器克隆一个数据库
db.commandHelp(name)           //返回帮助命令
db.copyDatabase(fromdb,todb,fromhost) //复制数据库，fromdb 为源数据库名称，todb 为目标数
                               //据库名称，fromhost 为源数据库服务器地址
db.createCollection(name,{size:3333,capped:333,max:88888}) //创建一个数据集，相当于一个表
db.currentOp()                 //取消当前库的当前操作
db.dropDataBase()              //删除当前数据库
db.eval(func,args)             //运行服务器端代码
db.getCollection(cname)        //取得一个数据集合，其他类似的用法：db['cname'] or
db.getCollenctionNames()       //取得所有数据集合的名称列表
db.getLastError()              //返回最后一个错误的提示消息
```

```
db.getLastErrorObj()            //返回最后一个错误的对象
db.getMongo()                   //取得当前服务器的连接对象 get the server
db.getMondo().setSlaveOk()      //允许当前连接读取备库中的成员对象
db.getName()                    //返回当前操作数据库的名称
db.getPrevError()               //返回上一个错误对象
db.getProfilingLevel()
db.getReplicationInfo()         //获得重复的数据
db.getSisterDB(name)            //获取服务器上面的数据库
db.killOp()                     //停止（杀死）在当前库的当前操作
db.printCollectionStats()       //返回当前库的数据集状态
db.printReplicationInfo()
db.printSlaveReplicationInfo()
db.printShardingStatus()        //返回当前数据库是否为共享数据库
db.removeUser(username)         //删除用户
db.repairDatabase()             //修复当前数据库
db.resetError()
db.runCommand(cmdObj)           //run a database command. if cmdObj is a string,
                                //turns it into {cmdObj:1}
db.setProfilingLevel(level)     //0=off,1=slow,2=all
db.shutdownServer()             //关闭当前服务程序
db.version()                    //返回当前程序的版本信息

db.test.find({id:10})           //返回 test 数据集 ID=10的数据集
db.test.find({id:10}).count()   //返回 test 数据集 ID=10的数据总数
db.test.find({id:10}).limit(2)//返回 test 数据集 ID=10的数据集从第二条开始的数据集
db.test.find({id:10}).skip(8)   //返回 test 数据集 ID=10的数据集从0到第八条的数据集
db.test.find({id:10}).limit(2).skip(8)//返回 test 数据集 ID=10的数据集从第二条到第八条的数据
db.test.find({id:10}).sort()    //返回 test 数据集 ID=10的排序数据集
db.test.findOne([query])        //返回符合条件的一条数据
db.test.getDB()                 //返回此数据集所属的数据库名称
db.test.getIndexes()            //返回此数据集的索引信息
db.test.group({key:...,initial:...,reduce:...[,cond:...]})
db.test.mapReduce(mayFunction,reduceFunction,<optional params>)
db.test.remove(query)                       //在数据集中删除一条数据
db.test.renameCollection(newName)           //重命名数据集名称
db.test.save(obj)                           //往数据集中插入一条数据
db.test.stats()                             //返回此数据集的状态
db.test.storageSize()                       //返回此数据集的存储大小
db.test.totalIndexSize()                    //返回此数据集的索引文件大小
db.test.totalSize()                         //返回此数据集的总大小
db.test.update(query,object[,upsert_bool])//在此数据集中更新一条数据
db.test.validate()                          //验证此数据集
```

```
db.test.getShardVersion()                    //返回数据集共享版本号
```

9.6.5 MongoDB 语法与现有关系型数据库 SQL 语法比较

MongoDB 语法与 MySQL 语法比较如表 9-1 所示。

表 9-1 MongoDB 语法与 MySQL 语法比较

MongoDB 语法	MySQL 语法
db.test.find({'name':'foobar'})	select * from test where name='foobar'
db.test.find()	select * from test
db.test.find({'ID':10}).count()	select count(*) from test where ID=10
db.test.find().skip(10).limit(20)	select * from test limit 10,20
db.test.find({'ID':{$in:[25,35,45]}})	select * from test where ID in (25,35,45)
db.test.find().sort({'ID':-1})	select * from test order by ID desc
db.test.distinct('name',{'ID':{$lt:20}})	select distinct(name) from test where ID<20
db.test.group({key:{'name':true},cond:{'name':'foo'}, reduce:function(obj,prev) {prev.msum+=obj.marks;},initial:{msum:0}})	select name,sum(marks) from test group by name
db.test.find('this.ID<20',{name:1})	select name from test where ID<20
db.test.insert({'name':'foobar','age':25})	insert into test ('name','age') values('foobar',25)
db.test.remove({})	delete * from test
db.test.remove({'age':20})	delete test where age=20
db.test.remove({'age':{$lt:20}})	delete test where age<20
db.test.remove({'age':{$lte:20}})	delete test where age<=20
db.test.remove({'age':{$gt:20}})	delete test where age>20
db.test.remove({'age':{$gte:20}})	delete test where age>=20
db.test.remove({'age':{$ne:20}})	delete test where age!=20
db.test.update({'name':'foobar'},{$set:{'age':36}})	update test set age=36 where name='foobar'
db.test.update({'name':'foobar'},{$inc:{'age':3}})	update test set age=age+3 where name='foobar'

注意以上命令大小写敏感。

9.6.6 可视化的客户端管理工具 MongoVUE

使用 mongo.exe 管理数据库虽然可行、功能也很强大，但是每次都要敲命令，既烦琐枯燥又效率低下。MongoDB 在 Windows 下的可视化操作的管理工具非常多，因为 MongoVUE 界面看上去舒服些，所以这里就简单介绍一下此工具。

下载地址为 http://www.mongovue.com/downloads/。

注意：官方提供的是收费版，试用期 15 天。

MongoVUE 的运行界面如图 9-55 所示。

图 9-55

在添加数据库连接之前，要先运行 Mongo 服务，如图 9-56~图 9-58 所示。

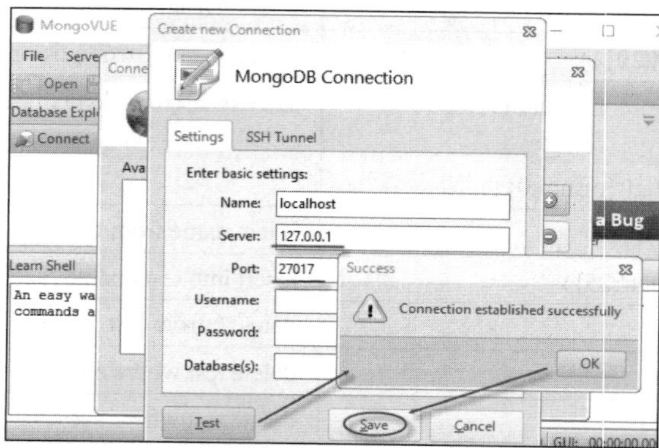

图 9-56

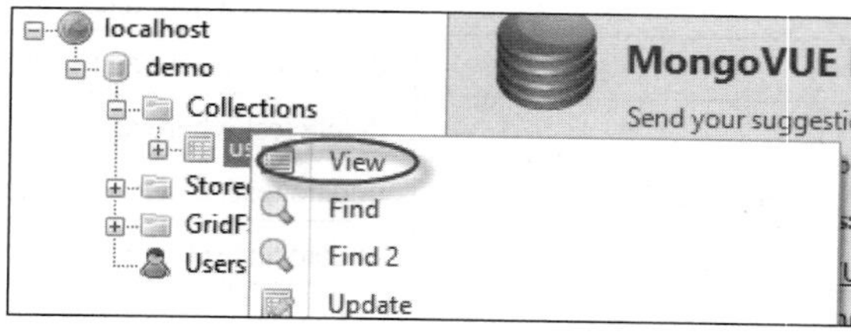

图 9-57

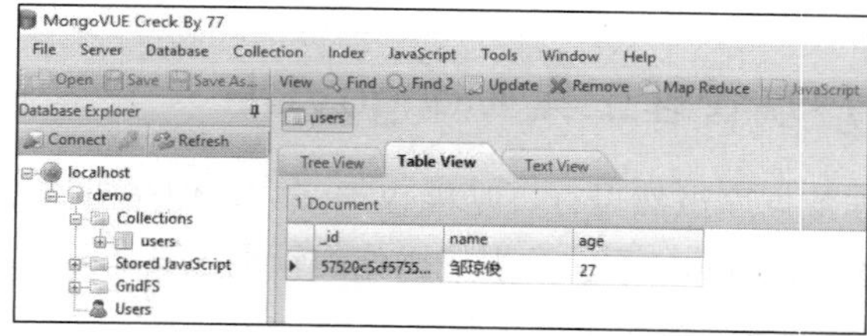

图 9-58

9.6.7 通过 C#的 samus 驱动进行操作

MongoDB 支持多种语言的驱动，这里我们只介绍 C#的驱动。C#驱动有很多种，每种驱动的形式大致相同，但是细节各有千秋，因此代码不能通用，比较常用的是官方驱动和 samus 驱动。samus 驱动除了支持一般形式的操作之外，还支持 linq 方式操纵数据。这里以 samus 驱动为例进行讲解。samus 驱动下载地址为 https://github.com/samus/mongodb-csharp。

这里下载的是驱动源码，下载后解压，再使用 VS 打开 MongoDB-CSharp-2010.sln 这个解决方案，编译其中的项目 MongoDB，在\bin\Debug 目录下面就能找到 MongoDB.dll。另外，还有一个项目 MongoDB.GridFS（主要用于存储大文件），不过这里我们只需要用到 MongoDB.dll。

（1）双击 mongodb_server.bat 运行 MongoDB 服务。

（2）新建控制台程序 ConsoleAppMongoDB，修改 Program.cs 代码：

```
using MongoDB;
using System;

namespace ConsoleAppMongoDB
{
    class Program
    {
        //集合名
        static string collectionName = "users";
        static void Main(string[] args)
        {
            //链接字符串
            string connectionString = "mongodb://127.0.0.1:27017";
            //数据库名
            string databaseName = "demo";

            //定义 Mongo 服务
            Mongo mongo = new Mongo(connectionString);
            //获取 databaseName 对应的数据库，不存在则自动创建
            MongoDatabase db = mongo.GetDatabase(databaseName) as MongoDatabase;
            //获取 collectionName 对应的集合，不存在则自动创建
            MongoCollection<Document> mongoCollection =
            db.GetCollection<Document>(collectionName) as MongoCollection<Document>;

            //链接数据库
            mongo.Connect();
            try
            {
```

```
            //定义一个文档对象，存入两个键值对
            Document doc = new Document();
            doc["ID"] = 1;
            doc["Msg"] = "Hello World!";

            //将这个文档对象插入集合
            mongoCollection.Insert(doc);
            //mongoCollection.Remove(doc);

            //在集合中查找键值对为 ID=1的文档对象
            Document docFind = mongoCollection.FindOne(new Document { { "ID", 1 } });

            //输出查找到的文档对象中“Msg”对应的值并输出
            Console.WriteLine(Convert.ToString(docFind["Msg"]));

            Insert(db);
            Update(db);
            Query(db);
            Delete(db);
            Console.ReadLine();
        }
        finally
        {
            //关闭链接
            mongo.Disconnect();
        }
    }

    /// <summary>
    /// 添加数据
    /// </summary>
    static void Insert(MongoDatabase db)
    {
        var col = db.GetCollection<Users>();
        //会自动创建一个名称为 Users 的集合，集合名称区分大小写
        //或者
        //var col = db.GetCollection("Users");
        Users users = new Users();
        users = new Users { name = "邹玉杰", age = 28 };
        col.Insert(users);
        users = new Users { name="邹琼俊",age=27};
        col.Insert(users);
```

```
    users = new Users { name = "邹宇峰", age =2 };
    col.Insert(users);
}
/// <summary>
/// 更新数据
/// </summary>
static void Update(MongoDatabase db)
{
    var col = db.GetCollection<Users>();
    //查出 Name 值为"邹琼俊"的第一条记录
    Users users = col.FindOne(x => x.name == "邹琼俊");
    //或者
    //Users users = col.FindOne(new Document { { "Name", "邹琼俊" } });
    if (users == null)
        return;
    users.age =26;
    col.Update(users, x => x.age == 27);
}

/// <summary>
/// 删除数据
/// </summary>
static void Delete(MongoDatabase db)
{
    var col = db.GetCollection<Users>();
    col.Remove(x => x.age == 27);
    //或者先查出 Age 值为27的第一条记录再移除
    //Users users = col.FindOne(x => x.age == 27);
    //col.Remove(users);
}

/// <summary>
/// 查询数据
/// </summary>
static void Query(MongoDatabase db)
{
    var col = db.GetCollection<Users>();
    var query = new Document { { "Name", "邹琼俊" } };
    //查询指定查询条件的全部数据
    var result1 = col.Find(query);
    //查询指定查询条件的第一条数据
    var result2 = col.FindOne(query);
```

```
            //查询全部集合里的数据
            var result3 = col.FindAll();
        }
    }
    public class Users
    {
        public string name { get; set; }
        public int age { get; set; }
    }
}
```

运行结果如下：

```
Hello World!
请按任意键继续. . .
```

我们来看一下数据库中的数据，如图 9-59 所示。

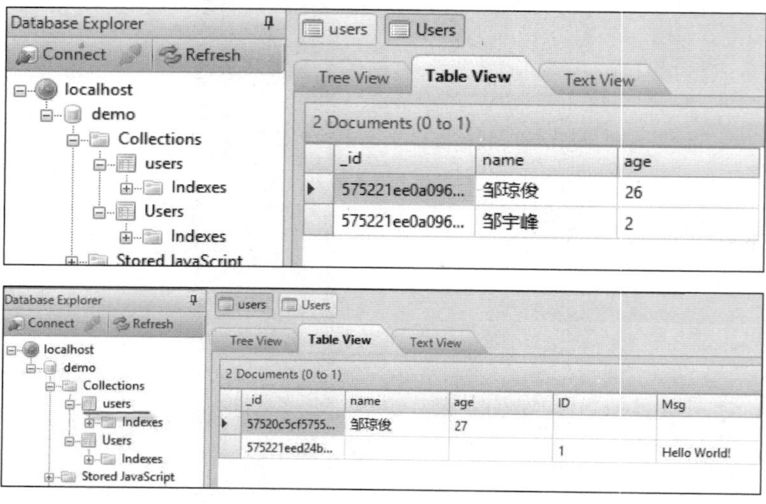

图 9-59

9.6.8 索引

索引在数据库中是非常重要的。MongoDB 自然也不例外，在 MongoDB 中创建索引十分方便。在 CMD 窗口输入如下命令查询集合中的所有索引：

```
> db.Users.getIndexes()
[
        {
                "v" : 1,
                "key" : {
```

```
                "_id" : 1
            },
            "name" : "_id_",
            "ns" : "demo.Users"
        }
]
>
```

从输出的结果我们可以看出，系统已经默认为 Users 集合创建了一个索引_id_。

MongoDB 在每个集合里都有一个默认的“_id”“字段”，它相当于“主键”。集合创建后系统会自动创建一个索引在“_id”键上，它是默认索引，索引名叫“_id_”，不允许删除。

1. 创建索引

在 MongoDB 中使用 ensureIndex()命令创建索引。

（1）创建单例索引。

```
> db.Users.ensureIndex({name:1})
{
        "createdCollectionAutomatically" : false,
        "numIndexesBefore" : 1,
        "numIndexesAfter" : 2,
        "ok" : 1
}
>
```

（2）创建联合索引（联合索引就是将多个字段作为一个索引）。

```
> db.Users.ensureIndex({name:1,age:-1})
{
        "createdCollectionAutomatically" : false,
        "numIndexesBefore" : 2,
        "numIndexesAfter" : 3,
        "ok" : 1
}
>
```

其中，关键字后面的数字表示索引的排序方向，1 表示升序，-1 表示倒序。

再次查看所有索引。

```
> db.Users.getIndexes()
[
        {
                "v" : 1,
```

```
                "key" : {
                        "_id" : 1
                },
                "name" : "_id_",
                "ns" : "demo.Users"
        },
        {

                "v" : 1,
                "key" : {
                        "name" : 1
                },
                "name" : "name_1",
                "ns" : "demo.Users"
        },
        {

                "v" : 1,
                "key" : {
                        "name" : 1,
                        "age" : -1
                },
                "name" : "name_1_age_-1",
                "ns" : "demo.Users"
        }
]
>
```

我们会发现索引的名称默认格式为“关键字_数字”，比如上面创建的两个索引的名称即为“name_1”和“name_1_age_-1”。

2. 删除索引

在 MongoDB 中可以使用 dropIndex()命令删除索引，例如：

```
> db.Users.dropIndex("name_1")
{ "nIndexesWas" : 3, "ok" : 1 }
>
```

第 10 章

站内搜索

在讲站内搜索之前，请先思考一个问题，我们的网站是怎么被客户访问到的呢？

- 直接在浏览器输入网址
- 通过搜索引擎搜索
- 通过外链跳转

事实上，很多人都是直接通过搜索引擎的方式（例如，在百度搜网站关键字）找到网站的。

10.1 SEO

10.1.1 SEO 简介

SEO（Search Engine Optimization，搜索引擎优化）的目的是让搜索引擎更多地收录网站的页面，让被收录页面的权重更靠前，让更多的人能够通过搜索引擎进入这个网站。这样 SEO 就显得非常重要，以致于有一些公司专门为其他企业提供 SEO 服务。但也会存在 SEO 做得再好，搜索排名也不在最前的情况。

例如，如图 10-1 所示，在百度中搜索“跨境电商”，我们会发现排在前面的记录后面都会带上“商业推广”字样，这是因为这些网站的运营商给百度交了钱，所以网站权重会靠前。

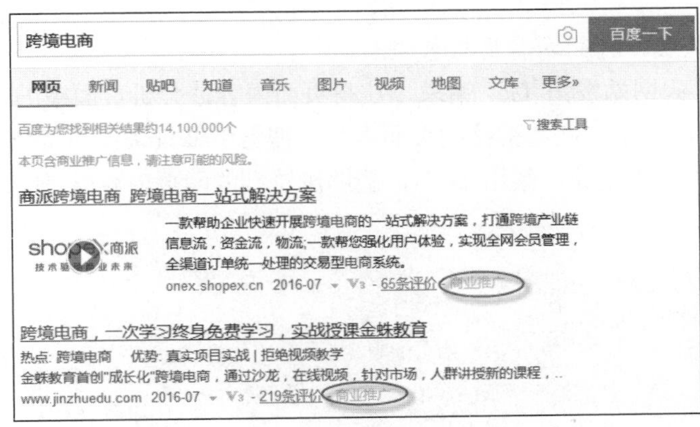

图 10-1

有人可能就要说了，对网站进行 SEO 优化岂不是没什么用了，还不如直接给百度交钱来得直接和高效。其实并不是那样的，我们可以从如下几点来考虑：

- 给百度交钱了，只能提升通过百度查找的排名，谷歌等其他搜索引擎还是原来的样子，如果均要提升，就要通通去交钱，而且这个钱通常都不在少数。
- 虽然说进行商业推广效果显著，但并不是所有的企业都会通过商业推广的方式来提升搜索排名，一句话：伤钱！
- 通过 SEO 手段，我们可以让网站的排名紧靠在进行了商业推广的同类网站之后，这样就可以为企业节省许多成本。

原理：蜘蛛会定时抓取网站的内容，发现网站内容变化、发现新增内容就反映到搜索引擎中。

蜘蛛（Spider） 爬网站：其实就是向网站发送 HTTP Get 请求的客户端。

SEO（搜索引擎优化）：让网站排名靠前，让网站更多的页面被搜索引擎收录。影响因素包括链接（外链、内链）、原创、关键词的数量、权重、小偷网站（站群）、降权（PR PageRank）、K 站（Kill 站，作弊）、伪原创等。

爬网站的过程：以百度为例，当百度蜘蛛发现网站“博客园”时，百度把“博客园”当成关键网站，然后顺着已知的网站链接找到新的网站或者新的页面。

SEO 的一个重要手段：建外链（外部链接），通过新网站的形式吸引蜘蛛。而对于非新网站，搜索引擎会考虑一个“权重”，重点考察外链数量。权重越高搜索结果越靠前，“权重”的重要因素就是“外链”数量、外链质量（外链网站的 PR（PageRank）值，决定一个网站的质量，PR 值越高网站越重要。PR 值受原创、创建时间等影响）和 Alexa 排名（全球网站排名）。

robots.txt 是公约，搜索引擎都建议遵守。它相当于一个指路牌：指出想让哪些搜索引擎搜索、想让哪些页面被搜索。

10.1.2 开发时要考虑 SEO

作为 Web 开发人员，我们均需要考虑 SEO。

搜索引擎蜘蛛：向网站发出 Get 请求，获得页面内容，分析页面中的超链接，进一步向页面中的超链接发出 Get 请求，获得链接的页面内容。搜索引擎只认识<a href="">，并且超链接不能是 js 动态生成的。也就是说，禁用 js 后，能够浏览到的网页内容就是蜘蛛能够爬到的内容。蜘蛛爬网站的原理如下：

```
WebClient wc = new WebClient();
string html = wc.DownloadString(" http://www.cnblogs.com");
//把 html 放入"数据库"
//分析 html 中的超链接 links
//foreach(string link in links)
//string html2 = wc.DownloadString(link);
```

网站的 SEO 基本原则：所有希望搜索引擎抓取的内容都应该是通过超链接（Get 请求）获得的，Post 获得的内容、js 脚本打开的页面是无法被搜索引擎抓取的，所以尽量不要通过 JS、Post 来进行页面导航和内容的生成。记住一点：搜索引擎一般不会执行 JS、不会发 Post 请求、不会提交表单。

动态菜单应该是把菜单内容静态地写到 HTML 中，然后通过隐藏、显示来切换，而不是动态绘制菜单内容。

使用 WebForm 开发的用户要注意：由于 LinkButton 是执行 JavaScript 向服务器发请求来进行 Redirect，而蜘蛛不会执行 JS，因此尽量不要用 LinkButton。

内嵌 JS 生成的超链接搜索引擎是看不到的。服务器端动态生成的内容是可以被蜘蛛看到的，蜘蛛只分析静态的 HTML 内容，不会去执行 JS。蜘蛛就相当于发出 Get 请求的 WebClient。禁用 JS 以后获得的内容就是蜘蛛能看到的内容。

这里通过一个 Demo 来演示 SEO 友好的 Ajax 操作。

（1）新建 MVC 项目 MvcAppSeoAjax。

（2）新建 Home 控制器，并添加 GetMsg 方法，代码如下：

```
[HttpPost]
public ContentResult GetMsg(int id)
{
        string strResult = string.Empty;
        switch (id)
        {
            case 1:
                strResult = @"搜索引擎蜘蛛（Spider）：向网站发出 Get 请求，获得页面内容，
                分析页面中的超链接，进一步地向页面中的超链接发 Get 请求，获得链接的页面内容。";
                break;
            case 2:
                strResult = @"因为 LinkButton 是执行 JavaScript 向服务器发请求来进行
                Redirect，而蜘蛛不会执行 JS，所以尽量不要用 LinkButton。";
                break;
            case 3:
                strResult = @"SEO 友好的 Ajax：做一个简单的根据参数来获得1、2、3三个文章的
                Ajax 效果";
                break;
            default:
                break;
        }
        return Content(strResult);
}
```

（3）添加 Index 视图：

```
@{
    ViewBag.Title = "Index";
}
@section Scripts{
    <script type="text/javascript">
        function getart(id) {
            $.post("/Home/GetMsg", { "id": id }, function (data) {
                $("#divMsg").html(data);
            });
        }
</script>
}
<ul>
    <li><a href="/Home/GetMsg/1" onclick="getart(1);return false;">1</a></li>
    <li><a href="/Home/GetMsg/2" onclick="getart(2);return false;">2</a></li>
    <li><a href="/Home/GetMsg/3" onclick="getart(3);return false;">3</a></li>
</ul>
<div id="divMsg">
</div>
```

注意，这里添加了 a 标签的 href 属性值，这是为了让爬虫能够爬到。实现 Ajax 时，同时制定 onclick 和 href，让普通用户走 onclick 的 Ajax 效果、搜索引擎蜘蛛走 href。对于开发人员来讲：搜索引擎只认 Get 请求获得的原始源代码；只认页面中的 a 标签，并且不能是 href="javascript:"。

比如，不能是这种的：

```
<a href="javascript:document.location='http://www.baidu.com'" >百度</a>
```

如果是使用 WebForm 开发，不要用 LinkButton 的 Onclick 中来做 Response.Redirect，因为生成的是 JS，请看如下代码：

```
<asp:LinkButton ID="LinkButton1" runat="server" onclick="LinkButton1_Click">LinkB
utton</asp:LinkButton>
```

生成的 HTML 源码如下：

```
<a id="LinkButton1" href="javascript:__doPostBack('LinkButton1','')">LinkButton</
a>
```

10.1.3 关于搜索

开发百度用的是通用搜索，这里主要做站内搜索。

有一定访问量的互联网站都有站内搜索功能，比如 verycd、优酷、豆瓣、cnblogs、mop、淘宝、大众点评网等。

REMARK like '%用户成功%'：模糊程度太低，无法匹配几个关键词不挨着的；全表扫描，效率低。

为什么不用数据库全文检索？数据库全文检索很傻瓜化，和普通 SQL 一样。数据全文检索灵活性不强。

注意：SQL Server 的专业版才支持全文检索功能。

使用 SQL Server 全文检索的步骤如下：

（1）设置需要检索的列属性，如图 10-2 所示。

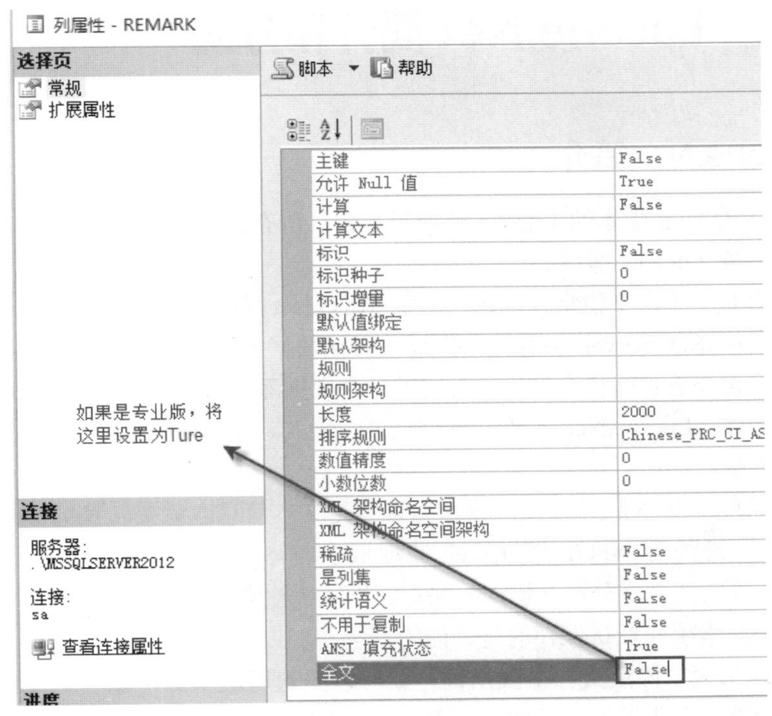

图 10-2

（2）全文检索的查询语法是：

```
select * from T where contains(REMARK,'用户成功')
```

为什么不用百度、Google 的站内搜索（site:cnblogs.com）：受制于人，会被 K；索引不及时、不全面、不精准；用户体验感差。

这里主要讲解最有广泛应用价值的站内搜索技术，像开发百度、Google 那种站外搜索相关的技术不讲（例如，在搜索引擎中输入： site:www.cnblogs.com 进行站内搜索）。

当我们新建一个网站后，要想让搜索引擎能够收录，就要先将网址提交到各搜索引擎入口，一般提交 2~3 天后就可以在各搜索引擎中输入关键字进行搜索了，如图 10-3 所示。

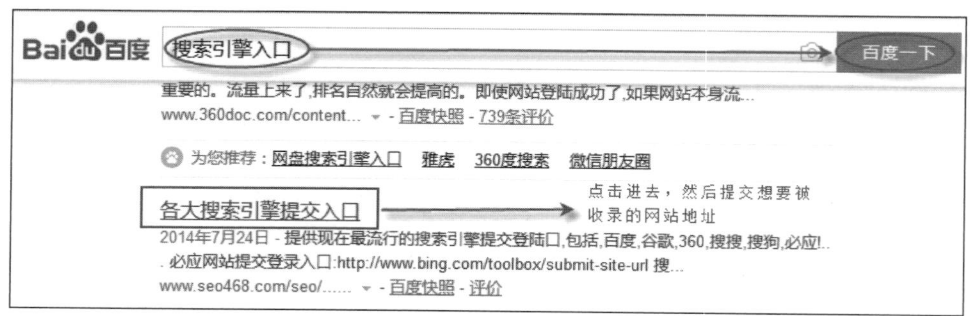

图 10-3

10.2 Lucene.Net 简介和分词

10.2.1 Lucene.Net 简介

Lucene.Net 是从 Java 版本的 Lucene 移植过来的，所有的类、方法都几乎和 Lucene 一模一样，因此使用时参考 Lucene 即可。

Lucene.Net 只是一个全文检索开发包（就像 ADO.Net 和管理系统的关系），而不是一个成型的搜索引擎，它往 SQLServer 中存完数据后，再把数据扔给 Lucene.Net，查询数据的时候从 Lucene.Net 查询，可以看作提供全文检索功能的一个数据库。SQL Server 中和 Lucene.Net 各存一份，但目的不一样。Lucene.Net 不管文本数据怎么来的，用户可以基于 Lucene.Net 开发满足自己需求的搜索引擎。

Lucene.Net 只能对文本信息进行检索。如果不是文本信息，就要转换为文本信息，比如要检索 Excel 文件，就要用 NPOI 把 Excel 读取成字符串，然后把字符串扔给 Lucene.Net。Lucene.Net 会把扔给它的文本分词（切词）保存，以加快检索速度。由于是保存的时候分词，因此搜索速度非常快！索引库默认保存的是“词的目录”。

假如要快速地从《三国演义》中找出词，就可以先遍历这本书，建一个词和页数的对应目录。第一次“找词”会非常慢，但是搜索会很快，如图 10-4 所示。

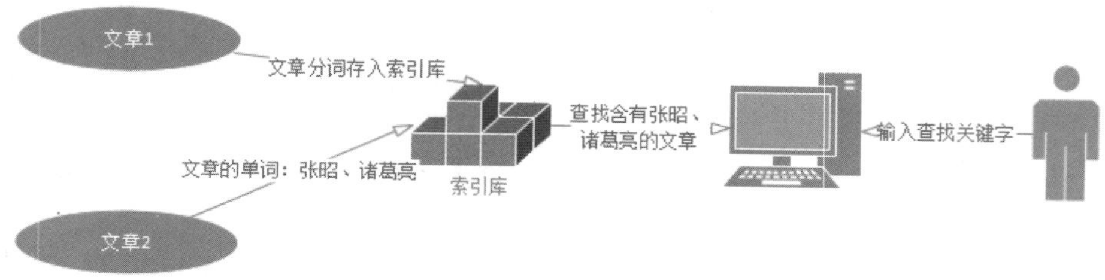

图 10-4

10.2.2　分词

分词是核心算法，搜索引擎内部保存的就是一个个的“词（Word）”。英文分词很简单，按照空格分隔就可以。中文则麻烦，例如把“诸葛亮舌战群儒”拆成“诸葛亮 舌战 群儒”。需要“the”“，”“和”“啊”“的”等对于搜索来说无意义的词一般都属于不参与分词的无意义单词(noise word)。

Lucene.Net 中不同的分词算法就是不同的类。所有分词算法类都从 Analyzer 类继承，不同的分词算法有不同的优缺点。

内置的 StandardAnalyzer 其实就是一元分词，是将英文按照空格、标点符号等进行分词，将中文按照单个字进行分词，一个汉字算一个词。如“诸葛亮舌战群儒”会分词为“诸 葛 亮 舌 战 群 儒”。

二元分词算法，每两个汉字算一个单词，“诸葛亮舌战群儒”会分词为“诸葛 葛亮 亮舌 舌战 战群 群儒”，例如 CJKAnalyzer.cs。

无论是一元分词还是二元分词，分词效率都比较高，但是会分出无用词，因此索引库大、查询效率低。

基于词库的分词算法基于一个词库进行分词，可以提高分词的成功率，例如 java 的庖丁解牛、.net 的盘古分词等，但是效率相对较低。

1. 一元分词 Demo

（1）新建 MVC 项目 SearchDemo。

（2）新建文件夹 lib，存放 dll 文件 Lucene.Net.dll。

（3）添加 Lucene.Net.dll 的引用。

（4）添加 Home 控制器和 Index 方法。

```
public ActionResult Index()
{
        return View();
}
[HttpPost]
public ActionResult Index(string txtBody)
{
        string strResult=AnalyzerResult(txtBody, new StandardAnalyzer());

        return Content(strResult);
}

private string AnalyzerResult(string txtBody, Analyzer analyzer)
{
        TokenStream tokenStream = analyzer.TokenStream("", new
        StringReader(txtBody));
```

```
            Lucene.Net.Analysis.Token token = null;
            StringBuilder sb = new StringBuilder();

            while ((token = tokenStream.Next()) != null)
            {
                sb.Append(token.TermText() + "\r\n");
            }
            return sb.ToString();
}
```

（5）异步查询。修改_Layout.cshtml 为：

```
<!DOCTYPE html>
<html>
<head>
    <meta charset="utf-8" />
    <meta name="viewport" content="width=device-width" />
    <title>@ViewBag.Title</title>
    @Styles.Render("~/Content/css")
    @Scripts.Render("~/bundles/jquery")
    @Scripts.Render("~/bundles/jqueryval")
</head>
<body>
    @RenderBody()
    @RenderSection("scripts", required: false)
</body>
</html>
```

在 Web.config 中的 appSettings 节点添加如下配置：

```
<add key="ClientValidationEnabled" value="true" />
<add key="UnobtrusiveJavaScriptEnabled" value="true" />
```

（6）添加 Index 视图：

```
@{
    ViewBag.Title = "一元分词";
}
@using (Ajax.BeginForm("Index", "Home", new AjaxOptions
{
    InsertionMode = InsertionMode.Replace,
    HttpMethod = "post",
    OnFailure = "fail",
    OnSuccess = "success",
```

```
    LoadingElementId = "lodeingmsg"
}, new { id="form1"}))
{
    <div> 输 入 内 容 :  @Html.TextArea("txtBody")<input  type="submit"  id="btnOne"
value="一元分词" />@Html.TextArea("txtResult")</div>
}
<div id="lodeingmsg" style="display: none;">加载中...</div>
<div id="msgDiv" style="color:red;"></div>
@section Scripts
{
<script type="text/javascript">
    function fail(txt) { $("#msgDiv").html("查询失败，失败信息：" + txt.responseText); }
    function success(txt) { $("#txtResult").html(txt); }
</script>
}
```

（7）查看运行结果，如图 10-5 所示。

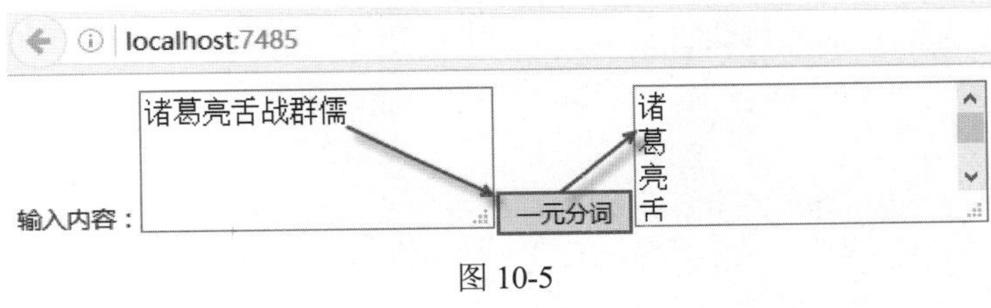

图 10-5

2. 二元分词 Demo

只需要将一元分词中的 StandardAnalyzer 替换为 CJKAnalyzer 即可。

（1）将 CJKAnalyzer.cs、CJKTokenizer.cs 这两个文件复制到项目 SearchDemo 根目录。

（2）在控制器中添加 TwoAnalyzer 方法：

```
public ActionResult TwoAnalyzer()
{
            return View();
}
[HttpPost]
public ActionResult TwoAnalyzer(string txtBody)
{
            string strResult = AnalyzerResult(txtBody, new CJKAnalyzer());

            return Content(strResult);
}
```

（3）添加 TwoAnalyzer 视图：

```
@{
    ViewBag.Title = "二元分词";
}
@using (Ajax.BeginForm("TwoAnalyzer", "Home", new AjaxOptions
{
    InsertionMode = InsertionMode.Replace,
    HttpMethod = "post",
    OnFailure = "fail",
    OnSuccess = "success",
    LoadingElementId = "lodeingmsg"
}, new { id="form1"}))
{
    <div> 输 入 内 容 ： @Html.TextArea("txtBody")<input  type="submit"  id="btnTwo"
value="二元分词" />@Html.TextArea("txtResult")</div>
}
<div id="lodeingmsg" style="display: none;">加载中...</div>
<div id="msgDiv" style="color:red;"></div>
@section Scripts
{
<script type="text/javascript">
    function fail(txt) { $("#msgDiv").html("查询失败，失败信息：" + txt.responseText); }
    function success(txt) { $("#txtResult").html(txt); }
</script>
}
```

（4）在浏览器地址栏中输入“http://localhost:7485/Home/TwoAnalyzer”，运行结果如图 10-6 所示。

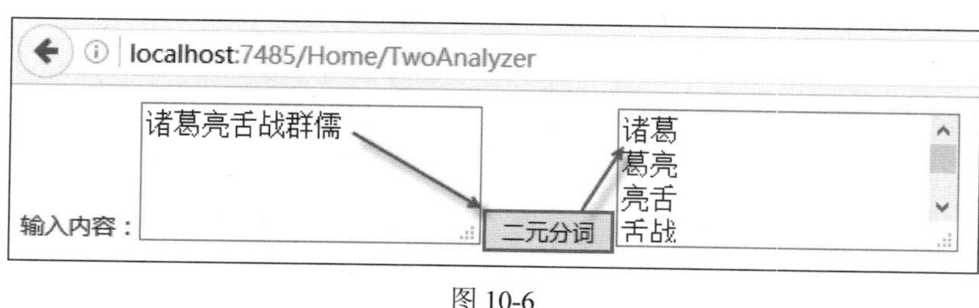

图 10-6

10.2.3 盘古分词算法的使用

官网：http://pangusegment.codeplex.com/releases/view/50811

http://pangusegment.codeplex.com/SourceControl/latest

具体用法请参考云盘中提供的 PanguMannual.pdf 或 Mannual.doc。

使用步骤：

（1）添加对 PanGu.dll（同目录下不要有 Pangu.xml，那个默认的配置文件的选项对于分词结果有很多无用信息）、PanGu.Lucene.Analyzer.dll 的引用。

（2）同二元分词 Demo 一样，只需将其中的 CJKAnalyzer 用 PanGuAnalyzer 代替。

（3）在项目根目录中添加 Dict 词库目录和词库文件。

（4）把 Dict 目录下的文件“复制到输出目录”设定为“如果较新则复制”，如图 10-7 所示。

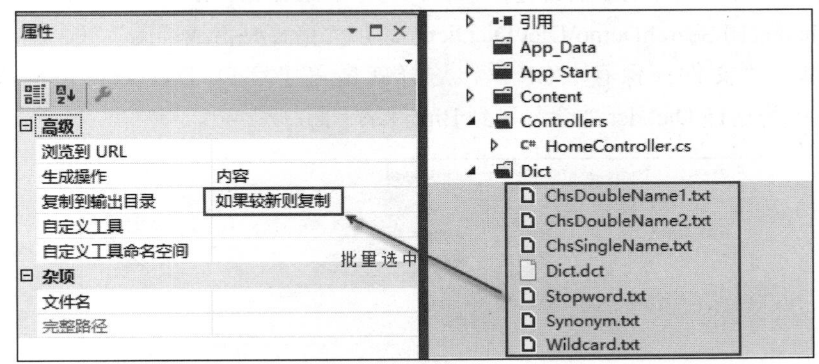

图 10-7

注意：如果此步骤不进行设置，运行时会发现提示需要 dct 文件，如图 10-8 所示。

图 10-8

由于不能把词库写死在 dll 中，因此需要提供单独的词库文件，根据报错放到合适的路径中。

通用技巧：把 Dict 目录下的文件“复制到输出目录”设定为“如果较新则复制”，每次生成的时候都会自动把文件复制到 bin\Debug 下，非常方便。（只有 Web 应用程序有那个选项，网站没有。）永远不要对 bing\debug 下的东西做直接修改，而是要改“源文件”。

词库的编辑：使用 DictManage.exe，对单词编辑的时候要先查找。在工作的项目中要将行业单词添加到词库中，比如荷兰牛栏、德国爱他美等。因为盘古分词的词库中只存在一些常用的单词，对于一些比较特殊的单词，需要手动添加到词库中才能够解析出来。

注：出现 Dict 路径的问题时，若没有找到配置文件则默认是 Dict 目录，设定 Pangu.xml 的复制到输出为“如果较新则复制”即可。或者将词典目录命名为 Dict，不要配置文件。

运行结果如图 10-9 所示。

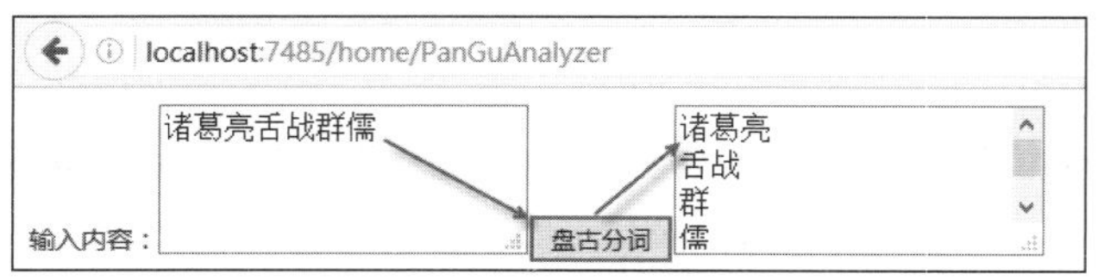

图 10-9

当遇到词库里面没有的一些词时，在分词时将查不到。如图 10-9 所示，“诸葛亮舌战群儒”这几个字被拆分了，如果想要把“诸葛亮舌战群儒”当成一个完整的词，就需要在词库中添加。可以通过 DictManage.exe 工具进行添加。运行 DictManage.exe，选择“文件→打开”，找到 D:\Study\Chapter10\SearchDemo\Dict\Dict.dct，添加“诸葛亮舌战群儒”（见图 10-10），添加完成之后，单击“文件→保存→OK”，这样就覆盖了字典 Dict.dct，然后重新生成项目 SearchDemo，因为这样 Dict.dct 才会复制到 Bin 目录下面。

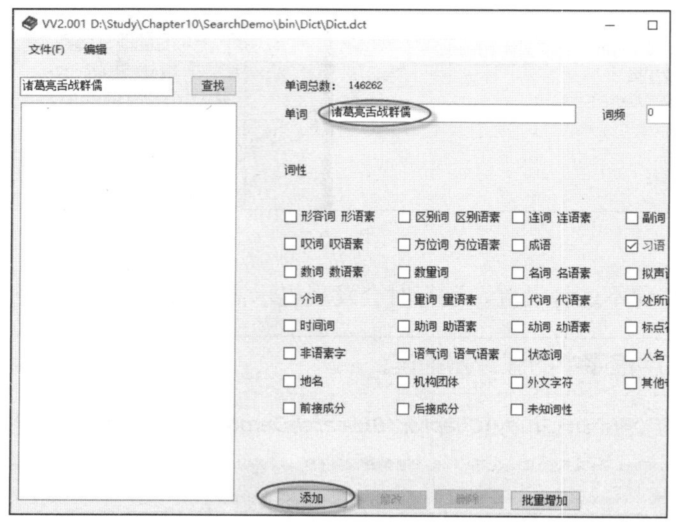

图 10-10

运行项目，在浏览器中输入地址“http://localhost:7485/Home/PanGuAnalyzer”，结果如图 10-11 所示。

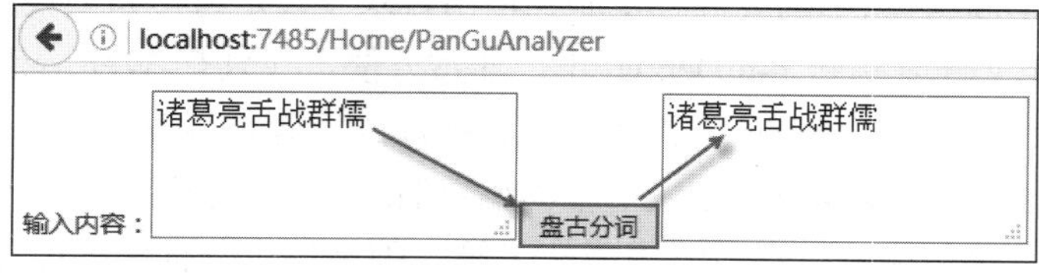

图 10-11

10.3 最简单的搜索引擎代码

发布文章的时候，先把文章数据存入数据库，然后在 Lucene.Net 中创建索引，最后搜索文章的时候直接从索引库进行搜索。在这里，为了方便，暂时没有建数据库，而是直接把“我的文章”目录中的文本文件作为数据源，然后把这些文本文件的数据在 Lucene.Net 中创建索引库。

这里通过一个 Demo 来演示。

（1）先创建一个文件夹 D:\Study\Chapter10\lucenedir，用于存放索引库数据，然后创建一个文件夹 D:\Study\Chapter10\我的文章，用于模拟存放真实的数据库记录。这里从网上找了《三国演义》中的几篇文章，以 txt 的形式存放在此目录中，如图 10-12 所示。

本地磁盘 (D:) › Study › Chapter10 › 我的文章

名称	修改日期	类型	大小
第二十一回 曹操煮酒论英雄 关公赚城斩车胄.txt	2016/7/3 18:27	文本文档	10 KB
第十六回 吕奉先射戟辕门 曹孟德败师淯水.txt	2016/7/3 18:29	文本文档	14 KB
第四十二回 张翼德大闹长坂桥 刘豫州败走汉津口.txt	2016/7/3 18:29	文本文档	8 KB
第四十三回 诸葛亮舌战群儒 鲁子敬力排众议.txt	2016/7/3 18:26	文本文档	11 KB
第四十五回 三江口曹操折兵 群英会蒋干中计.txt	2016/7/3 18:30	文本文档	11 KB
第五十二回 诸葛亮智辞鲁肃 赵子龙计取桂阳.txt	2016/7/3 18:30	文本文档	9 KB
第一回 宴桃园豪杰三结义 斩黄巾英雄首立功.txt	2016/7/3 18:28	文本文档	10 KB

图 10-12

（2）新建 SimpleSearch 控制器。

（3）添加 PanGu.dll、PanGu.Lucene.Analyzer.dll 的引用。

- PanGu.dll：盘古分词的核心组件。
- PanGu.Lucene.Analyzer.dll：盘古分词针对 Lucene.Net 的接口组件。

（4）添加名为 Index 的 Action 和两个方法。这里添加了足够详细的注释，相信大家都看得懂。这些代码不需要记住，事实上也很难记住，我们只需要拿过来极点五笔、看得懂、会修改和重构就可以了。在使用一些第三方开源框架时同样也是如此。要知道，大多数公司是不会给你太多时间去研究某一框架然后再来使用的，通常是知道 XX 框架能干什么、能解决什么样的问题，然后拿过来先用，在用的过程中有必要才去钻研。

```
using Lucene.Net.Analysis.PanGu;
using Lucene.Net.Documents;
using Lucene.Net.Index;
using Lucene.Net.Search;
using Lucene.Net.Store;
using SearchDemo.Models;
using System;
using System.Collections.Generic;
using System.IO;
using System.Web.Mvc;

namespace SearchDemo.Controllers
```

```
{
    public class SimpleSearchController : Controller
    {
        string indexPath = @"D:\Study\Chapter10\lucenedir";
        //注意和磁盘上文件夹的大小写一致，否则会报错。

        //表单中有多个 Submit，单击某个 Submit 均会提交表单，但是只会将用户所单击的表单元素的
        //value 值提交到服务端。
        public ActionResult Index(string txtSearch, string btnSearch,string
        btnCreate)
        {
            List<SearchResult> list = null;
            if (!string.IsNullOrEmpty(btnSearch))//如果单击的是查询按钮
            {
                list=Search(txtSearch);
            }
            else if(!string.IsNullOrEmpty(btnCreate)) //如果单击的是创建索引按钮
            {
                string msg = CreateIndex();
                ViewData["ShowInfo"] = string.IsNullOrEmpty(msg) ? "创建成功" : msg;
            }
            return View(list);
        }
        //查询
        private List<SearchResult> Search(string kw)
        {
            FSDirectory directory = FSDirectory.Open(new DirectoryInfo(indexPath),
            new NoLockFactory());
            IndexReader reader = IndexReader.Open(directory, true);
            IndexSearcher searcher = new IndexSearcher(reader);
            PhraseQuery query = new PhraseQuery();//查询条件
            query.Add(new Term("msg", kw));//where contains("msg",kw)
            //foreach (string word in kw.Split(' '))
            //先用空格，让用户去分词，空格分隔的就是词"诸葛亮"
            //{
            //    query.Add(new Term("msg", word));//contains("msg",word)
            //}
            query.SetSlop(100);
            //两个词的距离大于100（经验值）就不放入搜索结果，因为距离太远相关度就不高了
            TopScoreDocCollector collector = TopScoreDocCollector.create(1000,
            true);//盛放查询结果的容器
            searcher.Search(query, null, collector);
```

```
    //使用 query 这个查询条件进行搜索，搜索结果放入 collector
    //collector.GetTotalHits()总的结果条数
    ScoreDoc[] docs = collector.TopDocs(0,
    collector.GetTotalHits()).scoreDocs;//从查询结果中取出第 m 条到第 n 条的数据

    List<SearchResult> list = new List<SearchResult>();
    string msg = string.Empty;

    for (int i = 0; i < docs.Length; i++)//遍历查询结果
    {
        int docId = docs[i].doc;//拿到文档的 id，因为 Document 可能非常占内存
                                //（思考 DataSet 和 DataReader 的区别）
        //所以查询结果中只有 id，具体内容需要二次查询
        Document doc = searcher.Doc(docId);
        //根据 id 查询内容，放进去的是 Document，查出来的还是 Document
        SearchResult result = new SearchResult();
        result.Id = Convert.ToInt32(doc.Get("id"));
        msg = doc.Get("msg");//只有 Field.Store.YES 的字段才能用 Get 查出来
        result.Msg = msg.Length>200?msg.Substring(0,200)+"......":msg;
        result.Title = doc.Get("title");
        list.Add(result);
    }
    return list;
}
//创建索引库
private string CreateIndex()
{
    string msg = string.Empty;
    try
    {
        FSDirectory directory = FSDirectory.Open(new DirectoryInfo
         (indexPath), new NativeFSLockFactory());
        bool isUpdate = IndexReader.IndexExists(directory);//判断索引库是否存在
        if (isUpdate)
        {
            //如果索引目录被锁定（比如索引过程中程序异常退出），则首先解锁
            //Lucene.Net 在写索引库之前会自动加锁，在 close 的时候会自动解锁
            //不能多线程执行，只能处理意外被永远锁定的情况
            if (IndexWriter.IsLocked(directory))
            {
                IndexWriter.Unlock(directory);//un-否定。强制解锁
            }
```

```
        }
        IndexWriter writer = new IndexWriter(directory, new PanGuAnalyzer(),
        !isUpdate, Lucene.Net.Index.IndexWriter.MaxFieldLength.UNLIMITED);
        var files = System.IO.Directory.GetFiles(@"D:\Study\Chapter10\我的文
        章\", "*.txt");
        int i = 0;
        string title = string.Empty;

        foreach (var file in files)
        {
            i++;
            string txt =System.IO.File.ReadAllText(file,
            System.Text.Encoding.Default);
            title = file.Substring(file.LastIndexOf(@"\") + 1, file.Length -
            file.LastIndexOf(@"\") - 1);
            Document document = new Document();//一条 Document 相当于一条记录
            document.Add(new Field("id", i.ToString(), Field.Store.YES,
            Field.Index.NOT_ANALYZED));
            //每个 Document 可以有自己的属性（字段），所有字段名都是自定义的，
            //值都是 string 类型
            //Field.Store.YES 不仅要对文章进行分词记录，也要保存原文，就不用去数据库里
            //查一次了
            //需要进行全文检索的字段加 Field.Index. ANALYZED
            document.Add(new Field("title", title, Field.Store.YES,
            Field.Index.ANALYZED,
            Lucene.Net.Documents.Field.TermVector.WITH_POSITIONS_OFFSETS));
            document.Add(new Field("msg", txt, Field.Store.YES,
            Field.Index.ANALYZED, Lucene.Net.Documents.
            Field.TermVector.WITH_POSITIONS_OFFSETS));
            //防止重复索引，如果不存在则删除0条
            writer.DeleteDocuments(new Term("id", i.ToString()));
            //防止存在的数据//delete from t where id=i
            writer.AddDocument(document);//把文档写入索引库
        }
        writer.Close();
        directory.Close();//不要忘了 Close，否则索引结果搜不到
    }
    catch (Exception ex)
    {
        msg = ex.Message;
    }
    finally
```

```
            {
            }
            return msg;
        }
    }
}
```

（5）添加 Index 视图，这里引用了一些样式。关于样式，建议直接使用 Firebug 工具从网上抠代码。

```
@{
    ViewBag.Title = "Index";
}
@model List<SearchDemo.Models.SearchResult>
@using SearchDemo.Models;
<style type="text/css">
.search-list-con{width:640px;  background-color:#fff;  overflow:hidden;  margin-top:0px; padding-bottom:15px; padding-top:5px;}
.search-list{width:600px; overflow:hidden; margin:15px 20px 0px 20px;}
.search-list dt{font-family:'Microsoft Yahei'; font-size:16px; line-height:20px; margin-bottom:7px; font-weight:normal;}
.search-list dt a{color:#2981a9;}
.search-list dt a em{ font-style:normal; color:#cc0000;}
</style>
@using (@Html.BeginForm(null, null, FormMethod.Get))
{
    <div>@Html.TextBox("txtSearch")<input type="submit" value="搜索"
    name="btnSearch" /><input type="submit" name="btnCreate" value="创建索引" /></div>
    <div class="search-list-con">
        <dl class="search-list">
            @if (Model != null)
            {
                foreach (SearchResult viewModel in Model)
                {
                <dt><a href="javascript:void(0)">@viewModel.Title</a></dt>
                <dd>@MvcHtmlString.Create(viewModel.Msg)</dd>
                }
            }
        </dl>
    </div>
    <div>@ViewData["ShowInfo"]</div>
}
```

需要注意的是，这里把表单提交的方式设置为了 Get。这样做的目的是直接复制 URL 地址就能够查看到相应的查询结果。事实上，淘宝和百度的搜索也是采用的 Get 方式。

（6）编译运行，结果如图 10-13 所示。

图 10-13

单击“创建索引”按钮，运行结果如图 10-14 所示。

图 10-14

输入“孔明”，然后单击“搜索”按钮，运行结果如图 10-15 所示。

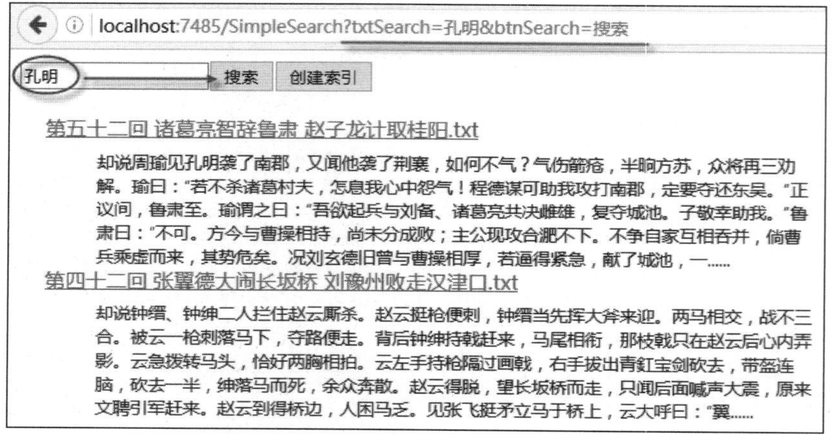

图 10-15

这样，最简单的搜索引擎代码就可以说是基本写好了，但是还有许多地方需要改进，接下来逐步进行优化。

10.4 搜索的第一个版本

在 10.3 节中只是简单地演示如何采用 Lucene.Net 来实现简单的搜索，数据源采用的是文本文件，而现实工作中的数据源一般是关系型数据库。另外，此种实现方式存在一个很大的缺陷，那就是没有考虑并发的情况。当并发出现的时候就会出现死锁，从而导致系统出现异常。

为了解决并发，我们可以考虑使用队列，通过生产者消费者模式来有效地解决并发问题。我们可以把所有产生的需要创建索引库的信息存放到一个队列里面，然后通过新建一个线程不断地从这个队列里面读取需要创建索引库的信息记录，再来进行创建索引库操作。

这里通过 EF 的 Code First 方式来实现 Demo。

（1）引入程序集 EntityFramework（System.Data.Entity）。

在项目 SearchDemo 中右击 Models 文件夹，选择“添加→新建项→数据→ADO.NET 实体数据模型→添加→空模型→完成”，然后删除 Model1.edmx，这样程序集 EntityFramework（System.Data.Entity）就被自动引用到项目中了。

（2）创建模型。在 Models 文件夹下面新建模型类 Article：

```
using System.ComponentModel.DataAnnotations;

namespace SearchDemo.Models
{
    public class Article
    {
        public int ArticleId { get; set; }
        [StringLength(100)]
        [Display(Name = "文章标题")]
        public string Title { get; set; }
        [StringLength(int.MaxValue)]
        [Display(Name = "文章内容")]
        public string Content { get; set; }
        [Display(Name = "创建时间")]
        public DateTime CreateTime { get; set; }
    }
}
```

（3）在配置文件中写连接字符串。“server=”后面的值就是我们用 SQL Server Management Studio 连接数据库时“服务器名称(S)”的值。

```
<add name="SearchDemoContext"connectionString="server= .\MSSQLSERVER2012;
database=SearchDemo;uid=sa;pwd=yujie1127" providerName="System.Data.SqlClient"/>
```

（4）创建上下文类 SearchDemoContext：

```
using System.Data.Entity;

namespace SearchDemo.Models
{
    public class SearchDemoContext : DbContext
    {
```

```
        public SearchDemoContext() : base("name=SearchDemoContext") { }
        public DbSet<Article> Article { get; set; }
    }
}
```

（5）新建控制器 Article，如图 10-16 所示。

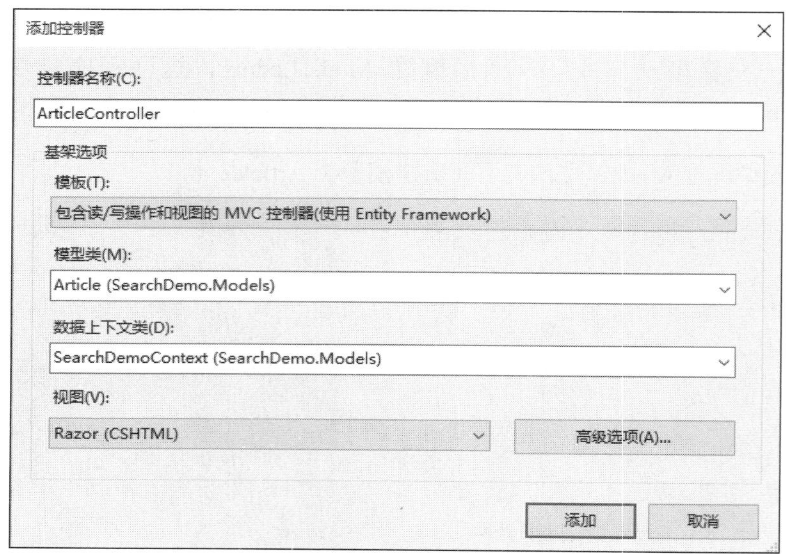

图 10-16

（6）在 Index Action 中添加如下代码：

```
//当操作的表存在时不创建，不存在时再创建
db.Database.CreateIfNotExists();
```

（7）修改 Create 视图。考虑到要添加文章内容，所以需要找一个富文本 UI 组件，这里使用 wysiwyg.js，下载地址为 http://www.htmleaf.com/jQuery/Form/201501221249.html。

在 lib 文件夹中新建一个文件夹 wysiwyg，把 css、js 这两个文件夹复制过来。修改 Create 视图代码：

```
@model SearchDemo.Models.Article

@{
    ViewBag.Title = "Create";
}
    <link rel="stylesheet" href="http://libs.useso.com/js/font-awesome/4.2.0/css/
font-awesome.min.css">
    <link rel="stylesheet" type="text/css" href="/lib/wysiwyg/css/normalize.css" />
    <link rel="stylesheet" type="text/css" href="/lib/wysiwyg/css/default.css">
    <link rel="stylesheet" type="text/css" href="/lib/wysiwyg/css/wysiwyg-editor.css"/>
```

```
<h2>Create</h2>
@using (Html.BeginForm()) {
    @Html.ValidationSummary(true)
    <fieldset>
        <legend>Article</legend>

        <div class="editor-label">
            @Html.LabelFor(model => model.Title)
        </div>
        <div class="editor-field">
            @Html.TextBoxFor(model => model.Title, new{style="width:300px;" })
            @Html.ValidationMessageFor(model => model.Title)
        </div>

        <div class="editor-label">
            @Html.LabelFor(model => model.Content)
        </div>
        <div class="editor-field">
             <textarea id="editor1" name="Content" placeholder="Type your text
             here..."></textarea>
            @Html.ValidationMessageFor(model => model.Content)
        </div>
        <p>
            <input type="submit" value="Create" />
        </p>
    </fieldset>
}
<div>
    @Html.ActionLink("Back to List", "Index")
</div>

@section Scripts {
    @Scripts.Render("~/bundles/jqueryval")
     <script type="text/javascript" src="/lib/wysiwyg/js/wysiwyg.js"></script>
     <script type="text/javascript" src="/lib/wysiwyg/js/wysiwyg-editor.js"></script>
     <script type="text/javascript" src="/lib/wysiwyg/js/demo.js"></script>
}
```

需要注意，由于要添加富文本格式的内容，因此需要在 Article 控制器中的 Create 方法上面添加：

```
[ValidateInput(false)]
```

并在 Web.config 中修改配置节点 httpRuntime，添加属性 requestValidationMode="2.0"：

```
<system.web>
<httpRuntime targetFramework="4.6" requestValidationMode="2.0"/>
```

（8）创建数据库和表

生成项目，然后运行。在浏览器中输入地址“http://localhost:7485/Article/Index”，就会在 SQLserver 中自动创建数据库并生成表，如图 10-17 所示。

图 10-17

往数据库中添加一条文章信息记录的时候，需要把这一条记录写入 Lucene.Net 中，进行创建索引库操作。考虑多个人同时添加文章的并发情况，我们需要把文章记录添加到数据库，然后存放到队列中，系统后台则额外开辟一个线程来监听这个队列，不断地从队列中取出数据并写入 Lucene.Net 中。

（9）添加队列。

这里通过添加一个类 SearchIndexManager 来封装索引库的操作，同时把队列也放置在这个类中统一管理，所以有必要把这个类的对象设置为单例。

```
public sealed class SearchIndexManager
{
       private static readonly SearchIndexManager searchIndexManager = new
       SearchIndexManager();
       private SearchIndexManager()
       {
       }
       public static SearchIndexManager GetInstance()
       {
          return searchIndexManager;
       }
```

```
Queue<IndexContent> queue = new Queue<IndexContent>();
/// <summary>
/// 向队列中添加数据
/// </summary>
/// <param name="Id"></param>
/// <param name="title"></param>
/// <param name="content"></param>
public void AddQueue(string Id, string title,string content, DateTime
createTime)
{
    IndexContent indexContent = new IndexContent();
    indexContent.Id = Id;
    indexContent.Title = title;
    indexContent.Content = content;
    indexContent.LuceneEnum = LuceneEnum.AddType;// 添加
    indexContent.CreateTime = createTime.ToString();
    queue.Enqueue(indexContent);
}
/// <summary>
/// 向队列中添加要删除数据
/// </summary>
/// <param name="Id"></param>
public void DeleteQueue(string Id)
{
    IndexContent indexContent = new IndexContent();
    indexContent.Id = Id;
    indexContent.LuceneEnum = LuceneEnum.DeleType;//删除
    queue.Enqueue(indexContent);
}

/// <summary>
/// 开启线程，扫描队列，从队列中获取数据
/// </summary>
public void StartThread()
{
    Thread myThread = new Thread(WriteIndexContent);
    myThread.IsBackground = true;
    myThread.Start();
}
private void WriteIndexContent()
{
    while (true)
```

```
        {
            if (queue.Count > 0)
            {
                CreateIndexContent();
            }
            else
            {
                Thread.Sleep(5000);//避免造成CPU空转
            }
        }
    }
    private void CreateIndexContent()
    {
        string indexPath = ConfigurationManager.AppSettings["lucenedir"];
        //注意和磁盘上文件夹的大小写一致，否则会报错。将创建的分词内容放在该目录下。
        FSDirectory directory = FSDirectory.Open(new DirectoryInfo(indexPath),
        new NativeFSLockFactory());//指定索引文件(打开索引目录)，FS指的是FileSystem
        bool isUpdate = IndexReader.IndexExists(directory);
        //IndexReader:对索引进行读取的类
        //该语句的作用：判断索引库文件夹是否存在以及索引特征文件是否存在
        if (isUpdate)
        {
            //同时只能有一段代码对索引库进行写操作。当使用IndexWriter打开directory时会
            //自动对索引库文件上锁。
            //如果索引目录被锁定（比如索引过程中程序异常退出），就首先解锁
            if (IndexWriter.IsLocked(directory))
            {
                IndexWriter.Unlock(directory);
            }
        }
        IndexWriter writer = new IndexWriter(directory, new PanGuAnalyzer(),
        !isUpdate, Lucene.Net.Index.IndexWriter.MaxFieldLength.UNLIMITED);
        //向索引库中写索引，并在这里加锁。

        while (queue.Count > 0)
        {
            IndexContent indexContent = queue.Dequeue();//将队列中的数据出队
            writer.DeleteDocuments(new Term("Id", indexContent.Id.ToString()));
            if (indexContent.LuceneEnum == LuceneEnum.DeleType)
            {
                continue;
            }
```

```
            Document document = new Document();//表示一篇文档
            //Field.Store.YES:表示是否存储原值。只有是 Field.Store.YES 时才能在后面用
            doc.Get("Id")取出值来。Field.Index. NOT_ANALYZED:不进行分词保存
            document.Add(new Field("Id", indexContent.Id, Field.Store.YES,
            Field.Index.NOT_ANALYZED));

            //Field.Index. ANALYZED:进行分词保存，也就是要为全文字段设置分词保存
            //（因为要进行模糊查询）
            //Lucene.Net.Documents.Field.TermVector.WITH_POSITIONS_OFFSETS:
            //不仅保存分词还保存分词的距离
            document.Add(new Field("Title", indexContent.Title, Field.Store.YES,
            Field.Index.ANALYZED, Lucene.Net.Documents.Field.TermVector.
            WITH_POSITIONS_OFFSETS));
            document.Add(new Field("Content", indexContent.Content,
            Field.Store.YES, Field.Index.ANALYZED, Lucene.Net.Documents.
            Field.TermVector.WITH_POSITIONS_OFFSETS));
            document.Add(new Field("CreateTime", indexContent.CreateTime,
            Field.Store.YES, Field.Index.NOT_ANALYZED));

            writer.AddDocument(document);
        }

        writer.Close();//会自动解锁
        directory.Close();//不要忘了 Close，否则索引结果搜不到
    }
}
```

（10）添加高亮显示。把查询关键字高亮显示，其实就是给查询到的内容中的关键字添加样式而已。

```
// 创建 HTMLFormatter,参数为高亮单词的前后缀
public static string CreateHightLight(string keywords, string Content)
{
        PanGu.HighLight.SimpleHTMLFormatter simpleHTMLFormatter =
         new PanGu.HighLight.SimpleHTMLFormatter("<font color=\"red\">",
         "</font>");
        //创建 Highlighter ，输入 HTMLFormatter 和盘古分词对象 Semgent
        PanGu.HighLight.Highlighter highlighter =
        new PanGu.HighLight.Highlighter(simpleHTMLFormatter,
        new Segment());
        //设置每个摘要段的字符数
        highlighter.FragmentSize = 150;
```

```
        //获取最匹配的摘要段
        return highlighter.GetBestFragment(keywords, Content);
}
```

（11）修改 Article 控制器中的 Create 方法，在文章保存之后再把记录存入队列中。

```
[HttpPost]
[ValidateInput(false)]
public ActionResult Create(Article article)
{
        if (ModelState.IsValid)
        {
            article.CreateTime = DateTime.Now;
            db.Article.Add(article);
            db.SaveChanges();

            SearchIndexManager.GetInstance().AddQueue(article.ArticleId.ToString
            (), article.Title, article.Content, article.CreateTime);
           return RedirectToAction("Index","Article");
        }

        return View(article);
}
```

（12）启动监听队列的线程。为了方便，这里直接把代码写到了 Global.asax 中，其实我们完全可以把这个监听队列的操作寄宿到 windows 服务中去。队列则不采用 C#自带的队列，而是使用分布式队列 Redis 代替。

```
protected void Application_Start()
{
        SearchIndexManager.GetInstance().StartThread();
        //开启线程扫描队列将数据取出来，写到 Lucene.Net 中
```

（13）添加控制器 NewSearch 来进行查询，代码和 10.3 节中的 SimpleSearch 类似，这里不再显示代码，大家可以查看源码。

（14）演示。生成项目后运行，在地址栏中输入“http://localhost:7485/Article/Create”，如图 10-18 所示。

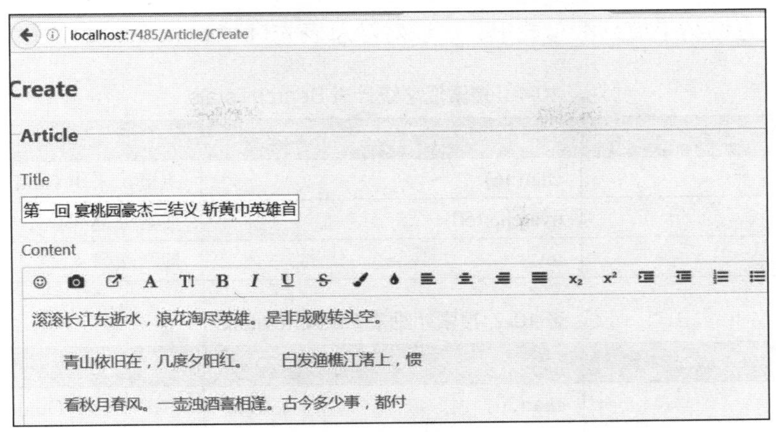

图 10-18

（15）添加内容。添加完成之后，在浏览器中输入“http://localhost:7485/NewSearch”，进行搜索展示，如图 10-19 所示。

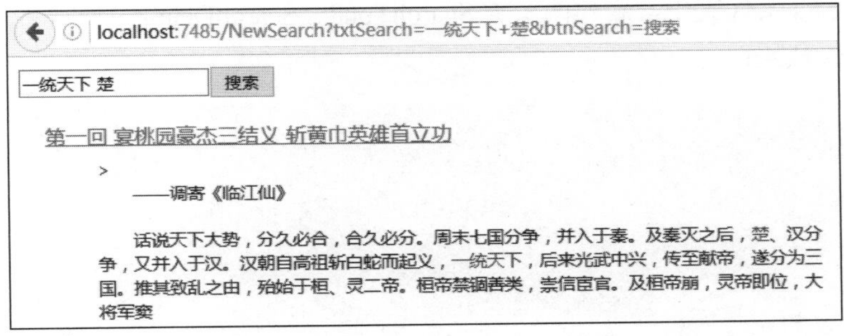

图 10-19

10.5 搜索的优化版

在 10.4 节已经完成了搜索的第一个版本，但是还有许多地方需要优化。比如说，要统计关键词搜索的频率高的词，即热词，以及像百度搜索那样在输入关键字后会自动把搜索相关的热词自动以下拉列表的形式呈现出来，还有搜索结果分页、查看文章明细等。

10.5.1 热词统计

1. 思路

（1）搜索关键字的统计实时性是不高的，也就是说我们可以定期地去进行统计。

（2）客户的每一次搜索记录都需要存起来，这样才能够统计得到。

通过第 1 点，我们脑海中会呈现一张汇总统计表（见表 10-1）。通过第 2 点，我们会想到使用一张搜索记录明细表（见表 10-2）。那方案就很明了了，只需要定期地从明细表中 Group by 查询，然后把查询结果放到汇总表中。可以对汇总表先进行 truncate，再进行 insert 操作。

表 10-1　搜索汇总统计表 SearchTotals

字段名称	字段类型	说明
Id	char(36)	主键，采用 Guid 方式存储
KeyWords	nvarchar(50)	搜索关键字
SearchCounts	int	搜索次数

表 10-2 搜索明细表 SearchDetails

字段名称	字段类型	说明
Id	char(36)	主键，采用 Guid 方式存储
KeyWords	nvarchar(50)	搜索关键字
SearchDateTime	datetime	搜索时间

2. 操作步骤

（1）在 Models 文件夹中，新建两个类 SearchTotal、SearchDetail。

SearchTotal.cs 代码：

```
using System;
using System.ComponentModel.DataAnnotations;

namespace SearchDemo.Models
{
    public class SearchTotal
    {
        public Guid Id { get; set; }
        [StringLength(50)]
        public string KeyWords { get; set; }
        public int SearchCounts { get; set; }
    }
}
```

SearchDetail.cs 代码：

```
using System;
using System.ComponentModel.DataAnnotations;

namespace SearchDemo.Models
{
    public class SearchDetail
    {
        public Guid Id { get; set; }
        [StringLength(50)]
```

```
        public string KeyWords { get; set; }
        public Nullable<DateTime> SearchDateTime { get; set; }
    }
}
```

（2）修改 SearchDemoContext 类，新增属性 SearchTotal、SearchDetail：

```
using System.Data.Entity;

namespace SearchDemo.Models
{
    public class SearchDemoContext : DbContext
    {
        public SearchDemoContext() : base("name=SearchDemoContext") { }
        public DbSet<Article> Article { get; set; }
        //下面两个属性是新增加的
        public DbSet<SearchTotal> SearchTotal { get; set; }
        public DbSet<SearchDetail> SearchDetail { get; set; }
    }
}
```

（3）更新数据库。由于修改了 EF 上下文，新增了两个模型类，因此需要进行迁移，更新数据库操作。

重新编译应用程序，然后选择“工具→库程序包管理器→程序包管理控制台”。打开控制台，输入“enable-migrations –force”，然后按回车键，在项目资源管理器中出现 Migrations 文件夹。打开 Configuration.cs 文件，将 AutomaticMigrationsEnabled 值改为 true，然后在控制台中输入“update-database”运行。操作完成之后，会在数据库 SearchDemo 中新建两张表，即 SearchTotals、SearchDetails，而原来的 Articles 表保持不变，如图 10-20 所示。

图 10-20

（4）保存搜索记录。用户在每次搜索的时候，都要把搜索记录存入 SearchDetails 表中。为了方便，这里在用户每次单击“搜索”之后就立即往 SearchDetails 表中插入记录了，也就是同步操作。实际上，为了提升搜索的效率，我们可以采用异步操作，即把搜索记录的数据先写入 redis 队列中，后台再开辟一个线程来监听 redis 队列，然后把队列中的搜索记录数据写入到数据表中。因为在每次单击“搜索”的时候，把记录写入 redis 和直接把记录写入关系型数据库的效率是相差很大的。

```
//先将搜索的词插入到明细表
        SearchDetail _SearchDetail = new SearchDetail { Id = Guid.NewGuid(),
        KeyWords = kw, SearchDateTime = DateTime.Now };
        db.SearchDetail.Add(_SearchDetail);
        int r = db.SaveChanges();
```

（5）定时更新 SearchTotals 表记录。这种定时任务操作可以采用 Quartz.Net 框架。为了方便，这里把 Quartz.Net 的 Job 寄宿在控制台程序中，而实际工作中，则更倾向于将其寄宿在 Windows 服务中。如果有必要，可以把这个定时更新 SearchTotals 表记录的程序部署到独立服务器，减轻 Web 服务器的压力。

3. Quartz.Net 的 Job 实现步骤

新建控制台程序 QuartzNet，添加 Quartz.dll 和 Common.Logging.dll 的程序集引用。这里采用 Database First 的方式添加 ADO.NET 实体数据模型，把表 SearchTotals、SearchDetails 添加进来。

（1）添加 KeyWordsTotalService.cs 类，里面封装两个方法。清空 SearchTotals 表，然后把 SearchDetails 表的分组查询结果插入 SearchTotals 表，这里只统计近 30 天内的搜索明细。

```
namespace QuartzNet
{
    public class KeyWordsTotalService
    {
        private SearchDemoEntities db = new SearchDemoEntities();
        /// <summary>
        /// 将统计的明细表的数据插入
        /// </summary>
        /// <returns></returns>
        public bool InsertKeyWordsRank()
        {
            string sql = "insert into SearchTotals(Id,KeyWords,SearchCounts) select
             newid(),KeyWords,count(*)  from SearchDetails where
             DateDiff(day,SearchDetails.SearchDateTime,getdate())<=30 group by
             SearchDetails.KeyWords";
            return this.db.Database.ExecuteSqlCommand(sql) > 0;
        }
        /// <summary>
        /// 删除汇总中的数据
        /// </summary>
        /// <returns></returns>
        public bool DeleteAllKeyWordsRank()
```

```
        {
            string sql = "truncate table SearchTotals";
            return this.db.Database.ExecuteSqlCommand(sql) > 0;
        }
    }
}
```

（2）添加 TotalJob.cs 类，继承 Ijob 接口，并实现 Execute 方法。

```
namespace QuartzNet
{
    public class TotalJob : IJob
    {
        /// <summary>
        /// 将明细表中的数据插入到汇总表中
        /// </summary>
        /// <param name="context"></param>
        public void Execute(JobExecutionContext context)
        {
            KeyWordsTotalService bll = new KeyWordsTotalService();
            bll.DeleteAllKeyWordsRank();
            bll.InsertKeyWordsRank();
        }
    }
}
```

（3）修改 Program.cs 类:

```
using Quartz;
using Quartz.Impl;
using System;

namespace QuartzNet
{
    class Program
    {
        static void Main(string[] args)
        {
            IScheduler sched;
            ISchedulerFactory sf = new StdSchedulerFactory();
            sched = sf.GetScheduler();
            JobDetail job = new JobDetail("job1", "group1",
            typeof(TotalJob));//IndexJob 为实现了 IJob 接口的类
```

```
            DateTime ts = TriggerUtils.GetNextGivenSecondDate(null, 5);
            //5秒后开始第一次运行
            TimeSpan interval = TimeSpan.FromSeconds(50);//每隔50秒执行一次
            Trigger trigger = new SimpleTrigger("trigger1", "group1", "job1",
            "group1", ts, null,SimpleTrigger.RepeatIndefinitely,interval);
            //每若干时间运行一次，时间间隔可以放到配置文件中指定

            sched.AddJob(job, true);
            sched.ScheduleJob(trigger);
            sched.Start();
            Console.ReadKey();
        }
    }
}
```

这里直接把 Job 和计划都直接写到了代码中，理由还是因为方便。在实际工作中，我们应当把这些信息尽量写到配置文件中，这样后面改动起来会更方便，不需要修改代码，只需要修改配置文件即可。

为了尽快看到效果，这里设置每隔 50 秒就进行一次统计操作，而在实际应用中，我们的时间间隔可能是几小时甚至一天，因为像这样的大数据统计，对实时性的要求不高，我们可以尽量减少对数据库的 IO 读写次数。

保持运行控制台程序 QuartzNet，然后进行搜索操作，这样后台就定期地生成了搜索统计记录。

10.5.2 热门搜索

1. 展示热门搜索

展示热门搜索其实就是从表 SearchTotals 中按照搜索次数进行降序排列，然后取出数条记录。

在 LastSearch 控制器中的 Index 方法中添加如下代码：

```
var keyWords = db.SearchTotal.OrderByDescending(a => a.SearchCounts).Select(x
    => x.KeyWords).Skip(0).Take(6).ToList();
        ViewBag.KeyWords = keyWords;
```

在 View 视图中添加如下代码：

```
<div id="divKeyWords"><span>热门搜索：</span>
@if (ViewBag.KeyWords != null) {
        foreach (string v in ViewBag.KeyWords) {
         <a href="#">@v</a>
```

```
            }
}</div>
```

想要实现的效果如图 10-21 所示，即单击一个热词时会自动加载到文本框，然后单击“搜索”按钮。

图 10-21

在 View 中添加代码：

```
<script type="text/javascript">
    $(function () {
        $("#divKeyWords a").click(function () {
            $("#txtSearch").val($(this).html());
            $("#btnSearch").click();
        });
});
</script>
```

2. 搜索下拉框

这里引入一个第三方 js 框架 Autocomplete。它能在文本框中输入文字的时候自动从后台抓取数据下拉列表。

云盘中提供了 Autocomplete.rar，将其解压，然后复制到 SearchDemo 项目中的 lib 目录下。

在 SearchDemo 项目中的 KeyWordsTotalService.cs 类中添加如下方法：

```
using System;
using System.Collections.Generic;
using System.Data.SqlClient;
using System.Linq;

namespace SearchDemo.Common
{
    public class KeyWordsTotalService
    {
        private SearchDemoContext db = new SearchDemoContext();

        public List<string> GetSearchMsg(string term)
        {
            try
            {
```

```
            //存在 SQL 注入的安全隐患
            //string sql = "select KeyWords from SearchTotals where KeyWords
            //like '"+term.Trim()+"%'";
            //return db.Database.SqlQuery<string>(sql).ToList();
            string sql = "select KeyWords from SearchTotals where KeyWords like
            @term";
            return db.Database.SqlQuery<string>(sql, new SqlParameter("@term",
            term+"%")).ToList();
        }
        catch (Exception ex)
        {
            throw new Exception(ex.Message);
        }
    }
  }
}
```

然后在 LastSearch 控制器中添加方法：

```
/// <summary>
/// 获取客户列表 模糊查询
/// </summary>
/// <param name="term"></param>
/// <returns></returns>
public string GetKeyWordsList(string term)
{
        if (string.IsNullOrWhiteSpace(term))
            return null;

        var list = new KeyWordsTotalService().GetSearchMsg(term);
        //序列化对象
        //尽量不要用 JavaScriptSerializer，因为性能差，完全可以用 Newtonsoft.Json 来代替
        //System.Web.Script.Serialization.JavaScriptSerializer js = new
        //System.Web.Script.Serialization.JavaScriptSerializer();
        //return js.Serialize(list.ToArray());
        return JsonConvert.SerializeObject(list.ToArray());
}
```

View 代码如下：

```
<link href="~/lib/Autocomplete/css/ui-lightness/jquery-ui-1.8.17.custom.css"
rel="stylesheet" />
<script src="~/lib/Autocomplete/js/jquery-ui-1.8.17.custom.min.js"></script>
```

```
<script type="text/javascript">
    $(function () {
        $("#divKeyWords a").click(function () {
            $("#txtSearch").val($(this).html());
            $("#btnSearch").click();
        });
        getKeyWordsList("txtSearch");
    });
    //自动加载搜索列表
    function getKeyWordsList(txt) {
        if (txt == undefined || txt == "")
            return;
        $("#" + txt).autocomplete({
            source: "/LastSearch/GetKeyWordsList",
            minLength: 1
        });
    }
</script>
```

10.5.3　标题和内容都支持搜索并高亮展示

在 10.4 节中只支持在内容中对关键词进行搜索，而实际上，我们可能既要支持在标题中搜索，也要支持在内容中搜索。

这里引入了 BooleanQuery，我们的查询条件也添加了一个 titleQuery。

在搜索方法中，修改代码：

```
PhraseQuery query = new PhraseQuery();//查询条件
PhraseQuery titleQuery = new PhraseQuery();//标题查询条件
List<string> lstkw = LuceneHelper.PanGuSplitWord(kw);
//对用户输入的搜索条件进行拆分。

foreach (string word in lstkw){
            query.Add(new Term("Content", word));//contains("Content",word)
            titleQuery.Add(new Term("Title", word));
}
query.SetSlop(100);
//两个词的距离大于100（经验值）就不放入搜索结果，因为距离太远相关度就不高了

BooleanQuery bq = new BooleanQuery();
//Occur.Should 表示 Or , Must 表示 and 运算
bq.Add(query, BooleanClause.Occur.SHOULD);
```

```
bq.Add(titleQuery, BooleanClause.Occur.SHOULD);

TopScoreDocCollector collector = TopScoreDocCollector.create(1000,true);
//盛放查询结果的容器
searcher.Search(bq, null, collector);
//使用 query 这个查询条件进行搜索，搜索结果放入 collector
```

假设输入“诸葛亮”，只要标题或者内容中存在“诸葛亮”这 3 个字的记录就都会被查找出来。

10.5.4 与查询、或查询、分页

前面我们在搜索的时候其实采用的都是与查询，也就是说，输入“诸葛亮周瑜”，则只会查找出既存在诸葛亮又存在周瑜的记录。但是有时候我们只想查询存在诸葛亮或者存在周瑜的记录，也就是所谓的或查询。

在界面中添加一个“或查询”复选框，让用户决定采用何种方式进行查询。

至于分页，这里采用 MvcPager。关于 MvcPager 的使用方法可参见 4.6.3 小节。

View 的完整代码预览如下：

```
@{
    ViewBag.Title = "Index";
}
@model PagedList<SearchDemo.Models.SearchResult>
@using Webdiyer.WebControls.Mvc;
@using SearchDemo.Models;
<style type="text/css">
.search-text2{   display:block;   width:528px;   height:26px;   line-height:26px;
float:left;   margin:3px   5px;   border:1px   solid   gray;   outline:none;   font-
family:'Microsoft Yahei'; font-size:14px;}
.search-btn2{width:102px;    height:32px;    line-height:32px;    cursor:pointer;
border:0px;    background-color:#d6000f;font-family:'Microsoft    Yahei';    font-
size:16px;color:#f3f3f3;}
.search-list-con{width:640px;  background-color:#fff;  overflow:hidden;  margin-
top:0px; padding-bottom:15px; padding-top:5px;}
.search-list{width:600px; overflow:hidden; margin:15px 20px 0px 20px;}
.search-list dt{font-family:'Microsoft Yahei'; font-size:16px; line-height:20px;
margin-bottom:7px; font-weight:normal;}
.search-list dt a{color:#2981a9;}
.search-list dt a em{ font-style:normal; color:#cc0000;}
#divKeyWords {text-align:left;width:520px;padding-left:4px;}
#divKeyWords a {text-decoration:none;}
```

```
#divKeyWords a:hover {color:red;}
</style>
<link      href="~/lib/Autocomplete/css/ui-lightness/jquery-ui-1.8.17.custom.css"
rel="stylesheet" />
@using(@Html.BeginForm(null, null, FormMethod.Get))
{
   @Html.Hidden("hidfIsOr")
   <div>@Html.TextBox("txtSearch",  null,  new  {  @class="search-text2"})<input
type="submit"  value=" 搜 索 "  name="btnSearch"  id="btnSearch"   class="search-
btn2"/><input type="checkbox" id="isOr" value="false"/>或查询</div>
   <div id="divKeyWords"><span>热门搜索：</span>@if (ViewBag.KeyWords != null) {
         foreach (string v in ViewBag.KeyWords) {
          <a href="#">@v</a>
         }
      }</div>
   <div class="search-list-con">
      <dl class="search-list">
         @if (Model != null&& Model.Count > 0)
         {
            foreach (var viewModel in Model)
            {
            <dt><a href="@viewModel.Url" target="_blank">@MvcHtmlString.Create
(viewModel.Title)</a><span style="margin-left:50px;">@viewModel.CreateTime</span>
</dt>
            <dd>@MvcHtmlString.Create(viewModel.Msg)</dd>
            }
         }
          @Html.Pager(Model, new PagerOptions
 {
    PageIndexParameterName = "id",
    ShowPageIndexBox = true,
    FirstPageText = "首页",
    PrevPageText = "上一页",
    NextPageText = "下一页",
    LastPageText = "末页",
    PageIndexBoxType = PageIndexBoxType.TextBox,
    PageIndexBoxWrapperFormatString = "请输入页数{0}",
    GoButtonText = "转到"
 })
    <br />
    >>分页共有@(Model==null? 0: Model.TotalItemCount) 篇文章 @(Model==null?0:Model.
CurrentPageIndex)/@(Model==null?0:Model.TotalPageCount)
```

```
        </dl>
    </div>
    <div>@ViewData["ShowInfo"]</div>
}
<script type="text/javascript">
    $(function () {
        $("#divKeyWords a").click(function () {
            $("#txtSearch").val($(this).html());
            $("#btnSearch").click();
        });
        getKeyWordsList("txtSearch");
        $("#isOr").click(function () {
            if ($(this).attr("checked") == "checked") {
                $("#hidfIsOr").val(true);
            }
            else {
                $("#hidfIsOr").val(false);
            }
        });
        if ($("#hidfIsOr").val() == "true") {
            $("input[type='checkbox']").prop("checked", true);
        }
    });
    //自动加载搜索列表
    function getKeyWordsList(txt) {
        if (txt == undefined || txt == "")
            return;
        $("#" + txt).autocomplete({
            source: "/LastSearch/GetKeyWordsList",
            minLength: 1
        });
    }
</script>
<script src="~/lib/Autocomplete/js/jquery-ui-1.8.17.custom.min.js"></script>
```

LastSearch 控制器中的方法如下：

```
public class LastSearchController : Controller
{
        //
        // GET: /LastSearch/

        string indexPath = System.Configuration.ConfigurationManager.
```

```
AppSettings["lucenedir"];
private SearchDemoContext db = new SearchDemoContext();

    public ActionResult Index(string txtSearch, bool? hidfIsOr, int id=1)
{
    PagedList<SearchResult> list = null;
    if (!string.IsNullOrEmpty(txtSearch))//如果单击的是查询按钮
    {
        //list = Search(txtSearch);
        list = (hidfIsOr == null || hidfIsOr.Value == false) ? OrSearch
         (txtSearch, id) : AndSearch(txtSearch, id);
    }
    var keyWords = db.SearchTotal.OrderByDescending(a => a.SearchCounts)
    .Select(x => x.KeyWords).Skip(0).Take(6).ToList();
    ViewBag.KeyWords = keyWords;
    return View(list);
}
//与查询
PagedList<SearchResult> AndSearch(String kw, int pageNo, int pageLen = 4)
{
    FSDirectory directory = FSDirectory.Open(new DirectoryInfo(indexPath),
     new NoLockFactory());
    IndexReader reader = IndexReader.Open(directory, true);
    IndexSearcher searcher = new IndexSearcher(reader);
    PhraseQuery query = new PhraseQuery();//查询条件
    PhraseQuery titleQuery = new PhraseQuery();//标题查询条件
    List<string> lstkw = LuceneHelper.PanGuSplitWord(kw);
    //对用户输入的搜索条件进行拆分

    foreach (string word in lstkw)
    {
        query.Add(new Term("Content", word));//contains("Content",word)
        titleQuery.Add(new Term("Title", word));
    }
    query.SetSlop(100);
     //两个词的距离大于100（经验值）就不放入搜索结果，因为距离太远相关度就不高了

    BooleanQuery bq = new BooleanQuery();
    //Occur.Should 表示 Or , Must 表示 and 运算
    bq.Add(query, BooleanClause.Occur.SHOULD);
    bq.Add(titleQuery, BooleanClause.Occur.SHOULD);

    TopScoreDocCollector collector=TopScoreDocCollector.create(1000,true);
    //盛放查询结果的容器
    searcher.Search(bq, null, collector);
    //使用 query 这个查询条件进行搜索，搜索结果放入 collector
```

```
    int recCount=collector.GetTotalHits();//总的结果条数
    ScoreDoc[] docs = collector.TopDocs((pageNo - 1) * pageLen,
    pageNo*pageLen).scoreDocs;//从查询结果中取出第m条到第n条的数据

    List<SearchResult> list = new List<SearchResult>();
    string msg = string.Empty;
    string title = string.Empty;

    for (int i = 0; i < docs.Length; i++)//遍历查询结果
    {
        int docId = docs[i].doc;
        //拿到文档的id，因为Document可能非常占内存（思考DataSet和DataReader的区别）
        //所以查询结果中只有id，具体内容需要二次查询
        Document doc = searcher.Doc(docId);
        //根据id查询内容。放进去的是Document，查出来的还是Document
        SearchResult result = new SearchResult();
        result.Id = Convert.ToInt32(doc.Get("Id"));
        msg = doc.Get("Content");//只有 Field.Store.YES的字段才能用Get查出来
        result.Msg = LuceneHelper.CreateHightLight(kw, msg);
         //将搜索的关键字高亮显示
        title = doc.Get("Title");
        foreach (string word in lstkw)
        {
            title=title.Replace(word,"<span style='color:red;'>
             "+word+"</span>");
        }
        //result.Title=LuceneHelper.CreateHightLight(kw, title);
        result.Title = title;
        result.CreateTime = Convert.ToDateTime(doc.Get("CreateTime"));
        result.Url = "/Article/Details?Id=" + result.Id + "&kw=" + kw;
        list.Add(result);
    }
    //先将搜索的词插入明细表
    SearchDetail _SearchDetail = new SearchDetail { Id = Guid.NewGuid(),
     KeyWords = kw, SearchDateTime = DateTime.Now };
    db.SearchDetail.Add(_SearchDetail);
    int r = db.SaveChanges();

    PagedList<SearchResult> lst = new PagedList<SearchResult>(list, pageNo,
     pageLen, recCount);
    lst.TotalItemCount = recCount;
    lst.CurrentPageIndex = pageNo;

    return lst;
}
//或查询
PagedList<SearchResult> OrSearch(String kw, int pageNo, int pageLen = 4)
```

```
{
    FSDirectory directory = FSDirectory.Open(new DirectoryInfo(indexPath),
    new NoLockFactory());
    IndexReader reader = IndexReader.Open(directory, true);
    IndexSearcher searcher = new IndexSearcher(reader);
    List<PhraseQuery> lstQuery = new List<PhraseQuery>();
    List<string> lstkw = LuceneHelper.PanGuSplitWord(kw);
    //对用户输入的搜索条件进行拆分。

    foreach (string word in lstkw)
    {
        PhraseQuery query = new PhraseQuery();//查询条件
        query.SetSlop(100);
        //两个词的距离大于100（经验值）就不放入搜索结果，因为距离太远相关度就不高了
        query.Add(new Term("Content", word));//contains("Content",word)

        PhraseQuery titleQuery = new PhraseQuery();//查询条件
        titleQuery.Add(new Term("Title", word));

        lstQuery.Add(query);
        lstQuery.Add(titleQuery);
    }

    BooleanQuery bq = new BooleanQuery();
    foreach (var v in lstQuery)
    {
        //Occur.Should 表示 Or , Must 表示 and 运算
        bq.Add(v, BooleanClause.Occur.SHOULD);
    }
    TopScoreDocCollector collector=TopScoreDocCollector.create(1000,true);
    //盛放查询结果的容器
    searcher.Search(bq, null, collector);
    //使用 query 这个查询条件进行搜索，搜索结果放入 collector

    int recCount = collector.GetTotalHits();//总的结果条数
    ScoreDoc[] docs = collector.TopDocs((pageNo - 1) * pageLen, pageNo *
    pageLen).scoreDocs;//从查询结果中取出第 m 条到第 n 条的数据

    List<SearchResult> list = new List<SearchResult>();
    string msg = string.Empty;
    string title = string.Empty;

    for (int i = 0; i < docs.Length; i++)//遍历查询结果
    {
        int docId = docs[i].doc;//拿到文档的 id，因为 Document 可能非常占内存
        //（思考 DataSet 和 DataReader 的区别）
        //所以查询结果中只有 id，具体内容需要二次查询
```

```
            Document doc = searcher.Doc(docId);
            //根据 id 查询内容。放进去的是 Document，查出来的还是 Document
            SearchResult result = new SearchResult();
            result.Id = Convert.ToInt32(doc.Get("Id"));
            msg = doc.Get("Content");//只有 Field.Store.YES 的字段才能用 Get 查出来
            result.Msg = LuceneHelper.CreateHightLight(kw, msg);
             //将搜索的关键字高亮显示
            title = doc.Get("Title");
            foreach (string word in lstkw)
            {
                title = title.Replace(word, "<span style='color:red;'>"
                 + word + "</span>");
            }
            //result.Title=LuceneHelper.CreateHightLight(kw, title);
            result.Title = title;
            result.CreateTime = Convert.ToDateTime(doc.Get("CreateTime"));
            result.Url = "/Article/Details?Id=" + result.Id + "&kw=" + kw;
            list.Add(result);
        }
        //先将搜索的词插入到明细表
        SearchDetail _SearchDetail = new SearchDetail { Id = Guid.NewGuid(),
         KeyWords = kw, SearchDateTime = DateTime.Now };
        db.SearchDetail.Add(_SearchDetail);
        int r = db.SaveChanges();

        PagedList<SearchResult> lst = new PagedList<SearchResult>(list, pageNo,
        pageLen, recCount);
        lst.TotalItemCount = recCount;
        lst.CurrentPageIndex = pageNo;

        return lst;
}

/// <summary>
/// 获取客户列表 模糊查询
/// </summary>
/// <param name="term"></param>
/// <returns></returns>
public string GetKeyWordsList(string term)
{
        if (string.IsNullOrWhiteSpace(term))
            return null;

        var list = new KeyWordsTotalService().GetSearchMsg(term);
        //序列化对象
        //尽量不要用 JavaScriptSerializer，因为性能差，完全可以用 Newtonsoft.Json 来代替
        //System.Web.Script.Serialization.JavaScriptSerializer js = new System.
```

```
        Web.Script.Serialization.JavaScriptSerializer();
        //return js.Serialize(list.ToArray());
        return JsonConvert.SerializeObject(list.ToArray());
}
```

最终演示效果预览如图 10-22~图 10-24 所示。

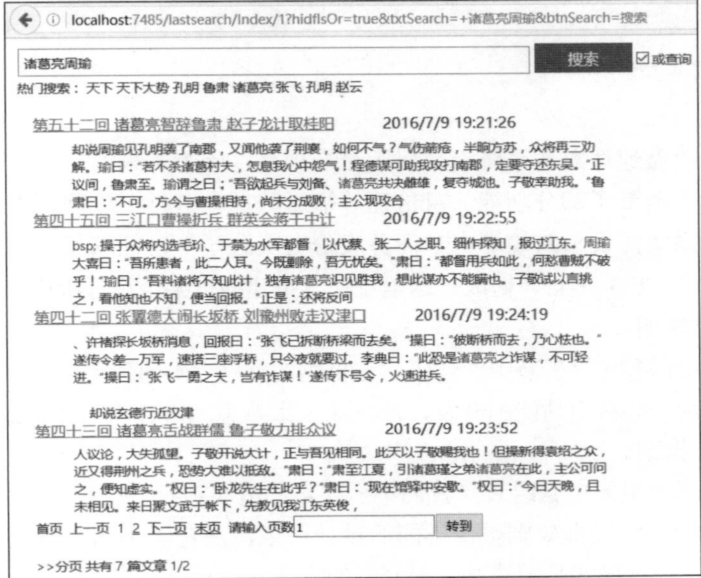

图 10-22

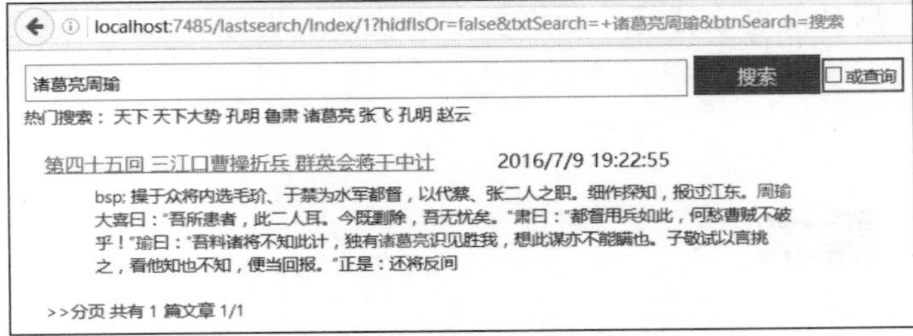

图 10-23

图 10-24

第 11 章 财务对账系统

本项目原名为“魔速达财务系统”，属于流产项目，因为并未商用，所以就在这里直接当示例教材了，本章中删减了部分功能，身份证等敏感数据也进行了处理，仅供大家学习使用，如果要商用，请联系“深圳市跨境翼电子商务股份有限公司”，征求其同意方可。项目后台框架的搭建者是该公司 CEO 李君、何成，这里仅在其搭建的框架上进行版本升级和开发，并自行搭建 UI 框架，特此注明。

项目后台框架可以参考 7.3 节中的代码，这里用的就是 7.3 的后台框架。本章侧重讲解如何搭建前台 UI 框架及一些常用功能的开发，旨在让大家熟悉 MVC 和 NHibernate 结合开发的方式，并自行组装 UI 框架。

软件开发工作真的很像电脑组装，我们不需要自己造 CPU、主板、硬盘等，只需要将各个组件进行组装。不同的开发框架则更像不同的设计图纸，决定了不同的组装方式。框架则是各种设计模式的集合。设计模式是解决某一类问题的方法。

显然，架构师就是设计电脑框架图纸的人，他能决定最终造造出来的电脑是否方便拆、是否利于散热、硬件扩展功能怎么样等，而许多自称软件工程师的人其实就是电脑组装工，也就是我们常说的码工。

11.1 需求

此项目的需求很简单，就是同步客户、订单、提单、身份证验证等数据信息，然后导入成本，最终生成对账单。此项目的目的就是对账。缩减版的功能列表如图 11-1 所示。

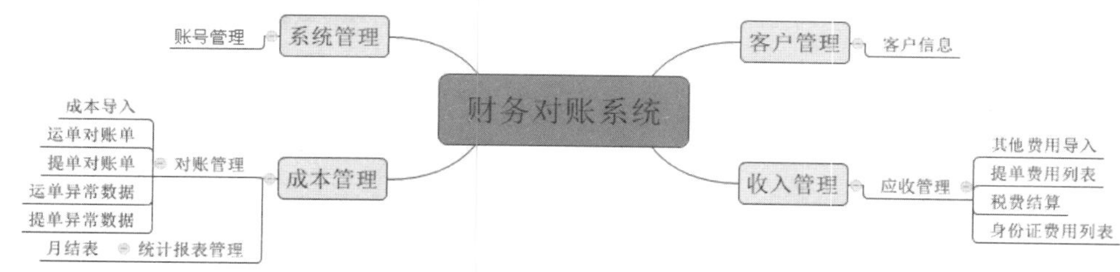

图 11-1

11.2 前台 UI 框架搭建

做此项目之前，从网页设计师手中拿到了 HTML 静态页面（没有一行 js），都是一个个零散的界面，这里需要做的事情是：

- 把零散的 HTML 界面连接起来。
- 自己编写 js 或者 jQuery 实现菜单效果。
- 把 HTML 页面集成在 MVC Razor 视图中。

在一些正规的软件公司中分工相对会比较细，会有专门的 UI 设计师、交互设计师、前端工程师，也有一些公司或者项目没有细分，所有的工作都直接交给.Net 程序员，这时程序员就要同时充当这些角色，也就是时下流行的全栈工程师（Full Stack developer）。

全栈工程师也叫全端工程师（同时具备前端和后台能力），是指掌握多种技能并能利用多种技能独立完成产品的人。

本想着使用第三方的 UI 框架（如 jQuery EasyUI、ExtJs、MiniUI 等）来搭建框架，但是公司要求必须做得和美工给的 HTML 页面样式一致，所以就不能用这些较复杂的 UI 框架了。这里主要使用 jQuery 和 jQuery 的一些 UI 插件。即便使用了第三方的 UI 插件也是非常痛苦的，因为需要修改 UI 插件的主题样式，要改得和美工给的界面差不多。

这里使用 UI 插件，一方面是为了提高用户体验，另一方面是为了减少编码，让 View 和 Controller 更好地结合。如果不添加 jQuery 的 UI 插件，直接在 View 中拼接美工所给的 HTML 页面，那么虽然看起来简单多了，但是界面复用性太差，需要更多的编码，所以这里通过框架来尽可能减少团队成员的编码量，提高开发效率。最终效果如图 11-2 所示。

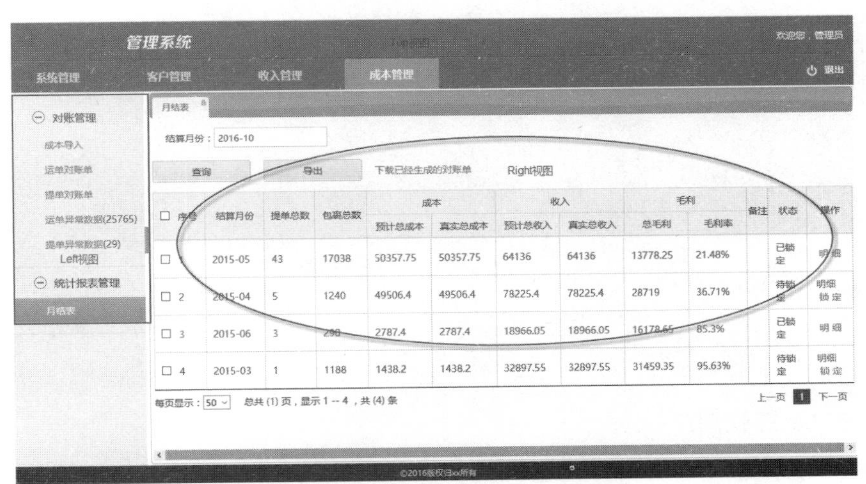

图 11-2

看到这样的后台界面，就会联想到使用 iframe 或者 frameset 来搭建，因为这样可以实现页面嵌入。项目组有同事说可以使用 MVC 里面的局部视图，跟以前 ASPX 视图里面的母版页差不多，但是这并不理想，不能每单击一个功能菜单都刷新整个界面，而且后面还要对菜单项做权限控制。园友“杰哥很忙”提议采用 Layout，通过@Html.Partial("_PartialHead")、@Html.Partial("_PartialMenu")、@Html.Partial("_PartialFoot")分别定义头部、左边菜单和尾部，然后通过 jQuery 进行处理可以达到局部刷新。

这里先不添加任何 js，一步一步来，然后持续改进，不断完善。

关于 iframe 和 frame 的区别大家可以从网上查找，这里不做过多解释。这里最终使用的是 frame。框架中用到的 js 和 css 为 jsCSSImg 文件夹。

（1）新建 ASP.NET MVC4 项目 MSD.WL.Site，把 jsCSSImg 目录中的所有文件夹复制到项目根目录，然后新建控制器 HomeController，这里用了 4 个 Action 方法，分别对应 4 个界面，Index 代表 frame 主界面。

```
public class HomeController : Controller
{
        public ActionResult Index()
        {
            ViewBag.Message = "欢迎使用财务模块";

            return View();
        }
        public ActionResult Top()
        {
            ViewBag.UserName = "超级管理员";
            ViewBag.AvailableBalance = "8888.00";
            return View();
        }
        public ActionResult Left()
        {
            return View();
        }
        public ActionResult Right()
        {
            return View();
        }
}
```

（2）新建 Index 视图：

```
@{
    ViewBag.Title = "";
    Layout = null;
}
```

```
<!DOCTYPE html>
<html lang="zh">
<head>
    <meta http-equiv="Content-Type" content="text/html; charset=utf-8" />
    <meta charset="utf-8" />
</head>
<frameset rows="104,*,30" cols="*" frameborder="no" border="0"
framespacing="0">@*顶部104px，底部30px，中间部分自适应*@
  <frame src="Home/Top" name="topFrame" scrolling="No" noresize="noresize"
  id="topFrame" title="topFrame" />
  <frameset cols="193,*" frameborder="no" border="0" framespacing="0"
  id="middenFram">@*左侧193px，右侧自适应*@
    <frame src="Home/Left" name="leftFrame" scrolling="No" noresize="noresize"
    id="leftFrame" title="leftFrame"/>
    <frame src="Home/Right" name="mainFrame" id="mainFrame" title="mainFrame" />
  </frameset>
  <frame src="/Content/Bootom.html" name="topFrame" scrolling="No"
  noresize="noresize" id="bootomFrame" title="topFrame" />
</frameset>
<noframes>
  <body> </body>
    </noframes>
</html>
```

（3）新建 Top 视图：

```
@{
    Layout = null;
}
<!DOCTYPE html>
<html lang="en">
<head>
    <title></title>
    <meta charset="utf-8">
    <meta name="viewport" content="width=device-width,initial-scale=1.0">
    <meta charset="utf-8" />
    <link href="~/Content/sharestyle.css" rel="stylesheet" />
    <style type="text/css">
        .hightCss
        {
            color: yellow;
        }
    </style>
```

```
</head>
<body>
    <div class="index_header">
        <div class="index_headertop">
            <div class="index_logo"><a href="#">
                <img src="/images/index_logo.png"></a></div>
            <div class="lgstatus">
                欢迎您，@ViewBag.UserName<i><em>20</em></i>可用余额：<span>￥@ViewBag.
                AvailableBalance</span>   
      <input type="submit" value="在线充值" class="btsty2">
            </div>
        </div>
        <div class="clear"></div>
        <div class="index_headerbot">
            <div class="nav_list">
                <ul>
                    <li><a href="#">业务管理</a>
                        <div class="nav_out" style="display: none;">
                            <i></i>
                            <p><a href="#">订单管理</a></p>
                            <p><a href="#">提单管理</a></p>
                            <p><a href="#">身份证管理</a></p>
                        </div>
                    </li>
                    <li class="slctd"><a href="#">财务管理</a>
                        <div class="nav_out" style="display: none;">
                            <i></i>
                            <p><a href="#">财务流水</a></p>
                            <p><a href="#">提单对账</a></p>
                            <p><a href="#">运单对账</a></p>
                            <p><a href="#">异常费用对账</a></p>
                            <p><a href="#">充值记录</a></p>
                        </div>
                    </li>
                    <li><a href="#">系统管理</a>
                        <div class="nav_out" style="display: none;">
                            <i></i>
                            <p><a href="#">基本信息管理</a></p>
                            <p><a href="#">认证管理</a></p>
                            <p><a href="#">修改密码</a></p>
                        </div>
                    </li>
```

```
            </ul>
        </div>
        <div class="fucnbx"><span><a href="#"><i class="ilChannel"></i>
        运单打印客户端下载</a></span> <span><a href="#"><i class="i2"></i>API 文档
        </a></span> <span><a href="#"><i class="i3"></i>退出</a></span> </div>
    </div>
  </div>
</body>
</html>
```

（4）新建 Right 视图：

```
@{
    Layout = null;
}
<!DOCTYPE html>
<html lang="zh">
<head>
    <title></title>
    <meta http-equiv="Content-Type" content="text/html; charset=utf-8" />
    <meta charset="utf-8" />
    <style type="text/css">
html, body
{
    margin: 0px;
    font-family: Arial, Sans-Serif; /*font-size: 62.5%;*/
    font-size: 12px;
    height: 100%;
    padding: 2px 4px 4px 0px;
    overflow:hidden;
}
    </style>
</head>
<body>
  <div class="rightcont">Hello,World
    </div>
    </body>
</html>
```

（5）新建 Left 视图：

```
@{
    Layout = null;
```

```
}
<!DOCTYPE html>
<html lang="zh">
<head>
    <title></title>
    <meta http-equiv="Content-Type" content="text/html; charset=utf-8" />
    <meta charset="utf-8" />
    <link href="~/Content/sharestyle.css" rel="stylesheet" />
    <link href="~/Content/main.css" rel="stylesheet" />
    <style type="text/css">
        body
        {
            margin:0px;
            padding:0px;
        }
    </style>
</head>
<body>
 <div class="leftbar" id="divOrder">
  <dl>
    <dt class="head2"id="dt_ulOrder"onclick='ShowMenuList("ulOrder")'>订单管理</dt>
    <ul class="box_n" id="ulOrder">
      <li><a href="#">批量新建订单</a></li>
      <li><a href="#">手工新建订单</a></li>
      <li><a class="nav_sub" href="#">订单草稿<span> (3) </span></a></li>
      <li><a href="#">已确认订单<span> (3) </span></a></li>
      <li><a href="#">待发货订单<span> (3) </span></a></li>
      <li><a href="#">已发货订单<span> (0) </span></a></li>
      <li><a href="#">订单回收站<span> (0) </span></a></li>
      <li><a href="#">退件<span> (0) </span></a></li>
      <li><a href="#">批量修改订单</a></li>
    </ul>
  </dl>
  <dl>
    <dt class="head1"id="dt_ulLading"onclick='ShowMenuList("ulLading")'>提单管理</dt>
    <ul class="box_n" id="ulLading" style="display:none;" >
      <li><a href="#">创建托盘</a></li>
      <li><a href="#">未交货托盘<span> (6) </span></a></li>
      <li><a href="#">已交货托盘</a></li>
      <li><a href="#">创建交货单</a></li>
      <li><a href="#">交货单列表</a></li>
      <li><a href="#">待预扣提单<span> (3) </span></a></li>
```

```
      <li><a href="#">已预扣提单</a></li>
    </ul>
  </dl>
  <dl>
    <dt class="head1" id="dt_ulIdentityCard"
    onclick='ShowMenuList("ulIdentityCard")'>身份证管理</dt>
    <ul class="box_n" id="ulIdentityCard" style="display:none;" >
      <li><a href="#">待验证身份证<span> (3) </span></a></li>
      <li><a href="#">无须验证身份证<span> (3) </span></a></li>
      <li><a href="#">已验证身份证<span> (3) </span></a></li>
    </ul>
  </dl>
</div>
 <div class="leftbar" id="divSysManage"></div>
    <div class="leftbar" id="divFinancial">
       <dl>
           <dt id="dt_ulChannel" class="head2"
            onclick='ShowMenuList("ulChannel")'>渠道费用管理</dt>
           <ul class="box_n" id="ulChannel">
               <li><a href="#" onclick="goNewPage('a.html','渠道分类');">
               渠道分类</a></li>
               <li><a target="mainFrame" id="channelManage"
               onclick="goNewPage('/Channel/Index','渠道管理');">渠道管理</a></li>
               <li><a href="#">分区管理</a></li>
               <li><a href="#">价格管理</a></li>
           </ul>
       </dl>
       <dl>
           <dt id="dt_ulFinancial" class="head1"
           onclick='ShowMenuList("ulFinancial")'>财务管理</dt>
           <ul class="box_n" id="ulFinancial" style="display: none;">
               <li><a href="#">财务流水</a></li>
               <li><a href="#">提单对账</a></li>
               <li><a href="#">运单对账</a></li>
               <li><a href="#">异常费用对账</a></li>
               <li><a href="#">充值记录</a></li>
           </ul>
       </dl>
    </div>
</body>
</html>
```

按 F5 键运行，界面如图 11-3 所示。

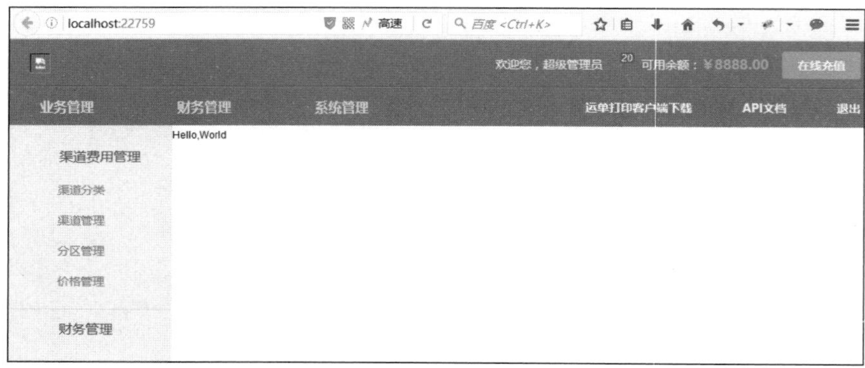

图 11-3

11.3 菜单特效

在 11.2 节已经把整个项目的框框搭建好了，但是还没有任何 js 效果实现。这一节就来讲解一下关于菜单的特效实现，需要的效果如图 11-4~图 11-6 所示。

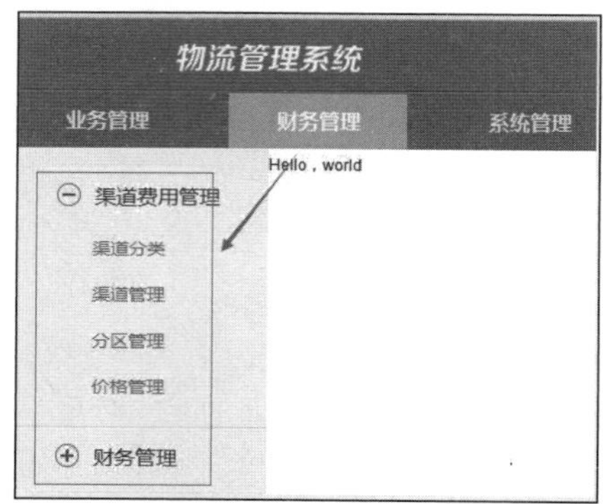

图 11-4

图 11-5

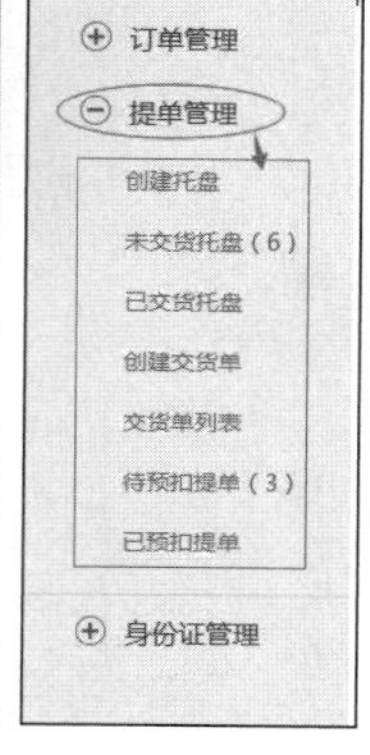

图 11-6

需求总结：

- 单击顶部菜单模块，左侧显示不同模块下面的菜单列表。
- 单击左侧菜单选项，展开下面的子菜单，并折叠其他菜单模块。菜单图标折叠显示为+，展开显示为-。

（1）在 Top 视图的 head 中添加 js，注意 js 代码添加在 css 引用之后。因为浏览器渲染是从上至下渲染的，我们一般让界面样式先加载，再去执行 js 代码。

```
<script src="~/Scripts/jquery-1.8.3.min.js"></script>
<script type="text/javascript">
        //控制 Left 视图中菜单模块的显示
        function showLeftList(divId) {
            self.parent.frames["leftFrame"].showDivMenu(divId);
        }
        //菜单单击高亮显示
        $(function () {
            $(".nav_list ul li a").click(function () {
                //$(".nav_list ul li a").css("color", "#ceebff");
                //$(this).css("color", "yellow");
                $(".nav_list ul li a").css("background-color", "");
                $(".nav_list ul li a").css("color", "#ceebff");
                $(this).css("background-color", "#66d354");
                $(this).css("color", "white");
            });
        });
</script>
```

（2）在 Top 视图中，向菜单项中添加 js 方法 showLeftList，传入一个参数，这个参数就是 Left 视图中菜单层的 id。

```
<li><a href="#" onclick="showLeftList('divOrder')">业务管理</a>
<li class="slctd"><a href="#" onclick="showLeftList('divFinancial')">财务管理</a>
<li><a href="#" onclick="showLeftList('divSysManage')">系统管理</a>
```

（3）查看 Left 视图代码，注意 id 的命名，因为这关系到 js 的调用，在 head 部分添加如下 js：

```
<script src="~/Scripts/jquery-1.7.1.min.js"></script>
<script type="text/javascript">
        //显示菜单下面的选项
        function ShowMenuList(id) {
            var objectobj = document.getElementById(id);
            var dtObj = document.getElementById("dt_" + id);
            if (objectobj.style.display == "none") {
                objectobj.style.display = "";
                dtObj.setAttribute("class", "head2");
                //其他菜单折叠
                $(dtObj).parent().parent().find("dt").not(dtObj).attr("class",
                 "head1"); //折叠
```

```
            $(objectobj).parent().parent().find("ul").not(objectobj).css("displ
            ay", "none"); //隐藏菜单子项
        }
        else {
            objectobj.style.display = "none";
            dtObj.setAttribute("class", "head1");
        }
    }
    //控制菜单模块的显示和隐藏
    function showDivMenu(divId) {
        $("#" + divId).css("visibility", "visible");
        $("#" + divId).siblings("div").css("visibility", "hidden");
    }
    $(function () {
        $(".box_n li a").click(function () {
            $(".box_n li a").removeClass("nav_sub");
            $(this).addClass("nav_sub");
        });
    });
    function goNewPage(url, name) {
        self.parent.frames["mainFrame"].addTab(url, name);
    }
</script>
```

至此，菜单特效就添加上去了，可以按 F5 键运行并查看效果。

11.4 面板折叠和展开

在 11.3 节已经添加了菜单特效，这一节来添加面板的折叠和展开功能，效果如图 11-7 所示。

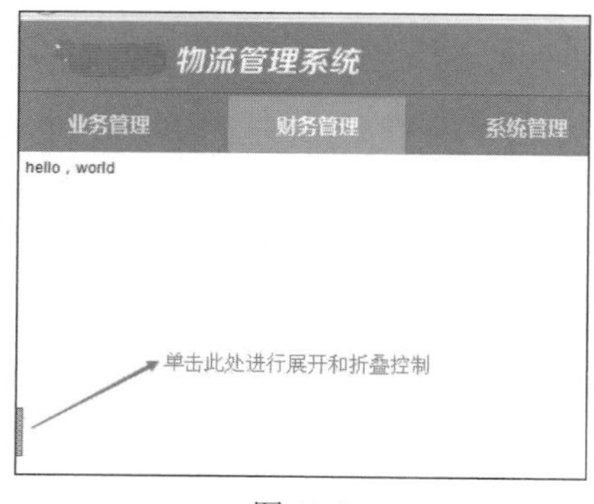

图 11-7

思路：在 Right 视图中添加一个 div，在这个 div 中存放一张图片，通过对这张图片的点击来控制 Left 视图的隐藏和显示，其实就是修改主框架 Index 视图中 frameset 的 cols 属性。

操作步骤如下：

（1）修改 Right 视图，在右侧添加一个 div，设置 float:left;，在里面存放一个图片按钮，作为面板折叠和展开的开关。需要注意的是，这里为了方便大家阅读，部分 css 样式直接写到了 View 中，然而在真实项目开发中，建议将 css 样式写入单独的样式文件中，方便复用、压缩合并，从而实现统一管理。

（2）添加一个 jquery 方法，调用父框架 Index 视图中的方法 hideShowFrame，修改 Index 视图中 frameset 的 cols 属性，从而控制界面的展示。

```
@{
    Layout = null;
}
<!DOCTYPE html>
<html lang="zh">
<head>
    <title></title>
    <meta http-equiv="Content-Type" content="text/html; charset=utf-8" />
    <meta charset="utf-8" />
    <style type="text/css">
        html, body
        {
            margin: 0px;
            font-family: Arial, Sans-Serif; /*font-size: 62.5%;*/
            font-size: 12px;
            height: 100%;
            padding: 2px 4px 4px 0px;
            overflow: hidden;
        }
    </style>
    <style type="text/css">
        .sidebar
        {
            width: 5px;
            height: 500px;
        }
            .sidebar .btn
            {
                width: 5px;
                height: 39px;
                background: url(/images/sidebar-on.gif);
                margin-top: 200px;
```

```
        }
            .sidebar .btn:hover
            {
                background-position: 0 -39px;
            }
        .fleft
        {
            float: left;
        }
    </style>
    <script src="~/Scripts/jquery-1.8.3.min.js"></script>
    <script type="text/javascript">
        $(function () {
            $("#divFolding").click(
                function () { self.parent.hideShowFrame(); }
                //hideShowFrame 是 Index 视图中的 js 方法
                );
        });
    </script>
</head>
<body>
    <div class="sidebar fleft">
        <div class="btn" id="divFolding"></div>
    </div>
    <div class="rightcont">
        hello, world
    </div>
</body>
</html>
```

（3）修改 Index 视图，添加如下 js：

```
<script type="text/javascript">
        //折叠展开面板
        function hideShowFrame() {
            if (document.getElementById("middenFram").cols == "193,*") {
                document.getElementById("middenFram").cols = "0,*";
            }
            else {
                document.getElementById("middenFram").cols = "193,*"
            }
        }
</script>
```

（4）按 F5 键运行并查看效果。

11.5 tab 多页签支持

在单击左侧菜单中的选项时，希望出现在 ExtJs、EasyUI 等中类似的 tab 页签功能，因为这样可以支持多个页面的浏览。有时候可能需要同时打开多个页面，如果不使用页签，那么每次查看某个页面都要重新调用并刷新，如果在网速慢或者该界面加载很耗时的情况下，简直会让人奔溃。因为不想引入整个 ExtJs 等的内容，就自然而然地想到了网上的 UI 插件。CleverTabs 比较适合这个项目，效果如图 11-8 所示。

图 11-8

操作步骤如下：

（1）修改 Right 视图，添加如下 js 和 css 引用：

```
<link href="~/Lib/CleverTabs/context/themes/base/jquery-ui.css"
    rel="stylesheet" />
<script src="~/Scripts/jquery-1.8.3.min.js"></script>
<script src="~/Lib/CleverTabs/scripts/jquery-ui.js"></script>
<script src="~/Lib/CleverTabs/scripts/jquery.cleverTabs.js"></script>
<script src="~/Lib/CleverTabs/scripts/jquery.contextMenu.js"></script>
```

（2）添加 js 方法：

```
var tabs;
$(function () {
            var h = $(document).height() - 35;
            $("#tabs").height(h);
            //如果不设置高度，默认就不是100%占满屏幕，所以这里使用计算的方式初始化界面高度
            tabs = $('#tabs').cleverTabs();
            $(window).bind('resize', function () {
                tabs.resizePanelContainer();
            });

            tabs.add({
                url: 'http://www.cnblogs.com/jiekzou/',
                label: '我的博客',
                //开启 Tab 后是否锁定(不允许关闭，默认: false)
```

```
            lock: false
        });
        $('input[type="button"]').button();

});
function addTab(url, name) {
        tabs.add({
            url: url,
            label: name
        });
}
```

（3）修改 Right 视图中的 body 主体：

```
<body>
    <div class="sidebar fleft">
        <div class="btn" id="divFolding"></div>
    </div>
    <div id="tabs" style="overflow: hidden; padding-top: 0px; height: 400px;">
        <ul>
        </ul>
    </div>
</body>
```

（4）在 Left 视图中添加如下 js 方法（单击 Left 中的菜单时调用 Right 视图中的添加页签方法 addTab）：

```
function goNewPage(url, name) {
        self.parent.frames["mainFrame"].addTab(url, name);
}
```

Left 视图中菜单的调用方法如下：

```
<li><a target="mainFrame" id="channelManage"
onclick="goNewPage('/Channel/Index','渠道管理');">渠道管理</a></li>
```

（5）按 F5 键运行，效果如图 11-9 所示。

图 11-9

11.6 Controller 和 View 的交互

这一节将用一个 Demo 来演示在此 UI 框架中，控制器和视图的交互。以渠道管理为例，效果如图 11-10~图 11-11 所示。

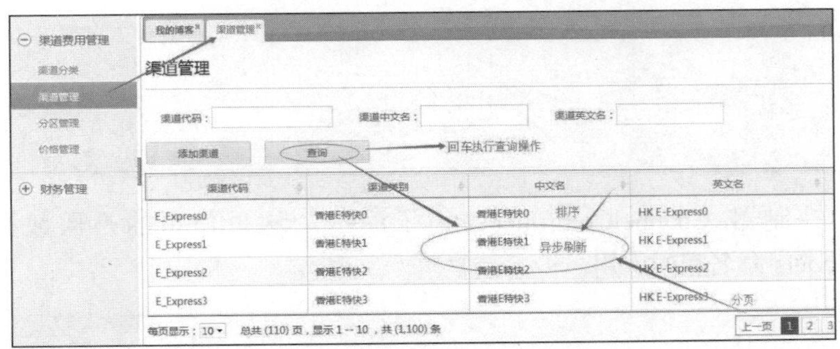

图 11-10

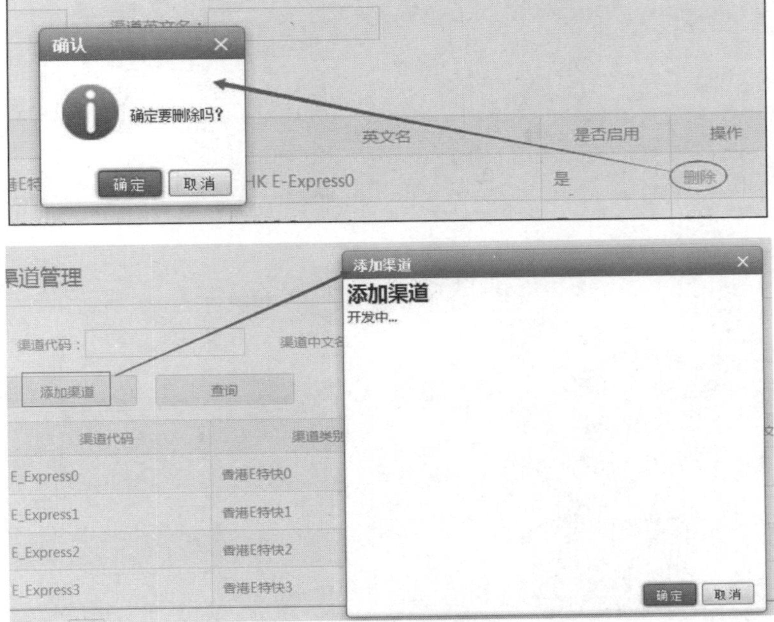

图 11-11

这里使用了基于 jQuery 的模态窗体组件 lhgdialog 和表格组件 dataTables。有关 dataTables 的更多资料请参考 http://dt.thxopen.com/example/，有关 lhgdialog 的更多资料请参考 http://www.lhgdialog.com/api/。

操作步骤如下：

（1）在 MVC 项目的 Models 文件夹中添加一个 model 类 ChannelInfo.cs，因为项目中的

ORM 框架使用的是 NHibernate，所以属性前面加了 virtual 。

```
public class ChannelInfo
{
    public virtual int ID { get; set; }
    public virtual string ChannelStyle { get; set; }
    public virtual string ChannelCode { get; set; }
    public virtual string CnName { get; set; }
    public virtual string EnName { get; set; }
    public virtual string Status { get; set; }
}
```

（2）添加控制器 ChannelController。为了演示，这里使用的是假数据。添加 using MSD.WL.Site.Models 命名空间引用。

```
public class ChannelController : Controller
{
    //
    // GET: /Channel/

    public ActionResult Index()
    {
        return View();
    }
    //添加渠道
    public ActionResult AddChannel()
    {
        return View();
    }

    [HttpPost]
    public JsonResult List(ChannelInfo filter)
    {
        List<ChannelInfo> list = new List<ChannelInfo>();
        for (int i = 0; i < 1100; i++)
        {
            list.Add(new ChannelInfo
            {
                ID = 1,
                ChannelCode = "E_Express" + i,
                ChannelStyle = "香港E特快" + i,
                CnName = "香港E特快" + i,
                EnName = "HK E-Express" + i,
```

```
                Status = "1"
            });
        }
        if (!string.IsNullOrEmpty(filter.ChannelCode))
        {
            list = list.Where(x => x.ChannelCode == filter.
            ChannelCode.Trim()).ToList();
        }
        if (!string.IsNullOrEmpty(filter.CnName))
        {
            list = list.Where(x => x.CnName == filter.CnName.Trim()).ToList();
        }
        if (!string.IsNullOrEmpty(filter.EnName))
        {
            list = list.Where(x => x.EnName == filter.EnName.Trim()).ToList();
        }

        //构造成 Json 的格式传递 iTotalRecords（总记录数）和 iTotalDisplayRecords
        //（每页显示的记录数）
        var result = new { iTotalRecords = 1100, iTotalDisplayRecords = 10,
        data = list };
        return Json(result, JsonRequestBehavior.AllowGet);
    }
}
```

（3）修改_Layout.cshtml，因为后面的 View 会用到：

```
<!DOCTYPE html>
<html lang="zh">
   <head>
   <meta http-equiv="Content-Type" content="text/html; charset=utf-8"/>
      <meta charset="utf-8" />
      <title>财务管理 @ViewBag.Title</title>
      <link href="~/favicon.ico" rel="shortcut icon" type="image/x-icon" />
      <meta name="viewport" content="width=device-width" />
      <link href="~/Content/sharestyle.css" rel="stylesheet" />
      <link href="~/Content/main.css" rel="stylesheet" />
      <script src="~/Scripts/jquery-1.8.3.min.js"></script>
      <script src="~/Lib/lhgdialog/lhgdialog.min.js?self=true&skin=iblue"></script>
   </head>
   <body>
         <section class="content-wrapper main-content clear-fix">
            @RenderBody()
```

```
        </section>
    </body>
</html>
```

（4）添加渠道管理的视图 Index，代码很简单，并有详细的注释，相信大家都看得懂。这里主要添加了列表展示、查询过滤和分页排序。datables 是支持服务器端分页排序的，但是这里只写了客户端排序，就是先一次性把所有的数据查出来再进行分页排序。在数据量小的情况下，体验还是非常不错的，也简单。如果数据量大，就要启用服务器分页，即每次按需取数据。关于 datables 服务器分页的内容，网上.NET 的例子非常少，不过摸索后也已经实现，只是此系列没有写出来。同时 datables 是支持缓存的，具体使用大家可以参考上面的网址内容，这里只做一个简单的引荐。

```
@{
    ViewBag.Title = "Index";

}
<style type="text/css">
    html, body
    {
        overflow:hidden;
    }
    #table_local tbody
    {
         height:50px;
    }
    table
    {
          overflow-y:auto;
          overflow-x:hidden;
    }
</style>
<link href="~/Lib/DataTables-1.10.6/media/css/jquery.dataTablesNew.css" rel="stylesheet" />
<script src="~/Lib/DataTables-1.10.6/media/js/jquery.dataTables.min.js"></script>
<script src="~/Content/DataTablesExt.js"></script>
<script type="text/javascript">
    //查询 刷新
    function reloadList() {
        var tables = $('#table_local').dataTable().api();
        //获取 DataTables 的 Api，详见 http://www.datatables.net/reference/api/
        tables.ajax.reload();
    }
    function deleteRecord(id) {
```

```
        $.dialog.confirm("确定要删除吗？", function () { $.dialog.alert("删除成功！"); }, null)
    }
    function successFun() {
        $.dialog.alert("渠道添加成功！");
    }
    //弹出框
    var dg;
    function showPublishWin() {
        dg = new $.dialog({
            id: "AddChannel",
            title: "添加渠道",
            content: "url:/Channel/AddChannel",
            width: 424,
            height: 320,
            max: false,
            min: false,
            lock: true,
            close: true,
            cancel: true, //X 按钮是否显示，如果设置了回调函数，就一定会显示
            //cancel: controlAllBtn,
            ok: successFun //单击确定执行的回调函数
        });
        dg.show();
    }
    document.onkeydown = function (event) {
        var e = event || window.event || arguments.callee.caller.arguments[0];
        if (e && e.keyCode == 27) { // 按 Esc 键
            //要做的事情
        }
        if (e && e.keyCode == 13) { // 按回车键
            //要做的事情
            reloadList();
        }
    };
</script>
<script type="text/javascript">
    $(function () {
        var h = $(document).height() - 258;
        $("#table_local").dataTable({
            //"iDisplayLength": 10,//每页显示10条数据
            //这里也可以设置分页，但是不能设置具体内容，只能是一维或二维数组的方式，所以推荐下面
            //language 里面的写法
```

```
 //"aLengthMenu": [[10, 15, 20, 25, 50, -1], [10, 15, 20, 25, 50, "All"]],
 bProcessing: true,
 //"dom": 'i,p',//l - Length changing, 选择每页显示行数下拉框的控件; f -
//Filtering input, 搜索过滤控件 t; - The Tabletools, 导出 excel、csv 的按钮
 //i - Information, 显示汇总信息（从 1 到 100 /共 1,288 条数据）;
 //p - Pagination, 分页控件; r - pRocessing, 显示加载时的进度条; C - copy,
 //显示复制, Excel 的控件
 //ajax: "/SendGoods/List",
 "scrollY": h, //垂直滚动
 "scrollCollapse": "true", //开启滚动
 "dom": 'tr<"bottom"lip><"clear">',     //这个是控制布局的, 不是很好理解
 "bServerSide": false,                  //指定从服务器端获取数据
 sServerMethod: "POST", //请求方式
 sAjaxSource: "@Url.Action("List", "Channel")", //数据源
 "fnServerParams": function (aoData) {  //查询条件
     aoData.push(
         { "name": "ChannelCode", "value": $("#ChannelCode").val() },
         { "name": "CnName", "value": $("#CnName").val() },
         { "name": "EnName", "value": $("#EnName").val() }
         );
 },
 columns: [
   {
       title: "1",
       "visible": false,
       "data": "ID", "sClass": "center",    //样式
       orderable: false,    //该列不排序
       "render": function (data, type, row) {    //列渲染
           return "<label class='position-relative'><input id='cbx" +
           data + "' type='checkbox' onclick='controlSelectAll(" + data
           + ")' class='cbx' value='" + data + "'/>";
       }
   },
   { "data": "ChannelCode", title: "渠道代码" },
   { "data": "ChannelStyle", title: "渠道类别" },
   { "data": "CnName", title: "中文名" },
   { "data": "EnName", title: "英文名" },
   {
      "data": "Status", title: "是否启用", orderable: false, "render":
       function (data, type, row, meta) { //自定义列
          if (data == "1") {
             return "是";
```

```
            }
            else {
                return "否";
            }
        }
    }
    , {
        "data": "ID", orderable: false, title: "操作", "render": function
        (data, type, row, meta) { //自定义列
            return "<a style='visibility:visible' onclick='deleteRecord
            (" + data + ")'>删除</a>";
        }
    }
],
paging: true,//分页
ordering: true,//是否启用排序
searching: false,//搜索
language: {
    lengthMenu: '每页显示：<select class="form-control input-xsmall">'
    + '<option value="5">5</option>' + '<option value="10">10</option>'
    + '<option value="15">15</option>'
      + '<option value="20">20</option>' + '<option
      value="25">25</option>' + '<option value="30">30</option>' +
      '<option value="35">35</option>' + '<option
      value="40">40</option>',//左上角的分页大小显示。
    search: '<span class="label label-success">搜索：</span>',
    //右上角的搜索文本，可以写 html 标签

    paginate: {//分页的样式内容
        previous: "上一页",
        next: "下一页",
        first: "",
        last: ""
    },

    zeroRecords: "暂无记录",//table tbody 内容为空时，tbody 的内容
    //下面三者构成了左下角的内容
    info: "总共 <span class='pagesStyle'>(_PAGES_) </span>页，显示 _START_
     -- _END_ ，共<span class='recordsStyle'> (_TOTAL_)</span> 条",
    //左下角的信息显示，大写的词为关键字，初始_MAX_ 条
    infoEmpty: "0条记录",//筛选为空时左下角的显示
    infoFiltered: ""//筛选之后的左下角筛选提示
```

```
            },
            pagingType: "full_numbers"//分页样式的类型

        });
        // $("#table_local_filter input[type=search]").css({ width: "auto" });
        //右上角的默认搜索文本框，不写这个就超出去了
    });
</script>
    <div class="areabx clear" style="margin-bottom:0px;padding-bottom:0px;">
        @using (Html.BeginForm("List", null, FormMethod.Get, new { @clase =
        "form-inline", @role = "form" }))
        {
            <div class="areabx_header">渠道管理</div>
            <ul class="formod mgt10">
                <li><span>渠道代码：</span>@Html.TextBox("ChannelCode","",
                new {@class="trade-time wid153" })</li>
                <li><span>渠道中文名：</span>@Html.TextBox("CnName", "",
                new {@class="trade-time" })</li>
                <li><span>渠道英文名：</span>@Html.TextBox("EnName", "",
                new {@class="trade-time" })</li>
            </ul>
            <div class="botbtbx pdb0">
                <input type="button" value="添加渠道" class="btn btn-primary"
                onclick="showPublishWin()"/>
                <input type="button" value="查询" onclick="reloadList();" class=
                "btn btn-primary">
            </div>
        }
        <div class="tob_box mgt15">
            <table id="table_local" class="display"  cellspacing="0"
            cellpadding="0" border="0" style="width:100%">
            </table>
        </div>
</div>
```

（5）添加视图 AddChannel：

```
@{
    ViewBag.Title = "添加渠道";
}
<style type="text/css">
    body {
        overflow:hidden;
```

```
    }
</style>
<h2>添加渠道</h2>
<div>开发中...</div>
```

由于要保持和美工给的样式风格一致，因此这里修改了 dataTables 的样式源码。

（6）按 F5 键运行。

11.7 增改查匹配

以客户信息界面为例，运行效果如图 11-12~图 11-13 所示。

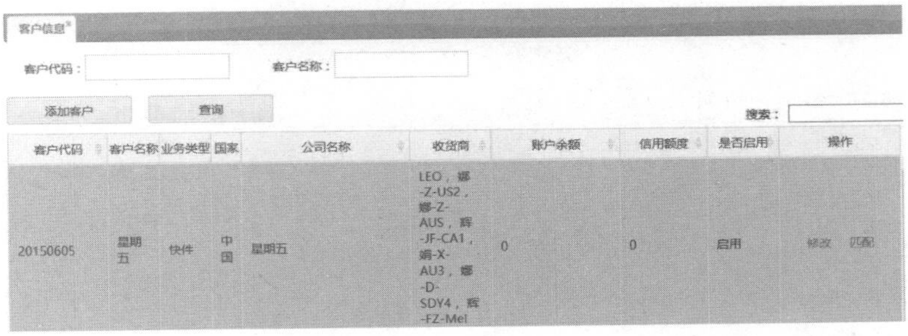

图 11-12

图 11-13

这里的添加和修改用了两个不同的视图，当然也可以把添加和修改放到同一个视图中，但是要写一些业务逻辑代码来区分当前调用的是修改还是添加，根据添加和修改的不同而对界面进行不同的操作。

添加控制器 Customer。需要注意的是，关于更新操作，NHibernate 每次都要先 load 一次，然后 update 一次，如果直接 save，就会把表中有但是界面上没有传过来的值全部更新为 null，相比之下 EF 就好多了。

```
/// <summary>
/// 客户控制器
/// </summary>
[Authorize]
public class CustomerController : Controller
{
    private string message = "<script>frameElement.api.opener.
    hidePublishWin('{0}', '{1}','{2}'); </script>";
    //消息，是否关闭弹出窗，是否停留在当前分页（0，1）

    #region 客户管理主页
    public ActionResult Index()
    {
        return View();
    }

    /// <summary>
    /// 客户列表
    /// </summary>
    /// <param name="filter"></param>
    /// <returns></returns>
    [HttpPost]
    public JsonResult List(CustomerFilter filter)
    {
        filter.PageSize = int.MaxValue;
        var dataSource = CustomerInfo.GetByFilter(filter);

        List<CustomerInfo> queryData = dataSource.ToList();

        var data = queryData.Select(u => new
        {
            ID = u.ID,
            CusCode = u.CusCode,
            CusName = u.CusName,
            BusssinessType = u.BusssinessType.GetDescription(false),
            Balance = u.Balance,
            CreditAmount = u.CreditAmount,
```

```
        Status = u.Status.GetDescription(false),
        Country = u.Country,
        CompanyName = u.CompanyName,
        Delivery = GetDeliveryList(u.ExpressCurInfoBy)

    });

    //构造成 Json 的格式传递
    var result = new { iTotalRecords = queryData.Count,
    iTotalDisplayRecords = 10, data = data };
    return Json(result, JsonRequestBehavior.AllowGet);
}

/// <summary>
/// 获取所有收货商名称
/// </summary>
/// <param name="list"></param>
/// <returns></returns>
private string GetDeliveryList(IList<ExpressCurInfo> list)
{
    StringBuilder result = new StringBuilder();
    if (list != null && list.Count > 0)
    {
        foreach (ExpressCurInfo ex in list)
        {
            if (result.Length > 0)
            {
                result.Append(",  ");
            }
            result.Append(ex.DeliveryName);
        }
    }
    return result.ToString();
}
#endregion

#region 添加客户
/// <summary>
/// 添加客户
/// </summary>
/// <param name="id"></param>
/// <returns></returns>
```

```
public ActionResult AddCustomer()
{
    ViewBag.Title = "添加客户";
    return View();
}

/// <summary>
/// 添加客户
/// </summary>
/// <param name="info"></param>
/// <returns></returns>
[HttpPost]
public ActionResult AddCustomer(CustomerInfo info)
{
    string msg = string.Empty;
    if (ModelState.IsValid)
    {
        try
        {
            info.Save();
            msg = "添加客户成功。";
        }
        catch (Exception ex)
        {
            msg = "添加客户失败！" + ex.Message;
            ViewBag.Msg = string.Format(message, msg, false, "1");
        }
        ViewBag.Msg = string.Format(message, msg, true, "0");
    }
    return View();
}
#endregion

#region 修改客户
/// <summary>
/// 修改客户
/// </summary>
/// <param name="id"></param>
/// <returns></returns>
public ActionResult UpdateCustomer(int id)
{
    ViewBag.Title = "修改客户";
```

```
    var result = CustomerInfo.Load(id);

    return View(result);
}

/// <summary>
/// 修改客户
/// </summary>
/// <param name="info"></param>
/// <returns></returns>
[HttpPost]
public ActionResult UpdateCustomer(CustomerInfo info)
{
    string msg = string.Empty;
    if (ModelState.IsValid)
    {
        try
        {
            CustomerInfo model = CustomerInfo.Load(info.ID);
            model.CusCode = info.CusCode;
            model.CusName = info.CusName;
            model.Phone = info.Phone;
            model.Tel = info.Tel;
            model.Email = info.Email;
            model.Fax = info.Fax;
            model.Country = info.Country;
            model.Address = info.Address;
            model.CompanyName = info.CompanyName;
            model.BusssinessType = info.BusssinessType;
            model.Status = info.Status;
            model.Update();
            msg = "修改客户成功。";
        }
        catch (Exception ex)
        {
            msg = "修改客户失败！" + ex.Message;
            ViewBag.Msg = string.Format(message, msg, false, "1");
        }
        ViewBag.Msg = string.Format(message, msg, true, "0");
    }
    return View();
}
```

```
#endregion

#region 客户匹配
public ActionResult DeliveryMatching(int id)
{
    ViewBag.Title = "收货商匹配";
    ViewBag.CustomerId = id;
    return View();
}

/// <summary>
/// 快件客户信息
/// </summary>
/// <param name="filter"></param>
/// <returns></returns>
[HttpPost]
public JsonResult DeliveryList()
{
    //filter.PageSize = int.MaxValue;
    var dataSource = ExpressCurInfo.GetAll();

    List<ExpressCurInfo> queryData = dataSource.Where(e => e.IsMatch ==
    false).ToList();
    var data = queryData.Select(u => new
    {
        ID = u.ID,
        DeliveryID = u.DeliveryID,
        DeliveryName = u.DeliveryName,
        AccountName = u.AccountName
    });

    //构造成Json的格式传递
    var result = new { iTotalRecords = queryData.Count,
    iTotalDisplayRecords = 10, data = data };
    return Json(result, JsonRequestBehavior.AllowGet);
}

[HttpPost]
public JsonResult AddMatching(int customerId, int deliveryId)
{
    string msg = string.Empty;
    bool result = true;
```

```
    try
    {
        CustomerInfo curCus = CustomerInfo.Load(customerId);
        ExpressCurInfo curExp = ExpressCurInfo.Load(deliveryId);
        curExp.MatchCustomer(curCus);
        msg = "匹配成功。";
        result = true;
    }
    catch (Exception ex)
    {
        msg = "匹配失败！" + ex.Message;
        result = false;
    }
    return Json(new { Msg = msg, Result = result ? "True" : "False" });
}
/// <summary>
/// 批量匹配
/// </summary>
/// <param name="customerId"></param>
/// <param name="deliveryId"></param>
/// <returns></returns>
[HttpPost]
public JsonResult AddMatchingList(int customerId, string ListID)
{
    string[] str = ListID.Split(new char[] { ',' }); //

    string msg = string.Empty;
    bool result = true;
    try
    {
        for (int i = 0; i < str.Length; i++)
        {
            CustomerInfo curCus = CustomerInfo.Load(customerId);
            ExpressCurInfo curExp =
            ExpressCurInfo.Load(Convert.ToInt32(str[i]));
            curExp.MatchCustomer(curCus);
        }
        msg = "匹配成功。";
        result = true;
    }
    catch (Exception ex)
    {
```

```
                msg = "匹配失败！" + ex.Message;
                result = false;
            }
            return Json(new { Msg = msg, Result = result ? "True" : "False" });
        }
        #endregion
}
```

客户信息列表视图 Index.cshtml：

```
@{
    ViewBag.Title = "客户信息";
}
<link href="~/libs/DataTables-1.10.6/media/css/jquery.dataTablesNew.css"rel="stylesheet" />
<script src="~/libs/lhgdialog/lhgdialog.min.js?self=true"></script>
<script src="~/libs/DataTables-1.10.6/media/js/jquery.dataTables.min.js"></script>
<script src="~/Scripts/DataTablesExt.js"></script>
<script type="text/javascript">
    //弹出框
    var addDG, updateDG, matchDG;
    var w = 424, h = 560; //宽，高
    //添加记录
    function showPublishWin() {
        addDG = new $.dialog({
            id: "AddChannel",
            title: "添加客户",
            content: "url:/Customer/AddCustomer",
            width: w,
            height: h,
            max: false,
            min: false,
            lock: true,
            close: true,
            btnBar: false
        });
        addDG.show();
    }
    //修改记录
    function modifyRecord(id) {
        updateDG = new $.dialog({
            id: "UpdateCustomer",
            title: "修改客户",
            content: "url:/Customer/UpdateCustomer/" + id,
```

```
            width: w,
            height: h,
            max: false,
            min: false,
            lock: true,
            close: true,
            btnBar: false
        });
        updateDG.show();
    }
    //隐藏弹出框
    function hidePublishWin(msg, result, isStay) {
        var icon = "success.gif";
        if (result == "False") {
            icon = "error.gif";
        }
        $.dialog({
            title: "提示",
            icon: icon,
            titleIcon: 'lhgcore.gif',
            content: msg,
            lock: true,
            ok: function () {
                if (result != "False") {
                    if (addDG) {
                        addDG.close();
                        addDG = null;
                    }
                    if (updateDG) {
                        updateDG.close();
                        updateDG = null;
                    }
                    if (matchDG) {
                        matchDG.close();
                        matchDG = null;
                    }
                    if (isStay == 0) {
                        reloadList();
                    }
                    else {
                        reloadListNew();
                    }
```

```
                }
            }
        });
    }
    //客户匹配
    function matchDelivery(id) {
        matchDG = new $.dialog({
            id: "UpdateCustomer",
            title: "客户匹配",
            content: "url:/Customer/DeliveryMatching/" + id,
            width: 802,
            height: h,
            max: false,
            min: false,
            lock: true,
            close: true,
            btnBar: false
        });
        matchDG.show();
    }
    //刷新
    function reloadList() {
        var tables = $('#table_local').dataTable().api();
        //获取 DataTables 的 Api，详见 http://www.datatables.net/reference/api/
        tables.ajax.reload();
    }
    //刷新，但是停留在当前分页
    function reloadListNew() {
        var tables = $('#table_local').dataTable().api();
        //获取 DataTables 的 Api，详见 http://www.datatables.net/reference/api/
        tables.ajax.reload(null, false);
    }
</script>
<script type="text/javascript">
    $(function () {
        var table = $("#table_local").dataTable({
            bProcessing: true,
            "scrollY": table_h1,
            "scrollCollapse": "true",
            "dom": 'ftr<"bottom"lip><"clear">',
            "bServerSide": false,                        //指定从服务器端获取数据
            sServerMethod: "POST",
```

```
sAjaxSource: "@Url.Action("List", "Customer")",
"fnServerParams": function (aoData) {  //查询条件
    aoData.push(
        { "name": "CusCode", "value": $("#CusCode").val() },
        { "name": "CusName", "value": $("#CusName").val() }
        );
},
columns: [{ title: "1", "visible": false, "data": "ID" },
   { "data": "CusCode", title: "客户代码" },
   { "data": "CusName", title: "客户名称" },
   { "data": "BusssinessType", title: "业务类型" },
   { "data": "Country", title: "国家" },
   { "data": "CompanyName", title: "公司名称", width: "200" },
   { "data": "Delivery", title: "收货商" },
   { "data": "Balance", title: "账户余额", width: "150" },
   { "data": "CreditAmount", title: "信用额度", width: "100" },
   { "data": "Status", title: "是否启用", width: "80" },
   {
       "data": "ID", orderable: false, title: "操作", width: "140",
       "render": function (data, type, row, meta) { //自定义列
           var re = "<div style='text-align:center'><a style=
           'visibility:visible' onclick='modifyRecord(" + data + ")'>
           修改</a>     ";
           re = re + "<a style='visibility:visible'
           onclick='matchDelivery(" + data + ")'>匹配</a></div>";
           return re;
       }
   }
 ],
 paging: true,//分页
 ordering: true,//是否启用排序
 searching: true,//搜索
 language: {
    "sProcessing": "处理中...",
    lengthMenu: lengthMenuStr,//左上角的分页大小显示
    search: '<span class="label label-success">搜索：</span>',
    //右上角的搜索文本，可以写 html 标签

    paginate: {//分页的样式内容
        previous: "上一页",
        next: "下一页",
        first: "",
```

```
                last: ""
              },

            zeroRecords: "暂无记录",//table tbody内容为空时，tbody的内容。
            //下面三者构成了左下角的内容。
            info: infoStr,//左下角的信息显示，大写的词为关键字，初始_MAX_ 条
            infoEmpty: "0条记录",//筛选为空时左下角的显示
            infoFiltered: ""//筛选之后的左下角筛选提示
          },
          pagingType: "full_numbers"//分页样式的类型

      });
      //设置选中行样式
      $('#table_local tbody').on('click', 'tr', function () {
          if ($(this).hasClass('selected')) {
              $(this).removeClass('selected');
          }
          else {
              table.$('tr.selected').removeClass('selected');
              $(this).addClass('selected');
          }
      });
  });
  //查询 刷新
  function reloadList() {
      var tables = $('#table_local').dataTable().api();//获取DataTables的Api,
      //详见 http://www.datatables.net/reference/api/
      tables.ajax.reload();
  }
</script>
<div class="areabx clear">
  @using (Html.BeginForm("List", null, FormMethod.Get, new { @clase =
  "form-inline", @role = "form" }))
  {
      <div id="divSearch">
          @*  <div class="areabx_header">客户信息</div>*@
          <ul class="formod">
              <li><span>客户代码：</span>@Html.TextBox("CusCode", "", new
              { @class = "trade-time wid153" })</li>
              <li><span>客户名称：</span>@Html.TextBox("CusName", "", new
              { @class = "trade-time" })</li>
              <li></li>
```

```
            </ul>
            <div class="botbtbx pdb0" style="margin-bottom: -30px;">
                <input type="button" value="添加客户" class="btn btn-primary"
                 onclick="showPublishWin()" />
                <input type="button" value="查询" onclick="reloadList();" class=
                "btn btn-primary">
            </div>
        </div>
    }
    <div class="tob_box mgt15">
        <table id="table_local" class="display" cellspacing="0" cellpadding="0"
        border="0" style="width: 100%">
        </table>
    </div>
</div>
```

添加 AddCustomer 视图：

```
@model Core.Customer.CustomerInfo
@using ProjectBase.Utils
@Html.Raw(ViewBag.Msg)
<style type="text/css">
        html, body {
        overflow-x:hidden;
    }
</style>
<div class="areabx clear">
@*    <div class="areabx_header">@ViewBag.Title</div>*@
    <div class="tian_xi">
        @using (Html.BeginForm("AddCustomer", "Customer", FormMethod.Post, new
        { @clase = "form-inline", @role = "form", name = "from1" }))
        {
            <table width="100%" border="0" cellpadding="0" cellspacing="0">
                <tbody>
                    <tr style="height: 40px;">
                        <td style="width: 120px; text-align: right;">客户代码：</td>
                        <td>
                            @Html.TextBoxFor(x=>x.CusCode,new{@class = "trade-timen",
                            @id = "cusCode" })<span class="wtps">* @Html.
                            ValidationMessageFor(m => m.CusCode)</span></td>
                    </tr>
                    <tr style="height: 40px;">
                        <td align="right">客户名称：</td>
```

```
        <td>
            @Html.TextBoxFor(x => x.CusName, new { @class = "trade-
            timen", @id = "cusName" })<span class="wtps">* @Html.
            ValidationMessageFor(m => m.CusName)</span></td>
    </tr>
    <tr style="height: 40px;">
        <td align="right">手机：</td>
        <td>
            @Html.TextBoxFor(x => x.Phone, new { @class =
            "trade-timen" })</td>
    </tr>
    <tr style="height: 40px;">
        <td align="right">电话：</td>
        <td>
            @Html.TextBoxFor(x => x.Tel, new { @class =
            "trade-timen" })</td>
    </tr>
    <tr style="height: 40px;">
        <td align="right">邮箱：</td>
        <td>
            @Html.TextBoxFor(x => x.Email, new { @class = "trade-
            timen", @id = "email" })<span class="wtps">
            @Html.ValidationMessageFor(m => m.Email)</span></td>
    </tr>
    <tr style="height: 40px;">
        <td align="right">传真：</td>
        <td>
            @Html.TextBoxFor(x => x.Fax, new { @class =
            "trade-timen" })</td>
    </tr>
    <tr style="height: 40px;">
        <td align="right">国家：</td>
        <td>
            @Html.TextBoxFor(x => x.Country, new { @class =
            "trade-timen" })</td>
    </tr>
    <tr style="height: 40px;">
        <td align="right">地址：</td>
        <td>
            @Html.TextBoxFor(x => x.Address, new { @class =
            "trade-timen" })</td>
    </tr>
```

```
                <tr style="height: 40px;">
                    <td align="right">公司名称：</td>
                    <td>
                        @Html.TextBoxFor(x => x.CompanyName, new { @class =
                        "trade-timen" })</td>
                </tr>
                <tr style="height: 40px;">
                    <td align="right">业务类型：</td>
                    <td>
                    @Html.DropDownListFor(x => x.BusssinessType,
                    @Html.EnumToList(typeof(Core.Customer.Busssiness), false),
                    new { @class = "trade-timen", style = "width:180px" })
                </tr>
                <tr style="height: 40px;">
                    <td align="right">是否启用：</td>
                    <td>是 @Html.RadioButtonFor(x => x.Status, "0", new { Checked
                     = "checked", @name = "status" })     
                        <span class="radioMagin">否 @Html.RadioButtonFor(x =>
                        x.Status, "1", new { @name = "status" })</span></td>
                </tr>
            </tbody>
        </table>
        <input type="submit" value="确定" class="popbtn1 mg">
        <input type="button" value="关闭" class="popbtn3 mg2"
        onclick="frameElement.api.opener.addDG.close();" />
    }
  </div>
</div>
```

添加 UpdateCustomer 视图：

```
@model Core.Customer.CustomerInfo
@using ProjectBase.Utils
@Html.Raw(ViewBag.Msg)
<style type="text/css">
        html, body {
        overflow-x:hidden;
    }
</style>
<div class="areabx clear">
@*    <div class="areabx_header">@ViewBag.Title</div>*@
    <div class="tian_xi">
        @using (Html.BeginForm("UpdateCustomer", "Customer", FormMethod.Post, new
```

```
{ @clase = "form-inline", @role = "form", name = "from1" }))
        {
            <table width="100%" border="0" cellpadding="0" cellspacing="0">
                <tbody>
                    <tr style="height: 40px;">
                        <td style="width: 120px; text-align: right;">客户代码：</td>
                        <td>
                            @Html.TextBoxFor(x => x.CusCode, new { @class = "trade-
                            timen", @id = "cusCode", @readOnly = "readOnly" })<span
                            class="wtps">* @Html.ValidationMessageFor(m =>
                            m.CusCode)</span></td>
                        @Html.HiddenFor(x => x.ID)
                    </tr>
                    <tr style="height: 40px;">
                        <td align="right">客户名称：</td>
                        <td>
                            @Html.TextBoxFor(x => x.CusName, new { @class = "trade-
                            timen", @id = "cusName" })<span class="wtps">*
                            @Html.ValidationMessageFor(m => m.CusName)</span></td>
                    </tr>
                    <tr style="height: 40px;">
                        <td align="right">手机：</td>
                        <td>
                            @Html.TextBoxFor(x => x.Phone, new { @class =
                            "trade-timen" })</td>
                    </tr>
                    <tr style="height: 40px;">
                        <td align="right">电话：</td>
                        <td>
                            @Html.TextBoxFor(x => x.Tel, new { @class =
                            "trade-timen" })</td>
                    </tr>
                    <tr style="height: 40px;">
                        <td align="right">邮箱：</td>
                        <td>
                            @Html.TextBoxFor(x => x.Email, new { @class = "trade-
                            timen", @id = "email" }) <span class=
                            "wtps">@Html.ValidationMessageFor(m => m.Email)</span></td>
                    </tr>
                    <tr style="height: 40px;">
                        <td align="right">传真：</td>
                        <td>
```

```
                @Html.TextBoxFor(x  =>  x.Fax,  new  {  @class  =  "trade-
timen" })</td>
            </tr>
            <tr style="height: 40px;">
                <td align="right">国家：</td>
                <td>
                    @Html.TextBoxFor(x => x.Country, new { @class = "trade-
                    timen" })</td>
            </tr>
            <tr style="height: 40px;">
                <td align="right">地址：</td>
                <td>
                    @Html.TextBoxFor(x => x.Address, new { @class = "trade-
                    timen" })</td>
            </tr>
            <tr style="height: 40px;">
                <td align="right">公司名称：</td>
                <td>
                    @Html.TextBoxFor(x => x.CompanyName, new { @class =
                    "trade-timen" })</td>
            </tr>
            <tr style="height: 40px;">
                <td align="right">业务类型：</td>
                <td>
                @Html.DropDownListFor(x => x.BusssinessType,
                @Html.EnumToList(typeof(Core.Customer.Busssiness), false),
                new { @class = "trade-timen", style = "width:180px" })
            </tr>
            <tr style="height: 40px;">
                <td align="right">是否启用：</td>
                <td>是 @Html.RadioButtonFor(x => x.Status, "0", new { Checked
                "checked", @name = "status" })     
                    <span class="radioMagin">否 @Html.RadioButtonFor(x =>
                    x.Status, "1", new { @name = "status" })</span></td>
            </tr>
        </tbody>
    </table>
    <input type="submit" value="确定" class="popbtn1 mg">
    <input type="button" value="关闭" class="popbtn3 mg2"
     onclick="frameElement.api.opener.updateDG.close();" />
  }
</div>
```

```
</div>
```

查询类 CustomerFilter:

```
public class CustomerFilter : ParameterFilter
{
    /// <summary>
    /// 客户代码
    /// </summary>
    public virtual string CusCode { get; set; }
    /// <summary>
    /// 客户名称
    /// </summary>
    public virtual string CusName { get; set; }

    /// <summary>
    /// 生产 NHQL 查询语句
    /// </summary>
    /// <returns></returns>
    public override string ToHql()
    {
        string hql = "";
        if (!string.IsNullOrEmpty(CusCode))
        {
            hql += " and Cus_Code =:CusCode ";
        }
        if (!string.IsNullOrEmpty(CusName))
        {
            hql += " and Cus_Name =:CusName ";
        }

        return hql;
    }

    /// <summary>
    /// 构造查询参数
    /// </summary>
    /// <returns></returns>
    public override Dictionary<string, object> GetParameters()
    {
        var result = new Dictionary<string, object>();
        if (!string.IsNullOrEmpty(CusCode))
        {
            result["CusCode"] = CusCode.Trim();
        }
        if (!string.IsNullOrEmpty(CusName))
        {
```

```
            result["CusName"] = CusName.Trim();
        }
        return result;
    }
}
```

这里只演示了控制器和视图的交互，数据层和业务层的代码可以参考 7.3 节。

11.8 统计报表

本节将通过一个 Demo 演示 Datatables 和 ASP.NET MVC 的完美结合，可以这么说，如果这样的界面都能做出来，那么后台系统 90%的界面功能都可以开发出来。

用 jQuery Datatables 来开发确实是比较烦人的（和 jQuery EasyUI、MiniUI、ExtJs 相比），因为用其他第三方 UI 框架来实现相同的功能非常简单，使用 Datatables 却是那么吃力。在官网，Datatables 默认使用的是 bootstraps 样式，这里已经重写了一部分样式。

公司原有的系统同样是使用 ASP.NET MVC 做的，完全不存在用户体验，于是这里就重新设置了 UI 框架（也可以说是组装，但是重写了许多东西）。

技术要点：① 服务器端分页。② 查询（模糊查询）。③ 界面操作刷新后依旧保留当前分页。④ 固定表头、表尾。⑤ 动态控制列的隐藏和显示。⑥ 全选、反选（数据行中的复选框全部被选中时，全选按钮自动被选中）。⑦ 服务器排序。⑧ 特殊字段标红显示。⑨ 滑动变色。⑩ 单击行选中变色……

整体效果如图 11-14~图 11-16 所示。

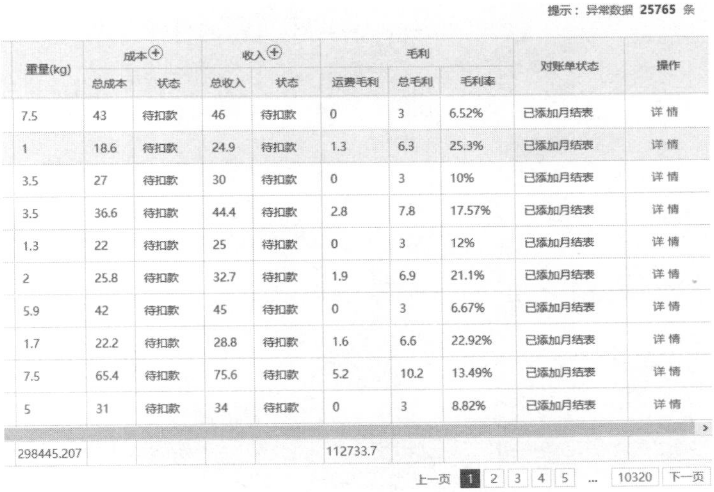

提示：异常数据 **25765** 条

重量(kg)	成本⊕		收入⊕		毛利			对账单状态	操作
	总成本	状态	总收入	状态	运费毛利	总毛利	毛利率		
7.5	43	待扣款	46	待扣款	0	3	6.52%	已添加月结表	详 情
1	18.6	待扣款	24.9	待扣款	1.3	6.3	25.3%	已添加月结表	详 情
3.5	27	待扣款	30	待扣款	0	3	10%	已添加月结表	详 情
3.5	36.6	待扣款	44.4	待扣款	2.8	7.8	17.57%	已添加月结表	详 情
1.3	22	待扣款	25	待扣款	0	3	12%	已添加月结表	详 情
2	25.8	待扣款	32.7	待扣款	1.9	6.9	21.1%	已添加月结表	详 情
5.9	42	待扣款	45	待扣款	0	3	6.67%	已添加月结表	详 情
1.7	22.2	待扣款	28.8	待扣款	1.6	6.6	22.92%	已添加月结表	详 情
7.5	65.4	待扣款	75.6	待扣款	5.2	10.2	13.49%	已添加月结表	详 情
5	31	待扣款	34	待扣款	0	3	8.82%	已添加月结表	详 情
298445.207					112733.7				

上一页 1 2 3 4 5 ... 10320 下一页

图 11-14

成本					收入				
邮政邮资	邮政邮件处理费	其他费用	总成本	状态	客户运费	客户操作费	其他费用	总收入	状态
36	7	0	43	待扣款	36	10	0	46	待扣款
15.6	3	0	18.6	待扣款	16.9	8	0	24.9	待扣款
20	7	0	27	待扣款	20	10	0	30	待扣款
33.6	3	0	36.6	待扣款	36.4	8	0	44.4	待扣款

图 11-15

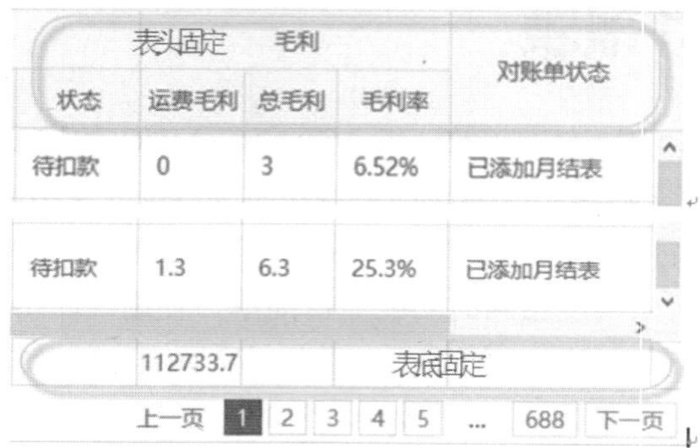

图 11-16

新建 Reconciliation 控制器：

```
[Authorize]
public class ReconciliationController : Controller
{
        private string message =
         "<script>frameElement.api.opener.hidePublishWin('{0}','{1}','{2}');</script>";
        //消息，是否关闭弹出窗，是否停留在当前分页（0，1）
        Dictionary<int, string> dicSort = new Dictionary<int, string>();
        //排序字段键值对列表 （列序号，列名称）

        #region 运单对账
        //运单对账
        public ActionResult WayBill(WayBillReconciliationFilter filter)
        {
            return View();
        }

        [HttpPost]
        public JsonResult WayBillList(WayBillReconciliationFilter filter)
        {
            WayBillExceptionFilter exceptionfilter = new WayBillExceptionFilter()
            { CusShortName = filter.CusShortName, LoadBillNum = filter.LoadBillNum,
```

```
ExpressNo = filter.ExpressNo, PostingTime = filter.PostingTime,
PostingTimeTo = filter.PostingTimeTo };
long counts =
Core.Reconciliation.WayBillException.GetExceptionCount(exceptionfilter);
//
DataTablesRequest parm = new DataTablesRequest(this.Request);
//处理对象
int pageIndex = parm.iDisplayLength == 0 ? 0 : parm.iDisplayStart /
 parm.iDisplayLength;
filter.PageIndex = pageIndex;    //页索引
filter.PageSize = parm.iDisplayLength;    //页行数
var DataSource = WayBillReconciliation.GetByFilter(filter) as
WRPageOfList<WayBillReconciliation>;

int i = parm.iDisplayLength * pageIndex;

List<WayBillReconciliation> queryData = DataSource.ToList();
var data = queryData.Select(u => new
{
    Index = ++i, //行号
    ID = u.ID,
    IsInputCost = u.IsInputCost,
    CusName = u.CusName, //客户简称
    PostingTime = u.PostingTime == null ? string.Empty :
    u.PostingTime.Value.ToStringDate(),//收寄日期
    ExpressNo = u.ExpressNo, //运单号
    BatchNO = u.LoadBillNum, //提单号
    Weight = u.Weight == null ? 0m : u.Weight / 1000, //重量
    WayBillFee = u.WayBillFee, //邮资
    ProcessingFee = u.ProcessingFee, //邮政邮件处理费
    InComeWayBillFee = u.ExpressFee, //客户运费
    InComeOprateFee = u.OperateFee, //客户操作费
    WayBillMargins = u.WayBillProfit, //运费毛利
    TotalMargins = u.ExpressFee + u.OperateFee + u.InComeOtherFee -
    (u.WayBillFee + u.ProcessingFee + u.CostOtherFee), //总毛利
    Margin = Math.Round((u.ExpressFee + u.OperateFee + u.InComeOtherFee
     == 0 ? 0m : (u.ExpressFee + u.OperateFee + u.InComeOtherFee -
     (u.WayBillFee + u.ProcessingFee + u.CostOtherFee)) / (u.ExpressFee
      + u.OperateFee + u.InComeOtherFee) * 100), 2),
      //毛利率，毛利率=(总收入-总的支出成本)/总收入*100%
    ReconcileDate = u.ReconcileDate.ToStringDate(), //对账日期
    CostOtherFee = u.CostOtherFee, //成本，其他费用
```

```
        CostTotalFee = u.WayBillFee + u.ProcessingFee + u.CostOtherFee,
        //成本，总费用
        CostStatus = u.CostStatus.ToChinese(),  //成本，状态
        InComeOtherFee = u.InComeOtherFee, //收入，其他费用
        InComeTotalFee = u.ExpressFee + u.OperateFee + u.InComeOtherFee,
        //收入，总费用
        InComeStatus = u.InComeStatus.ToChinese(),  //收入，状态
        Statement = u.Statement  //对账单状态
    });
    //构造成 Json 的格式传递
    var result = new
    {
        ExceptionCount = counts,
        iTotalRecords = DataSource.Count,
        iTotalDisplayRecords = DataSource.RecordTotal,
        data = data,
        TotalWeight = DataSource.StatModelBy.TotalWeight / 1000,
        TotalWayBillFee = DataSource.StatModelBy.TotalWayBillFee,
        TotalProcessingFee = DataSource.StatModelBy.TotalProcessingFee,
        TotalExpressFee = DataSource.StatModelBy.TotalExpressFee,
        TotalOperateFee = DataSource.StatModelBy.TotalOperateFee,
        SumWayBillProfit = DataSource.StatModelBy.TotalWayBillProfit,
        SumTotalProfit = 0m //总毛利求和
    };

    return Json(result, JsonRequestBehavior.AllowGet);
}

#endregion
/// <summary>
/// 运单详细页
/// </summary>
/// <param name="filter"></param>
/// <returns></returns>
[HttpGet]
public ActionResult WayBillDetailed(string ExpressNo)
{
    ViewBag.ExpressNo = ExpressNo;
    var list = Core.Reconciliation.WayBillReconciliation.GetByFilter(new
    WayBillReconciliationFilter() { ExpressNo = ExpressNo, PageIndex = 0,
     PageSize = 1 }).ToList();
```

```
            return View(list);  //.FirstOrDefault()
        }

}
```

新建 WayBill 视图：

```
@{
    ViewBag.Title = "运费对账";
}
<link href="~/libs/DataTables-1.10.6/media/css/jquery.dataTablesNew.css"
rel="stylesheet" />
<script src="~/libs/DataTables-
1.10.6/media/js/jquery.dataTables.min.js"></script>
<script src="~/Scripts/DataTablesExt.js"></script>
<script src="~/libs/My97DatePicker/WdatePicker.js"></script>
<script type="text/javascript">
    $(function () {
        var table = $("#table_local").dataTable({
            bProcessing: true,
            //"deferRender": true,//当处理大数据时，延迟渲染数据，有效提高 Datatables 处理能力
            "scrollY": table_h,
            "scrollX": $(document).width(),
            "scrollCollapse": "true",
            "dom": 'tr<"bottom"lip><"clear">',
            "bServerSide": true,                    //指定从服务器端获取数据
            sServerMethod: "POST",
            showRowNumber: true,
            sAjaxSource: "@Url.Action("WayBillList", "Reconciliation")",
            "initComplete": function (data, args) {
                getTotal(args);
                //var arr = new Array(7,8,9,10,11,14,15,16,17,18); //页面加载时隐藏的列
                //controlColumnShow(table, arr,false);
            },
            "fnServerParams": function (aoData) {  //查询条件
                aoData.push(
                    { "name": "CusShortName", "value": $("#CusShortName").val() },
                    { "name": "LoadBillNum", "value": $("#LoadBillNum").val() },
                    { "name": "ExpressNo", "value": $("#ExpressNo").val() },
                    { "name": "PostingTime", "value": $("#PostingTime").val() },
                    { "name": "PostingTimeTo", "value": $("#PostingTimeTo").val() },
                    { "name": "Margin", "value": $("#sltMargin").val() }
                 );
```

```
},
//跟数组下标一样，第一列从0开始，这里表格初始化时第三列默认降序
"order": [[2, "desc"]],
columns: [
   {
      "data": "IsInputCost", orderable: false, width: "60",
      "render": function (data, type, row, meta) {
         return " <input id='cbx" + data + "' type='checkbox'
         onclick='controlSelectAll(" + data + ")' class='cbx' value='"
         + data + "'/>  " + row.Index;
      }
   },
   //{ "data": "ReconcileDate",visible:false},//对账日期
   { "data": "CusName" }, //客户名称
   { "data": "PostingTime" },//收寄日期
   { "data": "ReconcileDate" },//对账日期
   { "data": "ExpressNo", orderable: false }, //运单号
   { "data": "BatchNO" },//提单号
   { "data": "Weight" },//重量
    {
       "data": "WayBillFee", visible: false, orderable: false, width:
       "80", "render": function (data, type, row, meta) { ///邮政邮资
          var re = "";
          (row.IsInputCost > 0) ? re = "<span>" : re = "<span
           class='preColor'>"
          re += +data + "</span>";
          return re;
       }
    },
    {
       "data": "ProcessingFee", visible: false, orderable: false,
       "render": function (data, type, row, meta) { //邮政邮件处理费
          var re = "";
          (row.IsInputCost > 0) ? re = "<span>" : re = "<span
           class='preColor'>"
          re += +data + "</span>";
          return re;
       }
    },
   {
      "data": "CostOtherFee", visible: false, orderable: false,
      "render": function (data, type, row, meta) { //其他费用
```

```
            var re = "";
            (row.IsInputCost > 0) ? re = "<span>" : re = "<span
             class='preColor'>"
            re += +data + "</span>";
            return re;
        }
    },
    {
        "data": "CostTotalFee", orderable: false, "render": function
        (data, type, row, meta) { //总成本
            var re = "";
            (row.IsInputCost > 0) ? re = "<span>" : re = "<span
             class='preColor'>"
            re += +data + "</span>";
            return re;
        }
    },
    { "data": "CostStatus", orderable: false, width: "80" },//状态
    { "data": "InComeWayBillFee", visible: false },//客户运费
    { "data": "InComeOprateFee", visible: false },//客户操作费
    { "data": "InComeOtherFee", visible: false },//其他费用
    { "data": "InComeTotalFee" },//总收入
    { "data": "InComeStatus", orderable: false, width: "80" },//状态
    {
        "data": "WayBillMargins", orderable: false, "render": function
        (data, type, row, meta) { //运费毛利
            var css = "";
            if (data < 0) {
                css = " class='numberColor'";
            }
            var re = "<div" + css + ">" + data + "</div>";
            return re;
        }
    },
    {
        "data": "TotalMargins", orderable: false, "render": function
        (data, type, row, meta) { //总毛利
            var css = "";
            if (data < 0) {
                css = " class='numberColor'";
            }
            var re = "<div" + css + ">" + data + "</div>";
```

```
                return re;
            }
        },
        {
            "data": "Margin", orderable: false, "render": function (data,
            type, row, meta) { //毛利率
                var css = "";
                if (data < 0) {
                    css = " class='numberColor'";
                }
                var re = "<div" + css + ">" + data + "%</div>";
                return re;
            }
        },
    { "data": "Statement", orderable: false, width: "130" },
    {
        "data": "ExpressNo", orderable: false, width: "100", "render":
        function (data, type, row, meta) { //操作
            var re = "<div style='text-align:center'><a
            style='visibility:visible' onclick='openDetail(\"" +
            row.ExpressNo + "\")'>详 情</a>  ";
            return re;
        }
    }
],
paging: true,//分页
ordering: false,//是否启用排序
searching: true,//搜索
language: {
    "sProcessing": "处理中...",
    lengthMenu: lengthMenuStr,//左上角的分页大小显示。
    search: '<span class="label label-success">搜索：</span>',
    //右上角的搜索文本，可以写html标签

    paginate: {//分页的样式内容
        previous: "上一页",
        next: "下一页",
        first: "",
        last: ""
    },

    zeroRecords: "暂无记录",//table tbody内容为空时，tbody的内容
```

```
        //下面三者构成了左下角的内容
        info: infoStr,//左下角的信息显示，大写的词为关键字，初始_MAX_ 条
        infoEmpty: "0条记录",//筛选为空时左下角的显示
        infoFiltered: ""//筛选之后的左下角筛选提示
    },
    pagingType: "full_numbers"//分页样式的类型
});
//设置选中行样式
$('#table_local tbody').on('click', 'tr', function () {
    if ($(this).hasClass('selected')) {
        $(this).removeClass('selected');
    }
    else {
        table.$('tr.selected').removeClass('selected');
        $(this).addClass('selected');
    }
});
//展开折叠列
$("#imgIncome").click(function () {
    var url = $("#imgIncome").attr("src");
    var arr = new Array(7, 8, 9);
    if (url == "/images/icon_9.png") {
        controlColumnShow(table, arr, true);
        $("#imgIncome").attr("src", "/images/icon_10.png");
    }
    else {
        controlColumnShow(table, arr, false);
        $("#imgIncome").attr("src", "/images/icon_9.png");
    }

});
//收入展开折叠
$("#imgCost").click(function () {
    var url = $("#imgCost").attr("src");
    var arr = new Array(12, 13, 14);
    if (url == "/images/icon_9.png") {
        controlColumnShow(table, arr, true);
        $("#imgCost").attr("src", "/images/icon_10.png");
    }
    else {
        controlColumnShow(table, arr, false);
        $("#imgCost").attr("src", "/images/icon_9.png");
```

```
            }
        });
    });
    function reloadList() {
        var tables = $('#table_local').dataTable().api();
        //获取 DataTables 的 Api，详见 http://www.datatables.net/reference/api/
        tables.ajax.reload(function () {
            var json = tables.context[0].json;
            getTotal(json);
        });
    }
    //统计
    function getTotal(json) {
        if (json) {
            //if (json.TotalWeight) {
                $("#spnTotalWeight").html(json.TotalWeight);
                $("#spnTotalWayBillFee").html(json.TotalWayBillFee);
                $("#spnTotalProcessingFee").html(json.TotalProcessingFee);
                $("#spnTotalExpressFee").html(json.TotalExpressFee);
                $("#spnTotalOperateFee").html(json.TotalOperateFee);
                $("#spnSumWayBillProfit").html(json.SumWayBillProfit);
                //$("#spnSumTotalProfit").html(json.SumTotalProfit);
            //}
        }
        if (json) {
            $("#UnusualCount").html("异常数据  <b style='color: red;'>" +
            json.ExceptionCount + "</b>  条");
        }
    }

    //打开异常界面
    function openUnusual() {
        //选中左侧菜单栏
        self.parent.parent.leftFrame.selectMenuItem("expressExData");
        var filter = '';
        if ($("#CusShortName").val().length > 0) {
            filter += filter.length > 0 ? "&CusShortName=" +
             $("#CusShortName").val() : "CusShortName=" + $("#CusShortName").val();
        }
        if ($("#LoadBillNum").val().length > 0) {
            filter += filter.length>0 ? "&LoadBillNum=" + ("#LoadBillNum").val() :
            "LoadBillNum=" + $("#LoadBillNum").val();
```

```
        }
        if ($("#ExpressNo").val().length > 0) {
            filter += filter.length > 0 ? "&ExpressNo=" + $("#ExpressNo").val() :
"ExpressNo=" + $("#ExpressNo").val();
        }
        if ($("#PostingTime").val().length > 0) {
            filter += filter.length > 0 ? "&PostingTime=" +
             $("#PostingTime").val() : "PostingTime=" + $("#PostingTime").val();
        }
        if ($("#PostingTimeTo").val().length > 0) {
            filter += filter.length > 0 ? "&PostingTimeTo=" +
            $("#PostingTimeTo").val() : "PostingTimeTo=" +
            $("#PostingTimeTo").val();
        }
        self.parent.addTab('@Url.Action("WayBillException",
        "Reconciliation")'.concat(filter.length > 0 ? '/?'.concat(filter) : ''),
        '运单异常数据');
    }
    //控制指定列的隐藏和显示（table，列索引数组，隐藏 or 显示:true,false)
    function controlColumnShow(table, arr, tag) {
        for (var i = 0; i < arr.length; i++) {
            table.fnSetColumnVis(arr[i], tag,false);
        }
    }
    //查看详情
    function openDetail(ExpressNo) {
        var url = "/Reconciliation/WayBillDetailed?ExpressNo=" + ExpressNo;
        goNewPage(url, "运单对账详情-" + ExpressNo);
    }

    //单击打开新页面
    function goNewPage(url, name) {
        self.parent.addTab(url, name);
    }
</script>
<div class="areabx clear">
    @using (Html.BeginForm("List", null, FormMethod.Get, new { @clase = "form-
    inline", @role = "form" }))
    {    <div id="divSearch">
        @*<div class="areabx_header">@ViewBag.Title</div>*@
        <ul class="formod">
            <li><span>客户简称：</span>@Html.TextBox("CusShortName", "", new { @class
```

```
            = "trade-time wid153" })</li>
            <li><span>提单号: </span>@Html.TextBox("LoadBillNum", "", new { @class =
            "trade-time" })</li>
        </ul>
        <ul class="formod">
            <li><span>运单号: </span>@Html.TextBox("ExpressNo", "", new { @class =
            "trade-time wid153" })</li>
            <li><span>收寄日期: </span>@Html.TextBox("PostingTime", "", new { @class
            = "trade-time wid153", @onClick = "WdatePicker
            ({maxDate:'#F{$dp.$D(\\'PostingTimeTo\\')}'})" })</li>
            <li><span class="css_Span">—</span> @Html.TextBox("PostingTimeTo", "",
            new { @class = "trade-time wid153", @onClick = "WdatePicker
            ({minDate:'#F{$dp.$D(\\'PostingTime\\')}'})" })</li>
            <li><span>毛利: </span><select class="trade-time" id="sltMargin"><option
            value="" selected="selected">全部</option>
                <option value="+">+</option>
                <option value="-">-</option>
            </select></li>
        </ul>
        <div class="botbtbx pdb0">
            <input type="button" value="查询" id="btnSearch" onclick="reloadList();"
            class="btn btn-primary" />
            <span class="spanMsg">提示:
                @*<a target="mainFrame"  id="UnusualCount"
                onclick="goNewPage('@Url.Action("WayBillException",
                "Reconciliation")','运单异常数据')"></a>*@
                <a target="mainFrame" id="UnusualCount" onclick="openUnusual()"></a>
            </span>
        </div>
    </div>
}
<div class="tob_box mgt15">
    <table id="table_local" class="display" cellspacing="0" cellpadding="0"
    border="0" style="width: 100%">
        <thead>
            <tr>
                <th rowspan="2"><input type='checkbox' id='chkAllColl'
                onclick='selectAll()' />序号</th>
                <th rowspan="2">客户简称</th>
                <th rowspan="2">收寄日期</th>
                <th rowspan="2">对账日期</th>
                <th rowspan="2">运单号</th>
```

```
        <th rowspan="2">提单号</th>
        <th rowspan="2">重量(kg)</th>
        <th colspan="5"><span>成本</span><span class="divIncome1"><img
        id="imgIncome" src="/images/icon_9.png" alt="收起/展开" title="
        收起/展开" /></span></th>
        <th colspan="5"><span>收入</span><span class="divIncome1"><img
        id="imgCost" src="/images/icon_9.png" alt="收起/展开" title="收起/
        展开" /></span></th>
        <th colspan="3">毛利</th>
        <th rowspan="2">对账单状态</th>
        <th rowspan="2">操作</th>
    </tr>
    <tr>
        <th>邮政邮资</th>
        <th>邮政邮件处理费</th>
        <th>其他费用</th>
        <th>总成本</th>
        <th>状态</th>
        <th>客户运费</th>
        <th>客户操作费</th>
        <th>其他费用</th>
        <th>总收入</th>
        <th>状态</th>
        <th>运费毛利</th>
        <th>总毛利</th>
        <th>毛利率</th>
    </tr>
</thead>
   <tfoot>
    <tr>
        <td>总计</td>
        <td></td>
        <td></td>
        <td></td>
        <td></td>
        <td></td>
        <td><span id="spnTotalWeight"></span></td>
        <td><span id="spnTotalWayBillFee"></span></td>
        <td><span id="spnTotalProcessingFee"></span></td>
        <td></td>
        <td></td>
        <td></td>
```

```
                <td><span id="spnTotalExpressFee"></span></td>
                <td><span id="spnTotalOperateFee"></span></td>
                <td></td>
                <td></td>
                <td></td>
                <td><span id="spnSumWayBillProfit"></span></td>
                <td><span id="spnSumTotalProfit"></span></td>
                <td></td>
                <td></td>
                <td></td>
            </tr>
        </tfoot>
    </table>
  </div>
</div>
```

"table.fnSetColumnVis(arr[i], tag);"这行代码控制列动态隐藏和展示的时候会重新加载数据，可以在后面加一个 false 参数，取消刷新。例如：

```
table.fnSetColumnVis(arr[i], tag,false);
```

请求参数封装类 DataTablesRequest 是从冠军的博客下载的，主要用于解析 datatables 的请求参数，由于 datatables 支持多列排序，因此比较复杂。下载的这个类有点问题，那就是获取的排序方式一直是 asc，这里进行了修改，修改后的代码如下：

```
using System;
using System.Collections.Generic;
using System.Linq;
using System.Web;

namespace ProjectBase.Utils
{
   // 排序的方向
   public enum SortDirection
   {
      Asc,    // 升序
      Desc    // 降序
   }

   // 排序列的定义
   public class SortColumn
   {
      public int Index { get; set; }                      // 列序号
      public SortDirection Direction { get; set; }    // 列的排序方向
   }
```

```
// 列定义
public class Column
{
    public string Name { get; set; }          // 列名
    public bool Sortable { get; set; }        // 是否可排序
    public bool Searchable { get; set; }      // 是否可搜索
    public string Search { get; set; }        // 搜索串
    public bool EscapeRegex { get; set; }     // 是否正则
}

public class DataTablesRequest
{
    private HttpRequestBase request;          // 内部使用的 Request 对象

    public DataTablesRequest(System.Web.HttpRequestBase request)
    // 用于 MVC 模式下的构造函数
    {
        this.request = request;

        this.echo = this.ParseStringParameter(sEchoParameter);
        this.displayStart = this.ParseIntParameter(iDisplayStartParameter);
        this.displayLength = this.ParseIntParameter(iDisplayLengthParameter);
        this.sortingCols = this.ParseIntParameter(iSortingColsParameter);

        this.search = this.ParseStringParameter(sSearchParameter);
        this.regex = this.ParseStringParameter(bRegexParameter) == "true";

        // 排序的列
        int count = this.iSortingCols;
        this.sortColumns = new SortColumn[count];
        for (int i = 0; i < count; i++)
        {
            SortColumn col = new SortColumn();
            col.Index = this.ParseIntParameter(string.Format("iSortCol_{0}",i));

            if(this.ParseStringParameter(string.Format("sSortDir_{0}",i))=="desc")
            {
                col.Direction = SortDirection.Desc;
            }
            else
            {
                col.Direction = SortDirection.Asc;
            }
            this.sortColumns[i] = col;
        }
```

```
        this.ColumnCount = this.ParseIntParameter(iColumnsParameter);

        count = this.ColumnCount;
        this.columns = new Column[count];

        if(this.ParseStringParameter(sColumnsParameter)==null||!this.
        ParseStringParameter(sColumnsParameter).Contains(','))
        {
            return;
        }
        string[] names = this.ParseStringParameter(sColumnsParameter)
        .Split(',');

        for (int i = 0; i < count; i++)
        {
            Column col = new Column();
            col.Name = names[i];
            col.Sortable = this.ParseStringParameter
            (string.Format("bSortable_{0}", i)) == "true";
            col.Searchable = this.ParseStringParameter
            (string.Format("bSearchable_{0}", i)) == "true";
            col.Search = this.ParseStringParameter(string.Format("sSearch_{0}", i));
            col.EscapeRegex = this.ParseStringParameter(string.Format("bRegex_{0}",
            i)) == "true";
            columns[i] = col;
        }
    }
    public DataTablesRequest(HttpRequest httpRequest)
    // 标准的 WinForm 方式下的构造函数
        : this(new HttpRequestWrapper(httpRequest))
    { }

    #region
    private const string sEchoParameter = "sEcho";

    // 起始索引和长度
    private const string iDisplayStartParameter = "iDisplayStart";
    private const string iDisplayLengthParameter = "iDisplayLength";

    // 列数
    private const string iColumnsParameter = "iColumns";
    private const string sColumnsParameter = "sColumns";

    // 参与排序列数
    private const string iSortingColsParameter = "iSortingCols";
    private const string iSortColPrefixParameter = "iSortCol_"; // 排序列的索引
    private const string sSortDirPrefixParameter = "sSortDir_";
```

```
// 排序的方向 asc, desc

// 每一列的可排序性
private const string bSortablePrefixParameter = "bSortable_";

// 全局搜索
private const string sSearchParameter = "sSearch";
private const string bRegexParameter = "bRegex";

// 每一列的搜索
private const string bSearchablePrefixParameter = "bSearchable_";
private const string sSearchPrefixParameter = "sSearch_";
private const string bEscapeRegexPrefixParameter = "bRegex_";
#endregion

private readonly string echo;
public string sEcho
{
   get { return echo; }
}

private readonly int displayStart;
public int iDisplayStart
{
   get { return this.displayStart; }
}

private readonly int displayLength;
public int iDisplayLength
{
   get { return this.displayLength; }
}

// 参与排序的列
private readonly int sortingCols;
public int iSortingCols
{
   get { return this.sortingCols; }
}

// 排序列
private readonly SortColumn[] sortColumns;
public SortColumn[] SortColumns
{
   get { return sortColumns; }
}
```

```
private readonly int ColumnCount;
public int iColumns
{
    get { return this.ColumnCount; }
}

private readonly Column[] columns;
public Column[] Columns
{
    get { return this.columns; }
}

private readonly string search;
public string Search
{
    get { return this.search; }
}

private readonly bool regex;
public bool Regex
{
    get { return this.regex; }
}

#region 常用的几个解析方法
private int ParseIntParameter(string name)            // 解析为整数
{
    int result = 0;
    string parameter = this.request[name];
    if (!string.IsNullOrEmpty(parameter))
    {
        int.TryParse(parameter, out result);
    }
    return result;
}

private string ParseStringParameter(string name)     // 解析为字符串
{
    return this.request[name];
}

private bool ParseBooleanParameter(string name)      // 解析为布尔类型
{
    bool result = false;
    string parameter = this.request[name];
    if (!string.IsNullOrEmpty(parameter))
    {
```

```
            bool.TryParse(parameter, out result);
        }
        return result;
    }
    #endregion
  }
}
```

11.9 服务器端排序

关于 jQuery datables 在服务器端的排序，在网上貌似没有.NET 的例子。事实上 datables 是支持多列排序的，本例只写了单列排序，运行效果如图 11-17 所示。

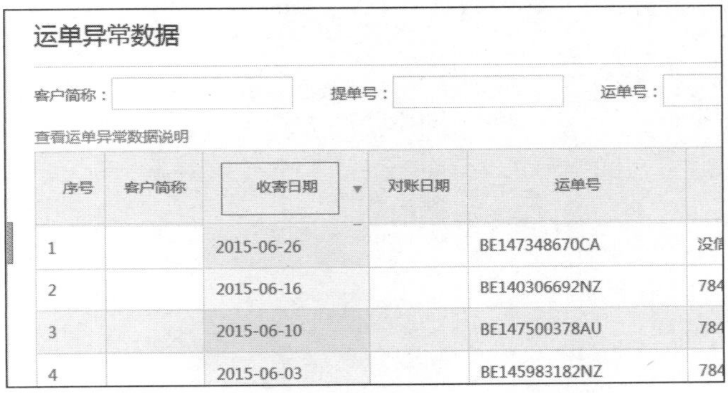

运单异常数据

客户简称：　　提单号：　　运单号：

查看运单异常数据说明

序号	客户简称	收寄日期	对账日期	运单号	
1		2015-06-26		BE147348670CA	没信
2		2015-06-16		BE140306692NZ	784
3		2015-06-10		BE147500378AU	784
4		2015-06-03		BE145983182NZ	784

图 11-17

在控制器中：

```
Dictionary<int, string> dicSort = new Dictionary<int, string>();
//排序字段键值对列表 （列序号，列名称）
/// <summary>
/// 运单异常数据
/// </summary>
/// <returns></returns>
public ActionResult WayBillException(WayBillExceptionFilter filter)
{
        return View(filter);
}

public JsonResult WayBillExceptionList(WayBillExceptionFilter filter)
{
        dicSort.Add(2, "w.PostingTime");

        DataTablesRequest parm = new DataTablesRequest(this.Request); //处理对象
```

```
int pageIndex = parm.iDisplayLength == 0 ? 0 : parm.iDisplayStart /
parm.iDisplayLength;
filter.PageIndex = pageIndex;    //页索引
filter.PageSize = parm.iDisplayLength;    //页行数

string strSortField = dicSort.Where(x => x.Key ==
parm.SortColumns[0].Index).Select(x => x.Value).FirstOrDefault();
string strSortDire = parm.SortColumns[0].Direction ==
SortDirection.Asc ? "asc" : "desc";

filter.OrderBy = " " + strSortField + " " + strSortDire;

var DataSource = Core.Reconciliation.
 WayBillException.GetByFilter(filter)as WRPageOfList<WayBillException>;

int i = parm.iDisplayLength * pageIndex;

List<WayBillException> queryData = DataSource.ToList();
var data = queryData.Select(u => new
{
    Index = ++i, //行号
    ID = u.ID,
    IsInputCost = u.IsInputCost,
    CusName = u.CusName, //客户简称
    PostingTime = u.PostingTime == null ? string.Empty :
    u.PostingTime.Value.ToStringDate(),//收寄日期
    ExpressNo = u.ExpressNo, //运单号
    BatchNO = u.LoadBillNum, //提单号
    Weight = u.Weight == null ? 0m : u.Weight / 1000, //重量
    WayBillFee = u.WayBillFee, //邮资
    ProcessingFee = u.ProcessingFee, //邮政邮件处理费
    InComeWayBillFee = u.ExpressFee, //客户运费
    InComeOprateFee = u.OperateFee, //客户操作费
    WayBillMargins = u.WayBillProfit, //运费毛利
    TotalMargins = u.ExpressFee + u.OperateFee + u.InComeOtherFee -
    (u.WayBillFee + u.ProcessingFee + u.CostOtherFee), //总毛利
    Margin = Math.Round((u.ExpressFee + u.OperateFee + u.InComeOtherFee
    == 0 ? 0m : (u.ExpressFee + u.OperateFee + u.InComeOtherFee -
    (u.WayBillFee + u.ProcessingFee + u.CostOtherFee)) / (u.ExpressFee
    + .OperateFee + u.InComeOtherFee) * 100), 2) + "%",
  //毛利率 毛利率=(总收入-总的支出的成本)/总收入*100%
    ReconcileDate = u.ReconcileDate.ToStringDate(), //对账日期
```

```
            CostOtherFee = u.CostOtherFee, //成本，其他费用
            CostTotalFee = u.WayBillFee + u.ProcessingFee + u.CostOtherFee,
             //成本，总费用
            CostStatus = u.CostStatus.ToChinese(),  //成本，状态
            InComeOtherFee = u.InComeOtherFee, //收入，其他费用
            InComeTotalFee = u.ExpressFee + u.OperateFee + u.InComeOtherFee,
            //收入，总费用
            InComeStatus = u.InComeStatus.ToChinese(),  //收入，状态
            ExceptionMsg = u.ExceptionMsg, //运单异常原因
            WayBillCostID = u.WayBillCostID //运单成本 ID
            // ExceptionType = u.ExceptionType  //运单异常状态
        });
        //decimal totalProfit = 0m;      //总毛利求和
        //构造成 Json 的格式传递
        var result = new
        {
            iTotalRecords = DataSource.Count,
            iTotalDisplayRecords = DataSource.RecordTotal,
            data = data
        };
        return Json(result, JsonRequestBehavior.AllowGet);
}
```

在 View 中，设置 datatables 的属性：

```
bServerSide: true,    //指定从服务器端获取数据
//跟数组下标一样，第一列从0开始，这里表格初始化时第三列默认降序
order: [[2, "desc"]],
```

当单击排序的时候，可以打开浏览器的 Firebug 查看数据，如图 11-18 所示。

如图 11-19 所示，第一列是排序字段的列索引，第二个字段标识有一个排序字段，因为这个控件是支持多列排序的。

图 11-18

iSortCol_0　2
iSortingCols　1

图 11-19

DataTablesRequest 类里面封装了对这些请求的处理：

```
string strSortField = dicSort.Where(x => x.Key == parm.SortColumns[0].
Index).Select(x => x.Value).FirstOrDefault();
```

```
            string strSortDire = parm.SortColumns[0].Direction ==
SortDirection.Asc ? "asc" : "desc";
```

11.10 从 ASP.NET MVC 中导出 Excel 文件

这里要在 ASP.NET MVC 站点上做 Excel 导出功能，但是要导出的 Excel 文件比较大，有几十兆字节，所以导出比较费时。为了不影响对界面的其他操作，采取异步的方式，后台开辟一个线程将 Excel 导出到指定目录，然后提供下载。导出的 Excel 涉及多个 sheet（工作簿）、表格合并、格式设置等，所以采用了 NPOI 组件。效果如图 11-20~图 11-22 所示。

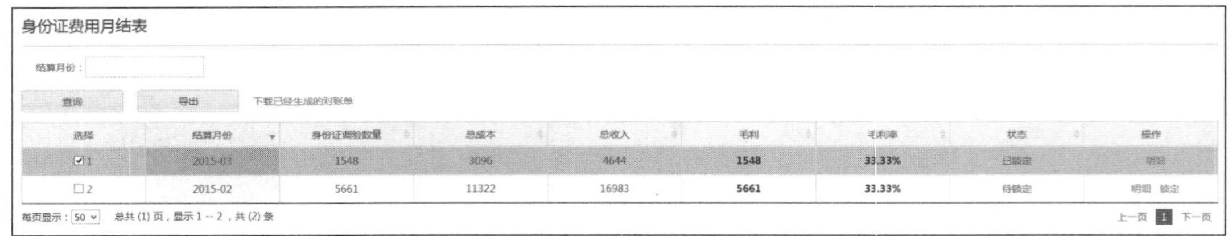

图 11-20

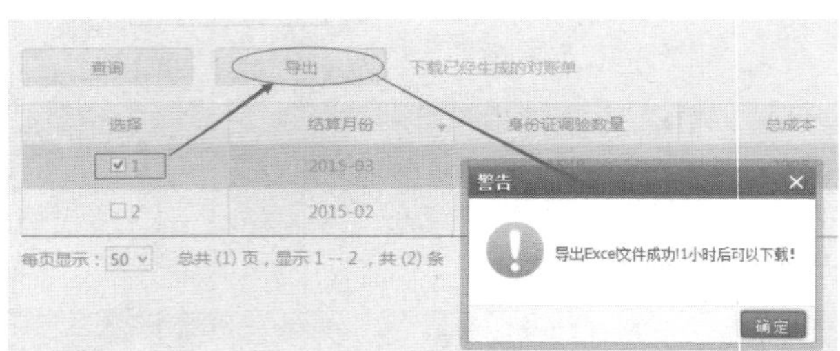

图 11-21

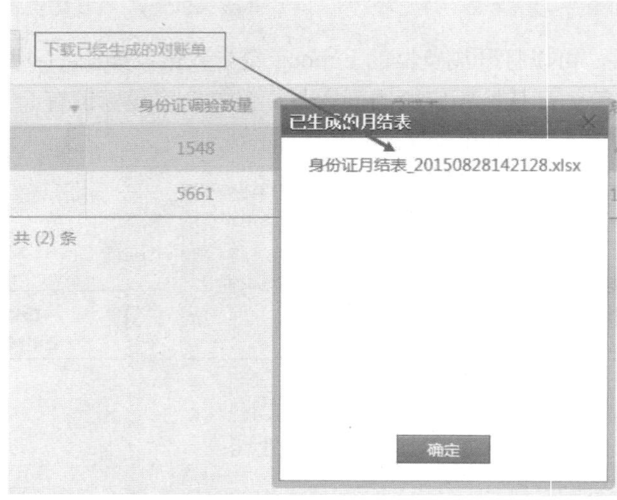

图 11-22

选中多行就会导出多个工作簿，一个是汇总，其他的都是明细数据，导出后运行结果如图 11-23~图 11-25 所示。

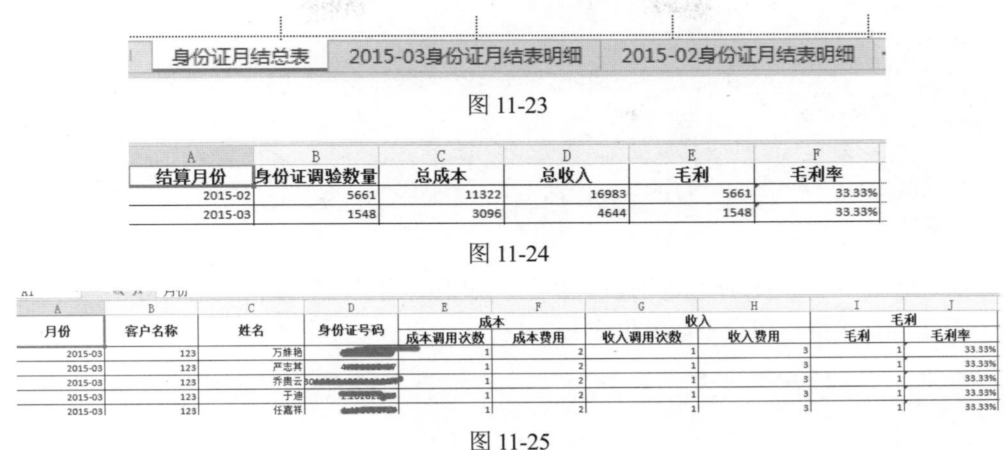

图 11-23

结算月份	身份证调验数量	总成本	总收入	毛利	毛利率
2015-02	5661	11322	16983	5661	33.33%
2015-03	1548	3096	4644	1548	33.33%

图 11-24

月份	客户名称	姓名	身份证号码	成本		收入		毛利	
				成本调用次数	成本费用	收入调用次数	收入费用	毛利	毛利率
2015-03	123	万姝艳	[illegible]	1	2	1	3	1	33.33%
2015-03	123	严志其	[illegible]	1	2	1	3	1	33.33%
2015-03	123	乔贵云	[illegible]	1	2	1	3	1	33.33%
2015-03	123	于迪	[illegible]	1	2	1	3	1	33.33%
2015-03	123	任嘉祥	[illegible]	1	2	1	3	1	33.33%

图 11-25

11.10.1 异步导出

下面是几个封装好的类，都是从网上找的，然后修改了一下。这几个类的很多方法都封装好了，十分利于复用。常见的 Excel 格式都可以导出，如果有特别的需求，也可以修改一下源代码进行扩展。

```
public JsonResult ExportExcel(MonthPayOffExportFilter filter)
{
        string excelPath = this.Server.MapPath(string.Format("/Excel/月结表
        _{0}.xls", DateTime.Now.ToString("yyyyMMddHHmmss") ));
        Core.Receivable.MonthPayOff.ExportExcel(excelPath, filter);

        var result = new { IsSuccess=true,Message="成功"};

        return Json(result);
    }

    /// <summary>
    /// 已生成的月结表列表
    /// </summary>
    /// <returns></returns>
    public ActionResult LoadExcelList()
    {
        string myDir = Server.MapPath("~/Excel");

        if (Directory.Exists(Server.MapPath("~/Excel")) == false)
```

```
        //如果不存在就创建 file 文件夹
        {
            Directory.CreateDirectory(Server.MapPath("~/Excel"));
        }
        DirectoryInfo dirInfo = new DirectoryInfo(myDir);
        List<LinkEntity> list = LinkEntityExt.ForFileLength(dirInfo);

        return View(list);
}
```

Global.asax.cs 在应用程序启动时监听队列，如果队列里面有数据就进行导出操作，即使操作人员离开了当前页面，也会不影响生成 Excel 操作，而且使用队列可以有效地解决并发问题。

```
public class MvcApplication : System.Web.HttpApplication
{
        protected void Application_Start()
        {
            AreaRegistration.RegisterAllAreas();

            WebApiConfig.Register(GlobalConfiguration.Configuration);
            FilterConfig.RegisterGlobalFilters(GlobalFilters.Filters);
            RouteConfig.RegisterRoutes(RouteTable.Routes);
            //BundleTable.EnableOptimizations = true;
            BundleConfig.RegisterBundles(BundleTable.Bundles);
            AuthConfig.RegisterAuth();
            RegisterContainer(ProjectBase.Data.IocContainer.Instance.Container);
            log4net.Config.XmlConfigurator.Configure();

            SysInfo.SysInfoModel = new SysInfo().GetSysConfigInfo();
        }

        public static void RegisterContainer(UnityContainer container)
        {
            var iRepositories = Assembly.Load("MSD.Finance.Core").GetTypes()
                .Where(p => p.FullName.Contains("Repositories")).ToList();
            var repositories = Assembly.Load("MSD.Finance.Data").GetTypes()
                .Where(p => p.FullName.Contains("Repository")).ToList();
            iRepositories.ForEach(
                p => container.RegisterType(p, repositories.Where(r => r.Name ==
p.Name.Substring(1, p.Name.Length - 1)).FirstOrDefault()));
        }
}
```

11.10.2 实时导出

实时导出有好几种方式，这里采用 FileResult 来进行导出，使用 FileResult 导出要求服务器上面必须存在 Excel 文件。如果没有选中任何行，就导出查询到的所有数据，否则导出选中行的数据。由于数据不是很多，因此这里采用实时导出的方式。

前台 js 代码：

```
//导出 Excel
function exportExcel(table) {
        var nTrs = table.fnGetNodes();
        //fnGetNodes 获取表格所有行，nTrs[i]表示第 i 行 tr 对象
        var row;
        var strdid = '';
        var selectCounts = 0;
        for (var i = 0; i < nTrs.length; i++) {
            if ($(nTrs[i])[0].cells[0].children[0].checked) {
                row = table.fnGetData(nTrs[i]);//fnGetData 获取一行的数据
                selectCounts++;
                strdid += "" + row.ID + ",";
            }
        }
        strdid = strdid.length > 0 ? strdid.substring(0, strdid.length - 1) :
        strdid;
        if (selectCounts < 1) { //按照查询结果进行导出
            window.location.href = '@Url.Action("ExportExcelByFilter",
            "Reconciliation")?' + "CusShortName=" + $("#CusShortName").val()
            +"&&LoadBillNum=" + $("#LoadBillNum").val() +"&&PostingTime=" +
            $("#PostingTime").val()+"&&PostingTimeTo="+$("PostingTimeTo").val() +
            "&&ExceptionType="+$("#ExceptionType").val();
        }
        else { //导出选中行
            //window.location.href = '@Url.Action("ExportExcelBySelect",
            //"Reconciliation")?' + "ListID=" + strdid; 地址栏太长会超出
            $.post('@Url.Action("ExportExcelBySelect", "Reconciliation")',
            { "ListID": strdid }, function (data) {
                window.location.href = data;
            });
        }
}
```

控制器中的代码如下：

```
/// <summary>
/// 导出选中的异常记录
/// </summary>
/// <param name="ListID"></param>
/// <returns></returns>
public JsonResult ExportExcelBySelect(string ListID)
{
          string url = "/Downloads/WayBillException/运单异常记录.xls";
          string excelUrl = Server.MapPath("~" + url);
          Core.Reconciliation.WayBillException.ExportExcel(excelUrl, ListID);
          return Json(url);
}
/// <summary>
/// 导出查询的异常记录
/// </summary>
/// <param name="filter"></param>
/// <returns></returns>
public FileResult ExportExcelByFilter(WayBillExceptionFilter filter)
{
          filter.PageSize = int.MaxValue;
          string excelUrl = Server.MapPath("~/Downloads/WayBillException/
          运单异常记录.xls");
          Core.Reconciliation.WayBillException.ExportExcel(filter,excelUrl);

          return File(excelUrl, "application/ms-excel", "运单异常记录.xls");
}
```

11.11 数据同步

代码所在文件夹为 DataSynchro，项目文件为 B2CSynchro.sln，项目结果如图 11-26 所示。

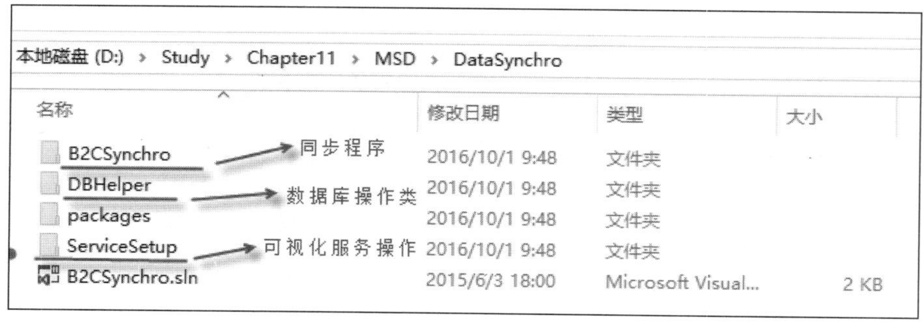

图 11-26

这个项目主要采用 Quartz.Net 定时任务框架来定时调用清关系统提供的 WebAPI 来抓取订单、提单、客户、身份证验证等数据信息。把定时任务寄宿在 Windows 服务中，然后通过一个可视化工具来操作服务的安装、卸载、启动、停用，以方便实施人员部署和操作。

关于清关系统提供的 WebAPI 暂时没有提供源码。

注意：本章示例项目所使用的数据库位于“MSD 项目完整版”目录下，分别是数据库脚本 msd_finacnce.sql 与数据库备份文件 160609142022.psc，如图 11-27 所示。

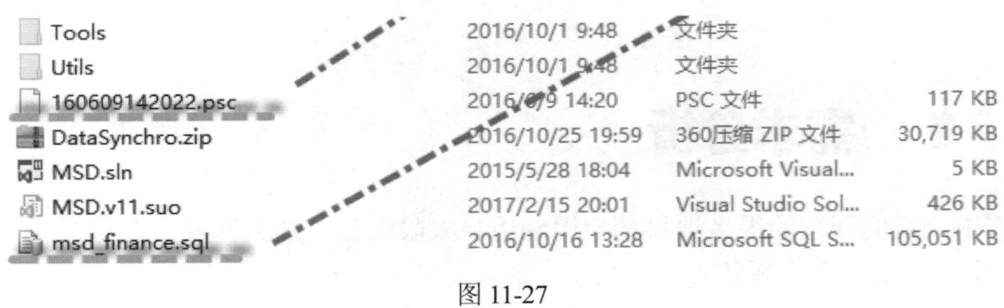

图 11-27

第 12 章 通用角色权限管理系统

12.1 需求分析

在工作中，我们开发的很多项目都会用到用户权限管理，那么开发一个相对比较通用的角色权限系统就可以实现各个系统复用，从而节省开发成本。

通用权限管理系统主要涉及以下几大常用功能模块：

- 系统管理：对系统信息进行统一管理式。
- 机构管理：不同的用户可能会在不同的组织机构下。
- 用户管理：维护用户账号及其他信息。
- 角色管理：维护系统角色信息、角色用户以及角色权限等。
- 菜单管理：维护系统菜单及功能模块。
- 字典管理：维护系统字典。
- 日志管理：管理系统访问日志。

系统所要实现的主要特点包括权限控制实现到按钮级权限控制；日志记录可记录客户端外网地址及操作；数据接口层支持多种常用数据库，可实现数据库移植。

机构管理和跨数据库支持留给大家去扩展开发，本项目中不做实现。同时，此角色权限系统的设计也存在一些不足的地方，大家可以自己优化。例如，添加一个临时用户，不得不新创建一个角色，然后给新角色赋权限，再把这个临时用户的角色配置为新角色。

扩展功能：

- 一键切换皮肤，并记录在客户端一段时间，比如 7 天。
- 自动登录功能，并实现 Session 共享。

12.2 技术选型

在清楚需求之后，我们首先要做的就是技术选型。

- ASP.NET MVC 4：本书讲解的就是这项技术。
- jQuery EasyUI：大众、传统化、简单易上手。

- EF：微软的 ORM 框架。
- SQL Server 2012：可与 EF 很好地集成。
- Redis：做日志队列。
- Log4Net：记录日志。
- Memcached：Session 共享。
- T4 模板：减少编码量。

这里采用的是 EF Database First 方式，因为：

- 数据库表结构是直接从以前的项目中拿过来的。
- 第 11 章 NHibernate 的代码映射方式和 EF 的 Code First 方式很像。
- 希望大家在学习完本章后可以尝试采用 Code First 方式来重写。

使用 EF Database First 方式需要注意的是：所有表必须设置主键，表和表之前的关系必须通过主外键进行关联。

12.3 数据库设计

由于以前公司项目使用的是 Oracle 数据库，因此转换为 SQL Server 表结构后字段都是大写的。

创建好数据库和数据表之后，PD 字典是可以从数据库中自动生成的，这里使用 PowerDesigner。之前经常有人会问 PD 字典如何导出图片，或者 PD 字典比较大如何滚动截图。其实，使用 PowerDesigner 的 Report 功能可以直接把 PD 字典导出到 Word 中，既可以是数据表模型图，也可以是 Word 表格。

以后如果要整理一个 Word 格式的数据库字典，就不要傻乎乎地手动去做了，可以编写一个 PowerDesigner 的 Report 模板，或者从网上下载一个模板，自己稍加修改，就可以自动导出成 Word 格式的字典。这里做了一个 wordTemplate.rtp 模板，大家以后如果要使用，可以直接把这个文件复制到 PowerDesigner 安装目录下面，这里安装在了 D 盘，所以目录是 D:\Program Files (x86)\Sybase\PowerDesigner 15\Resource Files\Report Templates。

关于“PowerDesigner 从 SQL Server 中反转为带注释的字典”的操作，大家可以参考 http://www.cnblogs.com/jiekzou/p/5721522.html 上的文章。“PowerDesigner 导出表到 word”的内容可以参考 http://blog.csdn.net/ferry_passion/article/details/8948456 上的文章。

数据表结构如图 12-1 所示。

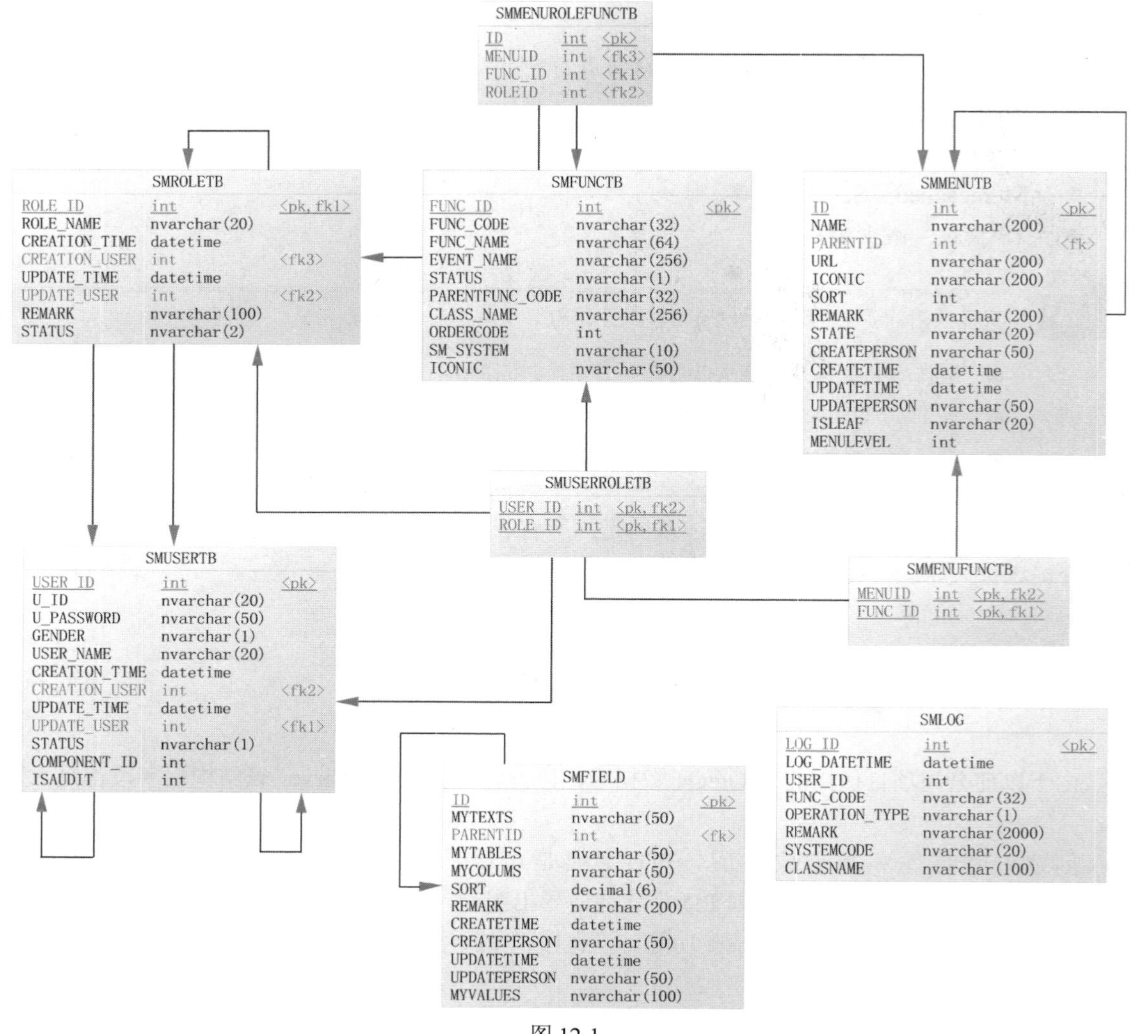

图 12-1

表清单如表 12-1 所示。

表 12-1　表清单

名称	说明
SMFIELD	字段表
SMFUNCTB	功能表
SMLOG	日志表
SMMENUFUNCTB	菜单功能表
SMMENUROLEFUNCTB	菜单角色功能表
SMMENUTB	菜单表
SMROLETB	角色表
SMUSERROLETB	用户角色表
SMUSERTB	用户表

表中的列清单如表 12-2~表 12-10 所示。

表 12-2　<SMFIELD> 的列清单

名称	数据类型	注释
ID	int	主键 ID
MYTEXTS	nvarchar(50)	名称
PARENTID	int	父节点
MYTABLES	nvarchar(50)	表名
MYCOLUMS	nvarchar(50)	字段名
SORT	decimal(6)	排序
REMARK	nvarchar(200)	备注
CREATETIME	datetime	创建时间
CREATEPERSON	nvarchar(50)	创建者
UPDATETIME	datetime	更新时间
UPDATEPERSON	nvarchar(50)	更新人
MYVALUES	nvarchar(100)	字段简称

表 12-3 <SMFUNCTB> 的列清单

名称	数据类型	注释
FUNC_ID	int	功能 ID，主键
FUNC_CODE	nvarchar(32)	功能编号
FUNC_NAME	nvarchar(64)	功能名称
EVENT_NAME	nvarchar(256)	事件名
STATUS	nvarchar(1)	状态（Y：启用，N：禁用）
PARENTFUNC_CODE	nvarchar(32)	父功能编号
CLASS_NAME	nvarchar(256)	类名
ORDERCODE	int	显示顺序
SM_SYSTEM	nvarchar(10)	系统名
ICONIC	nvarchar(50)	图标样式

表 12-4 <SMLOG> 的列清单

名称	数据类型	注释
LOG_ID	int	日志 ID
LOG_DATETIME	datetime	创建时间
USER_ID	int	用户 ID
FUNC_CODE	nvarchar(32)	功能编码
OPERATION_TYPE	nvarchar(1)	操作类型
REMARK	nvarchar(2000)	说明
SYSTEMCODE	nvarchar(20)	系统编码
CLASSNAME	nvarchar(100)	类名

表 12-5 <SMMENUFUNCTB> 的列清单

名称	数据类型	注释
MENUID	int	菜单 ID
FUNC_ID	int	功能 ID

表 12-6 <SMMENUROLEFUNCTB> 的列清单

名称	数据类型	注释
ID	int	主键 ID
MENUID	int	菜单 ID
FUNC_ID	int	功能 ID
ROLEID	int	角色 ID

表 12-7 <SMMENUTB> 的列清单

名称	数据类型	注释
ID	int	菜单 ID
NAME	nvarchar(200)	菜单名称
PARENTID	int	父菜单 ID
URL	nvarchar(200)	菜单地址
ICONIC	nvarchar(200)	菜单图标样式
SORT	int	显示顺序
REMARK	nvarchar(200)	备注
STATE	nvarchar(20)	状态（Y：启用，N：禁用）
CREATEPERSON	nvarchar(50)	创建者
CREATETIME	datetime	创建时间
UPDATETIME	datetime	更新时间
UPDATEPERSON	nvarchar(50)	更新者
ISLEAF	nvarchar(20)	是否叶子节点（Y：是，N：否）
MENULEVEL	int	层级

表 12-8 <SMROLETB> 的列清单

名称	数据类型	注释
ROLE_ID	int	角色 ID
ROLE_NAME	nvarchar(20)	角色名称
CREATION_TIME	datetime	创建时间
CREATION_USER	int	创建者
UPDATE_TIME	datetime	更新时间
UPDATE_USER	int	更新者
REMARK	nvarchar(100)	说明
STATUS	nvarchar(2)	状态（Y：是，N：否）

表 12-9 <SMUSERROLETB> 的列清单

名称	数据类型	注释
USER_ID	int	用户 ID
ROLE_ID	int	角色 ID

表 12-10 <SMUSERTB> 的列清单

名称	数据类型	注释
USER_ID	int	用户 ID
U_ID	nvarchar(20)	登录用户名
U_PASSWORD	nvarchar(50)	登录密码
GENDER	nvarchar(1)	性别（W：女，M：男）
USER_NAME	nvarchar(20)	用户姓名
CREATION_TIME	datetime	创建时间
CREATION_USER	int	创建者
UPDATE_TIME	datetime	更新时间
UPDATE_USER	int	更新者
STATUS	nvarchar(1)	状态（Y：启用，N：禁用）
COMPONENT_ID	int	机构 ID
ISAUDIT	int	审核通过（预留）

12.4 架构搭建

在新建项目、对项目进行命名时，一般都会遵循相应的命名规范，虽然不同的公司可能会有不同的命名规范，但是行业内用得比较多的命名规范是：以公司名开头，然后是项目名，最后是类库名称。例如，My.RolePermission.IBLL。其中，公司名称是 My，项目名称是 RolePermission，类库名称是 IBLL。

12.4.1 新建解决方案和项目

（1）新建空白解决方案。

（2）新建解决方案文件夹。新建解决方案文件夹可以分类打包一些项目，让整个项目变得更清晰，如图 12-2 所示。

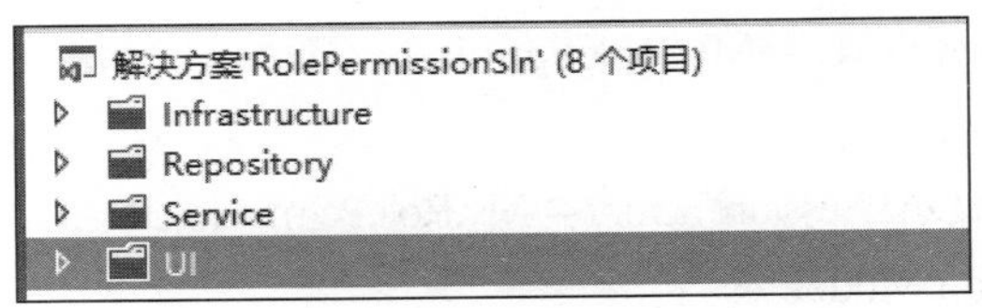

图 12-2

- Infrastructure：基础架构，包含一些通用的组件以及 IOC 容器。
- Repository：数据仓储。
- Service：服务层（业务逻辑和 API 接口），此层使用 WebService、WCF 或者 WebAPI 将业务封装，对外暴露成服务，也就是 SOA 化。在本项目中并没有 SOA 化，读者可以自行进行 SOA 化扩展。
- UI：界面（MVC 项目）。

（3）新建 MVC 项目。右击 UI 文件夹，选择“新建项目→Web→ASP.NET MVC 4 Web 应用程序”，并命名为“My.RolePermission.WebApp”，项目模板选择“基本”，视图引擎选择“Razor”。

（4）新建类库项目。

在解决方案文件夹 Service 下分别新建类库项目 My.RolePermission.IBLL、My.RolePermission.BLL。

在解决方案文件件 Repository 下分别新建类库项目 My.RolePermission.IDAL、My.RolePermission.DAL、My.RolePermission.DALSessionFactory。

在解决方案文件件 Infrastructure 下分别新建类库项目 My.RolePermission.Common、My.RolePermission.Model、My.RolePermission.Ioc。

最终的项目架构如图 12-3 所示。

图 12-3

添加项目之间的引用：

- My.RolePermission.Model -->My.RolePermission.Common。
- My.RolePermission.IDAL-->My.RolePermission.Model。
- My.RolePermission.DAL-->My.RolePermission.IDAL、My.RolePermission.Model、My .RolePermission.Ioc。
- My.RolePermission.DALSessionFactory-->My.RolePermission.IDAL、My.RolePermission.DAL、My.RolePermission.Model。
- My.RolePermission.IBLL--> My.RolePermission.Model、My.RolePermission.IDAL。

- My.RolePermission.BLL–> My.RolePermission.IBLL、My.RolePermission.IDAL、My.RolePermission.Model、My.RolePermission.DALSessionFactory、My.RolePermission.Common。
- My.RolePermission.WebApp--> My.RolePermission.IBLL、My.RolePermission.Model、My.RolePermission.BLL（为什么要引用这个 dll 呢？因为后面在使用 spring.net 注入的时候会用到）、My.RolePermission.Common、My.RolePermission.Ioc。

12.4.2 通用层搭建

这里是按照个人搭建框架的习惯进行的，一般顺序是通用层→Model→数据库访问层→业务逻辑层→UI 层，而且会先写接口后写具体实现类。

```
My.RolePermission.Ioc
```

这里选用 Ioc 框架是 Spring.net。

（1）添加 Spring.Core.dll 的引用。

（2）新建类 SpringHelper：

```
using Spring.Context;
using Spring.Context.Support;

namespace My.RolePermission.Ioc
{
    public class SpringHelper
    {
        #region Spring 容器上下文 +IApplicationContext SpringContext
        /// <summary>
        /// Spring 容器上下文
        /// </summary>
        private static IApplicationContext SpringContext
        {
            get
            {
                return ContextRegistry.GetContext();
            }
        }
        #endregion

        #region 获取配置文件配置的对象 +T GetObject<T>(string objName) where T : class
        /// <summary>
        /// 获取配置文件配置的对象
        /// </summary>
```

```
        /// <typeparam name="T"></typeparam>
        /// <param name="objName"></param>
        /// <returns></returns>
        public static T GetObject<T>(string objName) where T : class
        {
            return (T)SpringContext.GetObject(objName);
        }
        #endregion
    }
}
```

12.4.3 数据访问层搭建

1. My.RolePermission.Model

右击项目 My.RolePermission.Model，选择“添加→新建项→数据→ADO.NET 实体数据模型”，并命名为 RolePermission.edmx。

把 App.Config 中的如下配置节点复制到 My.RolePermission.WebApp 的 Web.config 中。

```
<add name="RolePermissionEntities"
connectionString="metadata=res://*/RolePermission.csdl|res://*/RolePermission.ssd
l|res://*/RolePermission.msl;provider=System.Data.SqlClient;provider connection
string=&quot;data source=.\MSSQLSERVER2012;initial catalog=RolePermission;user
id=sa;password=yujie1127;MultipleActiveResultSets=True;App=EntityFramework&quot;"
providerName="System.Data.EntityClient" />
```

2. My.RolePermission.IDAL

当我们使用不同的数据模型和领域模型时，仓储模式特别有用。仓储可以充当数据模型和领域模型之间的中介。在内部，仓储以数据模型的形式和数据库交互，然后给数据访问层之上的应用层返回领域模型。

在这里，因为使用数据模型作为领域模型，所以也会返回相同的模型。如果想要使用不同的数据模型和领域模型，就需要将数据模型的值映射到领域模型或使用任何映射库执行映射。

（1）新建文件 IBaseRepository.cs，定义仓储接口，在接口中定义一些通用数据库操作方法，即 CRUD 操作。

```
public interface IBaseRepository<T> where T : class,new()
{
        IQueryable<T> LoadEntities(Expression<Func<T, bool>> whereLambda);
        IQueryable<T> LoadPageEntities<model>(int pageIdex, int pageSize, out int
         toalCount, Expression<Func<T, bool>> whereLambda, string orderby, bool?
         isAsc);
```

```
        bool DeleteEntity(T entity);
        bool EditEntity(T entity);
        T AddEntity(T entity);

        IQueryable<T> LoadEntitiesAll(string entity);
}
```

（2）添加一个部分接口 IDBSession：

```
public partial interface IDBSession
{
DbContext Db { get; }
        bool SaveChanges();
        int ExecuteSql(string sql, params System.Data.SqlClient.SqlParameter[] pars);
        List<T> ExecuteSelectQuery<T>(string sql, params System.Data.SqlClient.
        SqlParameter[] pars);
}
```

（3）添加接口 IDBContextFactory，接口中只有一个方法，表示创建数据库上下文对象。

```
public interface IDBContextFactory
{
        DbContext CreateDbContext();
}
```

（4）使用 T4 模板自动生成具体的 IBaseRepository 接口。

之前我们新建了一个泛型接口 IBaseRepository<T>，所以有多少个数据库实体对象就要编写多少个类来继承这个泛型接口，然而这样的操作是乏味的，我们可以考虑使用 VS 中的 T4 模板自动生成这些代码。事实上，当我们使用 Database First 的方式自动生成 ADO.Net 实体模型的时候，也正是使用的 T4 模板自动生成了代码。通过文件名后缀.tt 就可以知道，如图 12-4 所示。

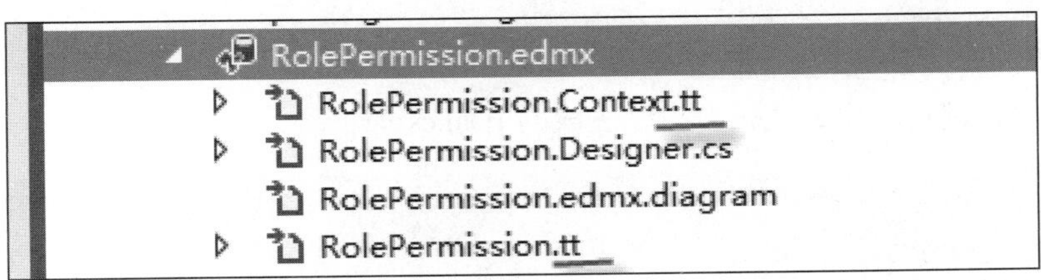

图 12-4

在 VS2012 中去掉了 T4 模板，只有一个文本模板，后缀同样是.tt。不过文本模板是可以直接当成 T4 模板来用的。

选择“添加→新建项→文件模板”，命名为 IDal.tt，然后把如下代码复制进去。

```
<#@ template language="C#" debug="false" hostspecific="true"#>
<#@ include file="EF.Utility.CS.ttinclude"#><#@
```

```
output extension=".cs"#>
<#
CodeGenerationTools code = new CodeGenerationTools(this);
MetadataLoader loader = new MetadataLoader(this);
CodeRegion region = new CodeRegion(this, 1);
MetadataTools ef = new MetadataTools(this);
string inputFile = @"..\My.RolePermission.Model\RolePermission.edmx";
EdmItemCollection ItemCollection = loader.CreateEdmItemCollection(inputFile);
string namespaceName = code.VsNamespaceSuggestion();
EntityFrameworkTemplateFileManager fileManager =
EntityFrameworkTemplateFileManager.Create(this);
#>
using My.RolePermission.Model;

namespace My.RolePermission.IDAL
{
<#
// Emit Entity Types
foreach (EntityType entity in ItemCollection.GetItems<EntityType>().OrderBy(e =>
e.Name))
{
    //fileManager.StartNewFile(entity.Name + "RepositoryExt.cs");
    //BeginNamespace(namespaceName, code);
#>
    public partial interface I<#=entity.Name#>Repository :
    IBaseRepository<<#=entity.Name#>>
    {
    }
<#}#>
}
```

按 Ctrl+S 组合键自动生成代码。自动生成的 IDal.cs 如下：

```
using My.RolePermission.Model;

namespace My.RolePermission.IDAL
{
    public partial interface ISMFIELDRepository :IBaseRepository<SMFIELD>
    {
    }
    public partial interface ISMFUNCTBRepository :IBaseRepository<SMFUNCTB>
    {
    }
```

```
    public partial interface ISMLOGRepository :IBaseRepository<SMLOG>
    {
    }
    public partial interface ISMMENUROLEFUNCTBRepository :IbaseRepository
    <SMMENUROLEFUNCTB>
    {
    }
    public partial interface ISMMENUTBRepository :IBaseRepository<SMMENUTB>
    {
    }
    public partial interface ISMROLETBRepository :IBaseRepository<SMROLETB>
    {
    }
    public partial interface ISMUSERTBRepository :IBaseRepository<SMUSERTB>
    {
    }
}
```

（5）创建文本模板 IDBSession.tt：

```
<#@ template language="C#" debug="false" hostspecific="true"#>
<#@ include file="EF.Utility.CS.ttinclude"#>
<#@ output extension=".cs"#>
<#
CodeGenerationTools code = new CodeGenerationTools(this);
MetadataLoader loader = new MetadataLoader(this);
CodeRegion region = new CodeRegion(this, 1);
MetadataTools ef = new MetadataTools(this);
string inputFile = @"..\My.RolePermission.Model\RolePermission.edmx";
EdmItemCollection ItemCollection = loader.CreateEdmItemCollection(inputFile);
string namespaceName = code.VsNamespaceSuggestion();
EntityFrameworkTemplateFileManager                    fileManager              =
EntityFrameworkTemplateFileManager.Create(this);
#>
namespace My.RolePermission.IDAL
{
    public partial interface IDBSession
    {
<#
// Emit Entity Types
foreach (EntityType entity in ItemCollection.GetItems<EntityType>().OrderBy(e =>
e.Name))
{
```

```
    //fileManager.StartNewFile(entity.Name + "RepositoryExt.cs");
    //BeginNamespace(namespaceName, code);

#>
        I<#=entity.Name#>Repository I<#=entity.Name#>Repository{get;set;}
<#}#>
    }
}
```

按 Ctrl+S 组合键自动生成代码。自动生成的 IDBSession1.cs 代码如下：

```
namespace My.RolePermission.IDAL
{
    public partial interface IDBSession
    {
        ISMFIELDRepository ISMFIELDRepository{get;set;}
        ISMFUNCTBRepository ISMFUNCTBRepository{get;set;}
        ISMLOGRepository ISMLOGRepository{get;set;}
        ISMMENUROLEFUNCTBRepository ISMMENUROLEFUNCTBRepository{get;set;}
        ISMMENUTBRepository ISMMENUTBRepository{get;set;}
        ISMROLETBRepository ISMROLETBRepository{get;set;}
        ISMUSERTBRepository ISMUSERTBRepository{get;set;}
    }
}
```

3. My.RolePermission.DAL

（1）在 My.RolePermission.DAL 中新建文件 BaseRepository.cs，申明一个泛型类实现数据层接口 IBaseRepository。我们不用为每个实体创建一个单独的仓储类，而是采用对所有实体类都有效的泛型仓储，这样一个类就可以对所有的实体类执行 CRUD 操作。

```
public class BaseRepository<T> : IBaseRepository<T> where T : class,new()
{
        /// <summary>
        /// EF 上下文对象
        /// </summary>
        DbContext Db = new DbContextFactory().CreateDbContext();
        /// <summary>
        /// 查询
        /// </summary>
        /// <param name="whereLambda"></param>
        /// <returns></returns>
        public IQueryable<T> LoadEntities(Expression<Func<T, bool>> whereLambda)
        {
```

```
    return Db.Set<T>().Where<T>(whereLambda);
}
/// <summary>
/// 查询所有
/// </summary>
/// <param name="entity">需要包含的实体</param>
/// <returns></returns>
public IQueryable<T> LoadEntitiesAll(string entity)
{
    return string.IsNullOrEmpty(entity)?Db.Set<T>().AsQueryable():
     Db.Set<T>().Include(entity).AsQueryable();
}
/// <summary>
/// 实现对数据的分页查询
/// </summary>
/// <typeparam name="model">实体对象</typeparam>
/// <param name="pageIdex">页码</param>
/// <param name="pageSize">每页显示记录数</param>
/// <param name="toalCount">总记录数</param>
/// <param name="whereLambda">条件表达式</param>
/// <param name="orderby">排序字段</param>
/// <param name="isAsc">是否升序（true:升序, false：降序）</param>
/// <returns></returns>
public IQueryable<T> LoadPageEntities<model>(int pageIdex, int pageSize,
out int toalCount, Expression<Func<T, bool>> whereLambda, string orderby,
bool ? isAsc)
{
    var temp = Db.Set<T>().Where<T>(whereLambda.Compile()).AsQueryable();
    //Compile()
    toalCount = temp.Count();
    if (isAsc.HasValue)//排序
    {
        temp = isAsc.Value ? temp.OrderBy<T>(orderby).Skip<T>((pageIdex - 1)
         * pageSize).Take<T>(pageSize) : temp = temp.OrderByDescending<T>
          (orderby).Skip<T>((pageIdex - 1) * pageSize).Take<T>(pageSize); ;
    }
    else
    {
        temp = temp.Skip<T>((pageIdex - 1) * pageSize).Take<T>(pageSize);
    }
    return temp;
}
```

```
        /// <summary>
        /// 删除
        /// </summary>
        /// <param name="entity"></param>
        /// <returns></returns>
        public bool DeleteEntity(T entity)
        {
            Db.Entry<T>(entity).State = System.Data.EntityState.Deleted;
            return true;
        }
        /// <summary>
        /// 修改
        /// </summary>
        /// <param name="entity"></param>
        /// <returns></returns>
        public bool EditEntity(T entity)
        {
            Db.Set<T>().Attach(entity);
            Db.Entry<T>(entity).State = System.Data.EntityState.Modified;
            return true;
        }
        /// <summary>
        /// 添加
        /// </summary>
        /// <param name="entity"></param>
        /// <returns></returns>
        public T AddEntity(T entity)
        {
            Db.Set<T>().Add(entity);
            return entity;
        }
}
```

（2）添加 EntityFramework、System.Data.Entity 的引用。此外，创建一个类 DbContextFactory，用于创建 EF 上下文实例，并保证是线程内唯一。DbContextFactory 代码如下：

```
public class DbContextFactory
{
        /// <summary>
        /// 保证 EF 上下文实例是线程内唯一
        /// </summary>
        /// <returns></returns>
```

```
    public static DbContext CreateDbContext()
    {
        DbContext dbContext = (DbContext)CallContext.GetData("dbContext");
        if (dbContext == null)
        {
            dbContext = new RolePermissionEntities();
            CallContext.SetData("dbContext", dbContext);
        }
        return dbContext;
    }
}
```

不能每次使用 EF 上下文都通过 new 来创建对象，因为在不同层中使用 EF 上下文时就不再是同一个上下文对象了。我们可能会考虑使用单例模式，但是在这里我们不能通过单例模式来解决 EF 上下文对象的问题，因为如果使用单例模式，那么大家都在共用一个 EF 上下文对象，一直不能释放，一旦释放，其他正在使用的用户用不了了。我们可以考虑在一次请求中使用一个 EF 上下文对象。这样就既不影响其他用户的操作，也减少了数据库连接，即线程内唯一对象，因为一个请求就是一个线程。

关于 CallContext

CallContext 是类似于方法调用的线程本地存储区的专用集合对象，并提供对每个逻辑执行线程都唯一的数据槽。数据槽不在其他逻辑线程上的调用上下文之间共享。当 CallContext 沿执行代码路径往返传播并且由该路径中的各个对象检查时，可将对象添加到其中。

当对另一个 AppDomain 中的对象进行远程方法调用时，CallContext 类将生成一个与该远程调用一起传播的 LogicalCallContext 实例。只有公开 ILogicalThreadAffinative 接口并存储在 CallContext 中的对象被在 LogicalCallContext 中传播到 AppDomain 外部。不支持此接口的对象不在 LogicalCallContext 实例中与远程方法调用一起传输。

CallContext 中的所有方法都是静态的，并且在当前线程中的调用上下文上操作。

层次结构： System.Runtime.Remoting.Messaging.CallContext。

需要注意的是，在 BaseDal 的实现中并没有直接调用 EF 上下文进行数据库操作，而是把 EF 的 SaveChanges 方法提取到一个独立的接口中。因为一个业务中有可能涉及对多张表的操作，可以将操作的数据传递到数据层中相应的方法并打上相应的标记，最后调用该方法将数据一次性提交到数据库中，以避免多次连接数据库。

（3）新建文本模板 DAL.tt。

```
<#@ template language="C#" debug="false" hostspecific="true"#>
<#@ include file="EF.Utility.CS.ttinclude"#><#@
 output extension=".cs"#>
```

```
<#
CodeGenerationTools code = new CodeGenerationTools(this);
MetadataLoader loader = new MetadataLoader(this);
CodeRegion region = new CodeRegion(this, 1);
MetadataTools ef = new MetadataTools(this);
string inputFile = @"..\My.RolePermission.Model\RolePermission.edmx";
EdmItemCollection ItemCollection = loader.CreateEdmItemCollection(inputFile);
string namespaceName = code.VsNamespaceSuggestion();
EntityFrameworkTemplateFileManager fileManager =
EntityFrameworkTemplateFileManager.Create(this);
#>
using My.RolePermission.Model;
using My.RolePermission.IDAL;

namespace My.RolePermission.DAL
{
<#

// Emit Entity Types
foreach (EntityType entity in ItemCollection.GetItems<EntityType>().OrderBy(e =>
e.Name))
{
   //fileManager.StartNewFile(entity.Name + "RepositoryExt.cs");
   //BeginNamespace(namespaceName, code);

#>
    public partial class <#=entity.Name#>Repository :
    BaseRepository<<#=entity.Name#>>,I<#=entity.Name#>Repository
   {
   }
<#}#>
}
```

按 Ctrl+S 组合键自动生成代码，生成的 DAL.cs 代码如下：

```
using My.RolePermission.Model;
using My.RolePermission.IDAL;

namespace My.RolePermission.DAL
{
    public partial class SMFIELDRepository : BaseRepository<SMFIELD>,
    ISMFIELDRepository
   {
```

```
    }
    public partial class SMFUNCTBRepository : BaseRepository<SMFUNCTB>,
    ISMFUNCTBRepository
    {
    }
    public partial class SMLOGRepository : BaseRepository<SMLOG>,ISMLOGRepository
    {
    }
    public partial class SMMENUROLEFUNCTBRepository :
    BaseRepository<SMMENUROLEFUNCTB>,ISMMENUROLEFUNCTBRepository
    {
    }
    public partial class SMMENUTBRepository :
    BaseRepository<SMMENUTB>,ISMMENUTBRepository
    {
    }
    public partial class SMROLETBRepository :
    BaseRepository<SMROLETB>,ISMROLETBRepository
    {
    }
    public partial class SMUSERTBRepository : BaseRepository<SMUSERTB>,
    ISMUSERTBRepository
    {
    }
}
```

（4）新建文本模板 DalFactory.tt：

```
<#@ template language="C#" debug="false" hostspecific="true"#>
<#@ include file="EF.Utility.CS.ttinclude"#><#@
 output extension=".cs"#>
<#
CodeGenerationTools code = new CodeGenerationTools(this);
MetadataLoader loader = new MetadataLoader(this);
CodeRegion region = new CodeRegion(this, 1);
MetadataTools ef = new MetadataTools(this);
string inputFile = @"..\My.RolePermission.Model\RolePermission.edmx";
EdmItemCollection ItemCollection = loader.CreateEdmItemCollection(inputFile);
string namespaceName = code.VsNamespaceSuggestion();
EntityFrameworkTemplateFileManager fileManager =
EntityFrameworkTemplateFileManager.Create(this);
#>
using My.RolePermission.IDAL;
```

```
using My.RolePermission.Ioc;

namespace My.RolePermission.DAL
{
  public partial class DalFactory
  {
<#
    foreach (EntityType entity in ItemCollection.GetItems<EntityType>().OrderBy
    (e => e.Name))
    {
#>
     public static I<#=entity.Name#>Repository Get<#=entity.Name#>Repository
     {
       get
        {
         return SpringHelper.GetObject<I<#=entity.Name#>Repository>("<#=
         entity.Name#>Repository");
        }
     }
  <#}#>
 }
}
```

按 Ctrl+S 组合键自动生成代码，生成的 DAL.cs 代码如下：

```
using My.RolePermission.IDAL;
using My.RolePermission.Ioc;

namespace My.RolePermission.DAL
{
  public partial class DalFactory
  {
     public static ISMFIELDRepository GetSMFIELDRepository
     {
       get
        {
         return SpringHelper.GetObject<ISMFIELDRepository>("SMFIELDRepository");
        }
     }
     public static ISMFUNCTBRepository GetSMFUNCTBRepository
     {
       get
        {
```

```
            return SpringHelper.GetObject<ISMFUNCTBRepository>("SMFUNCTBRepository");
        }
    }
    public static ISMLOGRepository GetSMLOGRepository
    {
        get
        {
            return SpringHelper.GetObject<ISMLOGRepository>("SMLOGRepository");
        }
    }
    public static ISMMENUROLEFUNCTBRepository GetSMMENUROLEFUNCTBRepository
    {
        get
        {
            return SpringHelper.GetObject<ISMMENUROLEFUNCTBRepository>
            ("SMMENUROLEFUNCTBRepository");
        }
    }
    public static ISMMENUTBRepository GetSMMENUTBRepository
    {
        get
        {
            return SpringHelper.GetObject<ISMMENUTBRepository>("SMMENUTBRepository");
        }
    }
    public static ISMROLETBRepository GetSMROLETBRepository
    {
        get
        {
            return SpringHelper.GetObject<ISMROLETBRepository>("SMROLETBRepository");
        }
    }
    public static ISMUSERTBRepository GetSMUSERTBRepository
    {
        get
        {
            return SpringHelper.GetObject<ISMUSERTBRepository>("SMUSERTBRepository");
        }
    }
  }
}
```

4. My.RolePermission.DALSessionFactory

（1）在 My.RolePermission.DALSessionFactory 中添加 DBSessionFactory 类，用于创建线程唯一的 DBSession 对象。

```
public class DBSessionFactory
{
        public static IDBSession CreateDbSession()
        {
            IDBSession DbSession = (IDBSession)CallContext.GetData("dbSession");
            if (DbSession == null)
            {
                DbSession = new DBSession();
                CallContext.SetData("dbSession", DbSession);
            }
            return DbSession;
        }
}
```

（2）新建类 DBSession：

```
/// <summary>
/// DBSession;工厂类（数据会话层），作用：创建数据操作类的实例。业务层通过 DBSession
/// 调用相应的数据操作类的实例，将业务层与数据层解耦
/// </summary>
public partial class DBSession : IDBSession
{
        public DbContext Db
        {
            get { return new DbContextFactory().CreateDbContext(); }
        }
        /// <summary>
        /// 执行 EF 上下文的 SaveChanges 方法
        /// </summary>
        /// <returns></returns>
        public bool SaveChanges()
        {
            Db.Configuration.ValidateOnSaveEnabled = false;
            return Db.SaveChanges() > 0;
        }
        public int ExecuteSql(string sql, params
        System.Data.SqlClient.SqlParameter[] pars)
        {
```

```
            return pars==null?Db.Database.ExecuteSqlCommand(sql):Db.Database.
             ExecuteSqlCommand(sql, pars);
        }
        public List<T> ExecuteSelectQuery<T>(string sql,
        params System.Data.SqlClient.SqlParameter[] pars)
        {
            return Db.Database.SqlQuery<T>(sql, pars).ToList();
        }
}
```

（3）新建 DbSession.tt：

```
<#@ template language="C#" debug="false" hostspecific="true"#>
<#@ include file="EF.Utility.CS.ttinclude"#><#@
 output extension=".cs"#>
<#
CodeGenerationTools code = new CodeGenerationTools(this);
MetadataLoader loader = new MetadataLoader(this);
CodeRegion region = new CodeRegion(this, 1);
MetadataTools ef = new MetadataTools(this);
string inputFile = @"..\My.RolePermission.Model\RolePermission.edmx";
EdmItemCollection ItemCollection = loader.CreateEdmItemCollection(inputFile);
string namespaceName = code.VsNamespaceSuggestion();
EntityFrameworkTemplateFileManager fileManager =
EntityFrameworkTemplateFileManager.Create(this);
#>
using My.RolePermission.DAL;
using My.RolePermission.IDAL;

namespace My.RolePermission.DALSessionFactory
{
    public partial class DBSession : IDBSession
    {
<#
foreach (EntityType entity in ItemCollection.GetItems<EntityType>().OrderBy
(e => e.Name))
{
#>
          private I<#=entity.Name#>Repository _<#=entity.Name#>Repository;
        public I<#=entity.Name#>Repository I<#=entity.Name#>Repository
        {
            get
            {
```

```
                if(_<#=entity.Name#>Repository == null)
                {
                   // _<#=entity.Name#>Repository = new <#=entity.Name#>Repository();
                   _<#=entity.Name#>Repository
=DalFactory.Get<#=entity.Name#>Repository;
                }
                return _<#=entity.Name#>Repository;
            }
            set { _<#=entity.Name#>Repository = value; }
        }
<#}#>
    }
}
```

按 Ctrl+S 组合键自动生成代码，生成的 DBSession1.cs 代码如下：

```
using My.RolePermission.DAL;
using My.RolePermission.IDAL;

namespace My.RolePermission.DALSessionFactory
{
    public partial class DBSession : IDBSession
    {

        private ISMFIELDRepository _SMFIELDRepository;
        public ISMFIELDRepository ISMFIELDRepository
        {
            get
            {
                if(_SMFIELDRepository == null)
                {
                   // _SMFIELDRepository = new SMFIELDRepository();
                     _SMFIELDRepository =DalFactory.GetSMFIELDRepository;
                }
                return _SMFIELDRepository;
            }
            set { _SMFIELDRepository = value; }
        }

        private ISMFUNCTBRepository _SMFUNCTBRepository;
        public ISMFUNCTBRepository ISMFUNCTBRepository
        {
            get
```

```
    {
        if(_SMFUNCTBRepository == null)
        {
          // _SMFUNCTBRepository = new SMFUNCTBRepository();
             _SMFUNCTBRepository =DalFactory.GetSMFUNCTBRepository;
        }
        return _SMFUNCTBRepository;
    }
    set { _SMFUNCTBRepository = value; }
}

 private ISMLOGRepository _SMLOGRepository;
public ISMLOGRepository ISMLOGRepository
{
    get
    {
        if(_SMLOGRepository == null)
        {
          // _SMLOGRepository = new SMLOGRepository();
             _SMLOGRepository =DalFactory.GetSMLOGRepository;
        }
        return _SMLOGRepository;
    }
    set { _SMLOGRepository = value; }
}

 private ISMMENUROLEFUNCTBRepository _SMMENUROLEFUNCTBRepository;
public ISMMENUROLEFUNCTBRepository ISMMENUROLEFUNCTBRepository
{
    get
    {
        if(_SMMENUROLEFUNCTBRepository == null)
        {
          // _SMMENUROLEFUNCTBRepository=new SMMENUROLEFUNCTBRepository();
             _SMMENUROLEFUNCTBRepository =DalFactory.
             GetSMMENUROLEFUNCTBRepository;
        }
        return _SMMENUROLEFUNCTBRepository;
    }
    set { _SMMENUROLEFUNCTBRepository = value; }
}
```

```
 private ISMMENUTBRepository _SMMENUTBRepository;
public ISMMENUTBRepository ISMMENUTBRepository
{
    get
    {
        if(_SMMENUTBRepository == null)
        {
          // _SMMENUTBRepository = new SMMENUTBRepository();
             _SMMENUTBRepository =DalFactory.GetSMMENUTBRepository;
        }
        return _SMMENUTBRepository;
    }
    set { _SMMENUTBRepository = value; }
}

 private ISMROLETBRepository _SMROLETBRepository;
public ISMROLETBRepository ISMROLETBRepository
{
    get
    {
        if(_SMROLETBRepository == null)
        {
          // _SMROLETBRepository = new SMROLETBRepository();
             _SMROLETBRepository =DalFactory.GetSMROLETBRepository;
        }
        return _SMROLETBRepository;
    }
    set { _SMROLETBRepository = value; }
}

 private ISMUSERTBRepository _SMUSERTBRepository;
public ISMUSERTBRepository ISMUSERTBRepository
{
    get
    {
        if(_SMUSERTBRepository == null)
        {
          // _SMUSERTBRepository = new SMUSERTBRepository();
             _SMUSERTBRepository =DalFactory.GetSMUSERTBRepository;
        }
        return _SMUSERTBRepository;
    }
```

```
            set { _SMUSERTBRepository = value; }
        }
    }
}
```

DbContext 默认支持事务，当实例化一个新的 DbContext 对象时会创建一个新的事务，当调用 SaveChanges 方法时事务会提交。如果我们需要使用相同的 DbContext 对象把多个代码模块的操作放到一个单独的事务中，就可以使用工作单元（Unit of Work）。

工作单元本质上是一个类，可以在一个事务中跟踪所有的操作，然后将所有的操作作为原子单元执行，这里将把类 DBSession 作为工作单元。前面创建的所有仓储类都没有调用 SaveChanges 方法，原因在于工作单元可以通过 IOC 注入的方式获取每个仓储对象。当想保存修改时，可以在工作单元上调用 SaveChanges 方法，也就相当于在 DbContext 类上调用了 SaveChanges 方法。这样就会使涉及多个仓储的所有操作成为单个事务的一部分。

每个工作单元都有自己的事务，每个事务都可以使用相应的仓储类处理多个实体，当调用 DBSession 类的 SaveChanges 方法时，事务就会提交。

12.4.4 业务逻辑层

1. My.RolePermission.IBLL

（1）在 My.RolePermission.IBLL 中定义一个业务层父接口 IbaseService：

```
public interface IBaseService<T> where T : class,new()
{
    IDBSession GetCurrentDbSession { get; }
    IBaseRepository<T> CurrentRepository { get; set; }
    IQueryable<T> LoadEntities(Expression<Func<T, bool>> whereLambda);
    IQueryable<T> LoadPageEntities<s>(int pageIdex, int pageSize, out int
    toalCount, Expression<Func<T, bool>> whereLambda, string orderby, bool?
    isAsc);
    bool DeleteEntity(T entity);
    bool EditEntity(T entity);
    T AddEntity(T entity);
}
```

（2）新建文本模板 IBLL.tt：

```
<#@ template language="C#" debug="false" hostspecific="true"#>
<#@ include file="EF.Utility.CS.ttinclude"#><#@
 output extension=".cs"#>
<#
CodeGenerationTools code = new CodeGenerationTools(this);
```

```
MetadataLoader loader = new MetadataLoader(this);
CodeRegion region = new CodeRegion(this, 1);
MetadataTools ef = new MetadataTools(this);
string inputFile = @"..\My.RolePermission.Model\RolePermission.edmx";
EdmItemCollection ItemCollection = loader.CreateEdmItemCollection(inputFile);
string namespaceName = code.VsNamespaceSuggestion();
EntityFrameworkTemplateFileManager fileManager =
EntityFrameworkTemplateFileManager.Create(this);
#>
using My.RolePermission.Model;

namespace My.RolePermission.IBLL
{
<#
foreach (EntityType entity in ItemCollection.GetItems<EntityType>().OrderBy(e =>
e.Name))
{
#>
    public partial interface I<#=entity.Name#>Service : IBaseService<<#=entity.Name#>>
    {
    }
<#}#>
}
```

按 Ctrl+S 组合键自动生成代码，生成的 IBLL.cs 代码如下：

```
using My.RolePermission.Model;

namespace My.RolePermission.IBLL
{
    public partial interface ISMFIELDService : IBaseService<SMFIELD>
    {
    }
    public partial interface ISMFUNCTBService : IBaseService<SMFUNCTB>
    {
    }
    public partial interface ISMLOGService : IBaseService<SMLOG>
    {
    }
    public partial interface ISMMENUROLEFUNCTBService :
    IBaseService<SMMENUROLEFUNCTB>
    {
    }
```

```
    public partial interface ISMMENUTBService : IBaseService<SMMENUTB>
    {
    }
    public partial interface ISMROLETBService : IBaseService<SMROLETB>
    {
    }
    public partial interface ISMUSERTBService : IBaseService<SMUSERTB>
    {
    }
}
```

2. My.RolePermission.BLL

（1）在 My.RolePermission.BLL 中定义一个业务层父类 BaseService：

```
public abstract class BaseService<T> where T : class,new()
{
       public IDBSession GetCurrentDbSession
       {
          get
          {
             return DBSessionFactory.CreateDbSession();
          }

       }
       public IDAL.IBaseRepository<T> CurrentRepository { get; set; }
       public abstract void SetCurretnRepository();
       public BaseService()
       {
          SetCurretnRepository();
       }
       /// <summary>
       /// 查询
       /// </summary>
       /// <param name="whereLambda"></param>
       /// <returns></returns>
       public IQueryable<T> LoadEntities(Expression<Func<T, bool>> whereLambda)
       {
          return CurrentRepository.LoadEntities(whereLambda);
       }

       /// <summary>
       /// 实现对数据的分页查询
       /// </summary>
```

```
        /// <typeparam name="model">实体对象</typeparam>
        /// <param name="pageIdex">页码</param>
        /// <param name="pageSize">每页显示记录数</param>
        /// <param name="toalCount">总记录数</param>
        /// <param name="whereLambda">条件表达式</param>
        /// <param name="orderby">排序字段</param>
        /// <param name="isAsc">是否升序（true:升序, false:降序）</param>
        /// <returns></returns>
        public IQueryable<T> LoadPageEntities<s>(int pageIdex, int pageSize, out
        int toalCount, Expression<Func<T, bool>> whereLambda, string orderby,bool?
        isAsc)
        {
            return CurrentRepository.LoadPageEntities<s>(pageIdex, pageSize, out
            toalCount, whereLambda, orderby, isAsc);
        }
        public bool DeleteEntity(T entity)
        {
            CurrentRepository.DeleteEntity(entity);
            return this.GetCurrentDbSession.SaveChanges();
        }
        public bool EditEntity(T entity)
        {
            CurrentRepository.EditEntity(entity);
            return this.GetCurrentDbSession.SaveChanges();
        }
        public T AddEntity(T entity)
        {
            CurrentRepository.AddEntity(entity);
            this.GetCurrentDbSession.SaveChanges();
            return entity;
        }
}
```

（2）新建文本模板 BLL.tt：

```
<#@ template language="C#" debug="false" hostspecific="true"#>
<#@ include file="EF.Utility.CS.ttinclude"#><#@
 output extension=".cs"#>
<#
CodeGenerationTools code = new CodeGenerationTools(this);
MetadataLoader loader = new MetadataLoader(this);
CodeRegion region = new CodeRegion(this, 1);
MetadataTools ef = new MetadataTools(this);
```

```
string inputFile = @"..\My.RolePermission.Model\RolePermission.edmx";
EdmItemCollection ItemCollection = loader.CreateEdmItemCollection(inputFile);
string namespaceName = code.VsNamespaceSuggestion();
EntityFrameworkTemplateFileManager fileManager =
EntityFrameworkTemplateFileManager.Create(this);
#>
using My.RolePermission.IBLL;
using My.RolePermission.Model;

namespace My.RolePermission.BLL
{
<#
foreach (EntityType entity in ItemCollection.GetItems<EntityType>().OrderBy(e =>
e.Name))
{
#>
    public partial class <#=entity.Name#>Service :
    BaseService<<#=entity.Name#>>,I<#=entity.Name#>Service
    {
        public override void SetCurretnRepository()
        {
            CurrentRepository = this.GetCurrentDbSession.
            I<#=entity.Name#>Repository;
        }
    }
<#}#>
}
```

按 Ctrl+S 组合键自动生成代码，生成的 BLL.cs 代码如下：

```
using My.RolePermission.IBLL;
using My.RolePermission.Model;

namespace My.RolePermission.BLL
{
    public partial class SMFIELDService :BaseService<SMFIELD>,ISMFIELDService
    {
        public override void SetCurretnRepository()
        {
            CurrentRepository = this.GetCurrentDbSession.ISMFIELDRepository;
        }
    }
    public partial class SMFUNCTBService :BaseService<SMFUNCTB>,ISMFUNCTBService
```

```
{
    public override void SetCurretnRepository()
    {
        CurrentRepository = this.GetCurrentDbSession.ISMFUNCTBRepository;
    }
}
public partial class SMLOGService :BaseService<SMLOG>,ISMLOGService
{
    public override void SetCurretnRepository()
    {
        CurrentRepository = this.GetCurrentDbSession.ISMLOGRepository;
    }
}
public partial class SMMENUROLEFUNCTBService :BaseService<SMMENUROLEFUNCTB>,
ISMMENUROLEFUNCTBService
{
    public override void SetCurretnRepository()
    {
        CurrentRepository = this.GetCurrentDbSession.
         ISMMENUROLEFUNCTBRepository;
    }
}
public partial class SMMENUTBService :BaseService<SMMENUTB>,ISMMENUTBService
{
    public override void SetCurretnRepository()
    {
        CurrentRepository = this.GetCurrentDbSession.ISMMENUTBRepository;
    }
}
public partial class SMROLETBService :BaseService<SMROLETB>,ISMROLETBService
{
    public override void SetCurretnRepository()
    {
        CurrentRepository = this.GetCurrentDbSession.ISMROLETBRepository;
    }
}
public partial class SMUSERTBService :BaseService<SMUSERTB>,ISMUSERTBService
{
    public override void SetCurretnRepository()
    {
        CurrentRepository = this.GetCurrentDbSession.ISMUSERTBRepository;
    }
```

```
    }
}
```

最终代码结构如图 12-5 所示。

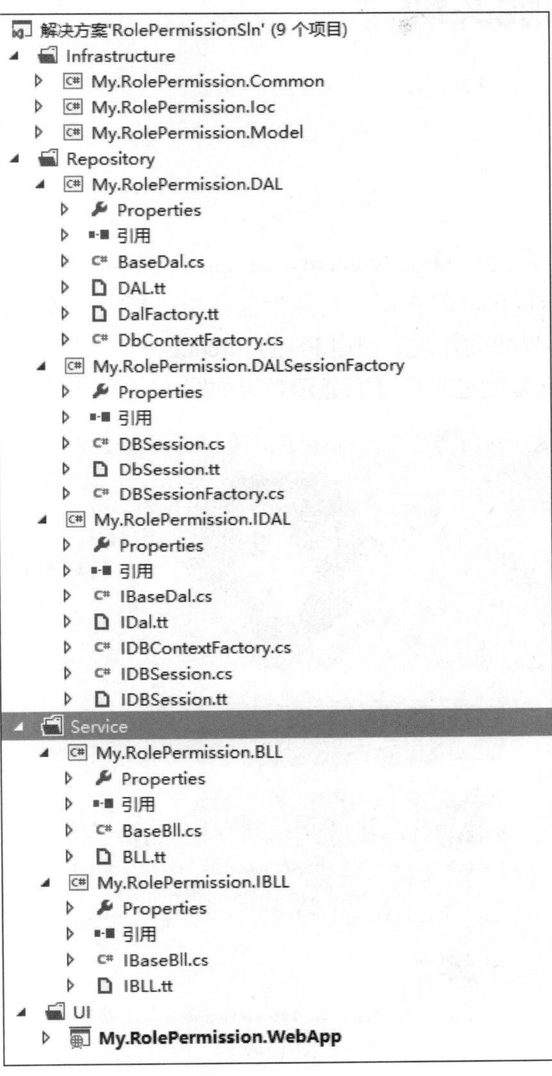

图 12-5

12.4.5 UI 层

jQuery EasyUI 是一组基于 jQuery 的 UI 插件集合，而 jQuery EasyUI 的目标就是帮助 Web 开发者更轻松地打造出功能丰富并且美观的 UI 界面。开发者不需要编写复杂的 JavaScript，也不需要对 css 样式有深入的了解，需要了解的只有一些简单的 html 标签。

到 JQuery EasyUI 官网 http://www.jeasyui.com/download/index.php 中下载该组件，并把 jquery-easyui-1.4.5 引入到项目中。

12.5 功能实现

12.5.1 用户登录

1. 制作登录界面

要制作登录界面了，直接去 http://www.mycodes.net/154/8456.htm 下载 HTML 的界面模板，当然一般在项目中美工会提供，但在部分创业型公司中可能没有美工，这样的工作就必须由开发人员来完成了。调前端界面推荐大家使用 FF 的 Firebug。

把静态的 HTML 界面复制过来后界面显示效果如图 12-6 所示。

图 12-6

这里没有验证码，可以去 http://www.easyicon.net/下载验证码图标，但是大小对不上，用 PS 进行一些图片处理就可以了。改造后的界面如图 12-7 所示。

图 12-7

引入过来的 css 和 js 文件可能是直接嵌套在 HTML 页面中的，我们可以单独提取出来，方便后面统一进行合并和压缩，这也属于前端优化技巧之一。

在引入这些 css 样式的时候要注意“../”相对路径和“/”绝对路径的区别。一般建议使用“/”这种绝对路径，“/”表示 Web 项目根目录。

2. 实现验证码

关于右侧验证码的图片生成代码，大家可以自行设计，也可以在网上搜索一些比较有趣的。验证码部分的 HTML 代码如下：

```
<div class="col-sm-2" style="padding-left:0px;"><img id="imgCode" alt="验证码" title="刷新验证码" onclick="RefreshValidateCode(this);" style="cursor:pointer;width:50px;height:45px;" src="/Login/ShowValidateCode"/></div>
<script type="text/javascript">
    function RefreshValidateCode(obj) {
        obj.src = "/Login/ShowValidateCode/" + Math.floor(Math.random() * 10000);
    }
</script>
```

刷新验证码这里，在图片的请求地址后面加了随机数，因为 URL 请求地址不变，浏览器就会认为是同一请求，直接从浏览器缓存中读取，不再发送 Web 请求。具体的验证码实现代码，大家可以参见源代码。

3. 实现登录

实现登录的步骤如下：

（1）检查验证码是否正确。

（2）检查用户名和密码是否正确。

（3）将登录信息存储到 Memcache，过去都是直接存到 Session 中。

（4）将登录信息存储到 Cookie。

（5）跳转到登录后台。

登录验证流程图如图 12-8 所示。

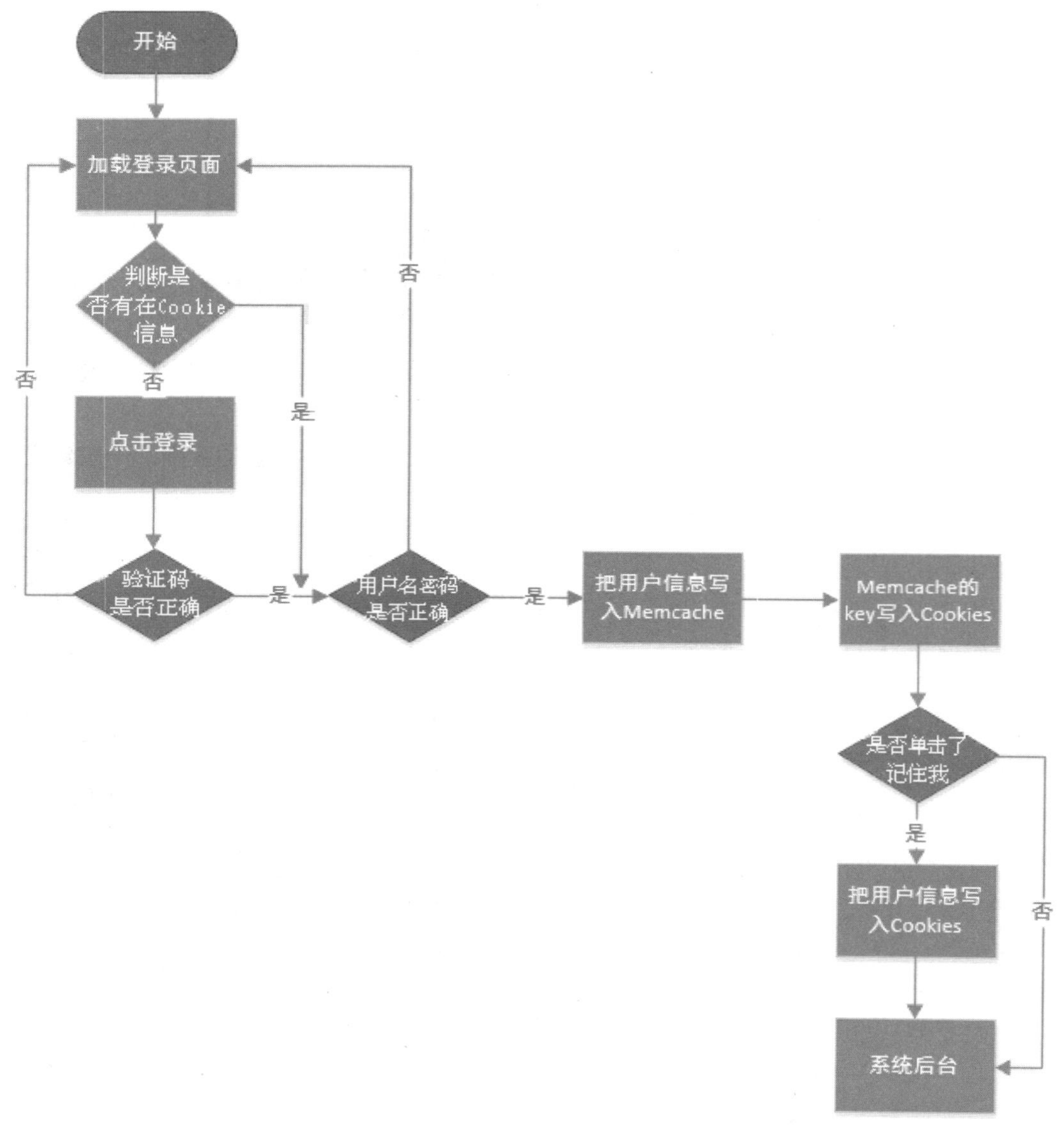

图 12-8

```
public ActionResult Index()
{
```

```
            CheckCookieInfo();
            return View();
}
/// <summary>
/// 单击登录系统
/// </summary>
/// <param name="model">登录信息</param>
/// <returns></returns>
[HttpPost]
[ValidateAntiForgeryToken]//用防伪造令牌来避免 CSRF 攻击
public ActionResult Index(LogOnModel model)
{
            #region 验证码验证

            if (Session["ValidateCode"] != null && model.ValidateCode != null &&
            model.ValidateCode.ToLower() != Session["ValidateCode"].ToString())
            {
                ModelState.AddModelError("Error_PersonLogin", "验证码错误！");
                return View();
            }
            Session["ValidateCode"] = null;
            #endregion

            if (ModelState.IsValid)
            {
                SMUSERTB person = SMUSERTBService.ValidateUser(model.PersonName,
                xEncrypt.EncryptText(model.Password));
                if (person != null) //登录成功
                {
                    Account account = person.ToAccount();

                    string sessionId = Guid.NewGuid().ToString();//作为 Memcache 的 key
                    try
                    {
                        MemcacheHelper.Set(sessionId, Common.SerializerHelper.
                        SerializeToString(account), DateTime.Now.AddMinutes(20));
                        //使用 Memcache 代替 Session 解决数据在不同 Web 服务器之间共享的问题
                    }
                    catch (Exception ex)
                    {
                        throw new Exception(ex.Message);
                    }
```

```
                Response.Cookies["sessionId"].Value = sessionId;
                //将 Memcache 的 key 以 Cookie 的形式返回到浏览器端的内存中，当用户再次请求其
                //他页面时请求报文中会以 Cookie 将该值再次发送到服务端

                if (model.RememberMe)
                {
                    HttpCookie ckUid = new HttpCookie("ckUid", model.PersonName);
                    HttpCookie ckPwd=new HttpCookie("ckPwd", xEncrypt.EncryptText
                    (model.Password));
                    ckUid.Expires = DateTime.Now.AddDays(3);
                    ckPwd.Expires = DateTime.Now.AddDays(3);
                    Response.Cookies["sessionId"].Expires =
                    DateTime.Now.AddDays(3);
                    Response.Cookies.Add(ckUid);
                    Response.Cookies.Add(ckPwd);
                }
                return RedirectToAction("Index", "Home");
            }
        }
        ModelState.AddModelError("Error_PersonLogin", "用户名或者密码出错。");
        return View();
}
```

4. 实现记住密码

实现记住密码一般是直接把用户登录信息存入 Cookie，用户名直接明文存，登录密码加密后存储。

勾选“记住我”复选框后，就把用户登录信息存入 Cookie，然后保存特定时间。当下次再浏览登录界面的时候，就从本地查找 Cookie 中是否存在登录信息，如果存在，就直接从 Cookie 中读取登录信息，然后验证登录用户信息的正确性，如果正确就把登录信息存入 Memcache 实现 Session 共享，然后直接跳转到登录后台。

```
/// <summary>
/// 判断 Cookie 信息
/// </summary>
private void CheckCookieInfo()
{
        if (Request.Cookies["ckUid"]!=null && Request.Cookies["ckPwd"]!=null)
        {
            string userName = Request.Cookies["ckUid"].Value;
            string userPwd = Request.Cookies["ckPwd"].Value;
            //判断 Cookie 中存储的用户密码和用户名是否正确.
            SMUSERTB person = SMUSERTBService.ValidateUser(userName, userPwd);
```

```
            if (person != null)
            {
                string sessionId = Guid.NewGuid().ToString();//作为Memcache的key
                var account = person.ToAccount();
                Common.MemcacheHelper.Set(sessionId, Common.SerializerHelper.
               SerializeToString(account), DateTime.Now.AddMinutes(20));
                //使用Memcache代替Session解决数据在不同Web服务器之间共享的问题
                Response.Cookies["sessionId"].Value = sessionId;
                //将Memcache的key以Cookie的形式返回到浏览器端的内存中，当用户再次请求其
                //他页面时请求报文中会以Cookie将该值再次发送到服务端
                //Response.Redirect("/Home/Index"); 尽量不要用这样的写法
                RedirectToAction("Index", "Home");
            }
            else
            {
                //如果说账号密码是错误的，就没必要再把登录用户名和密码存在Cookie中了
                Response.Cookies["ckUid"].Expires = DateTime.Now.AddDays(-1);
                Response.Cookies["ckPwd"].Expires = DateTime.Now.AddDays(-1);
            }
        }
}
```

5. 验证用户名和密码

在 ISMUSERTBService 接口中添加方法：

```
SMUSERTB ValidateUser(string userName, string password);
```

在 SMUSERTBService 中实现方法：

```
/// <summary>
/// 验证用户名和密码是否正确
/// </summary>
/// <param name="userName">用户名</param>
/// <param name="password">密码</param>
/// <returns>登录成功后的用户信息</returns>
public SMUSERTB ValidateUser(string userName, string password)
{
            if (String.IsNullOrWhiteSpace(userName) || String.
             IsNullOrWhiteSpace(password))
                return null;

            //获取用户信息,请确定web.config中的连接字符串正确
            return LoadEntities(x => x.U_ID == userName &&x.U_PASSWORD == password
```

```
&& x.STATUS == "Y").FirstOrDefault();
}
```

12.5.2 采用分布式的方式记录异常日志

多线程操作同一个文件时会出现并发问题。解决的一个办法就是给文件加锁（lock），但是这样的话，一个线程操作文件时，其他的都得等待，性能非常差。另外一个解决方案就是先将数据放在队列中，然后开启一个线程，负责从队列中取出数据，再写到文件中。

思路：把所有产生的日志信息存放到一个队列里面，然后通过新建一个线程不断地从这个队列里面读取异常信息，然后往日志里面写，也就是所谓的生产者、消费者模式。这里的队列不是直接使用 C#中的 Queue，而是使用 Redis 分布式队列。使用 C#中的 Queue 的弊端是队列和 Web 应用程序都是在同一台 Web 服务器上面的，Queue 会占用 Web 服务器内存，而且不易扩展。

（1）在 Filter 文件夹中新建类 MyExceptionAttribute，继承类 HandleErrorAttribute，代码如下：

```
using System.Web.Mvc;
using ServiceStack.Redis;

namespace My.RolePermission.WebApp.Filter
{
    public class MyExceptionAttribute: HandleErrorAttribute
    {
        public static RedisClient client = new RedisClient("127.0.0.1", 6379);
        //发布到正式环境时，记得更改 IP 地址和默认端口，并且设置密码
        #region 如果设置密码
        //static string host = "127.0.0.1";/*访问 host 地址*/
        //static string password = "2016@Test.88210_yujie";/*实例 id:密码*/
      //static readonly RedisClient client = new RedisClient(host, 6379, password);
        #endregion
        //public static IRedisTypedClient<string> redis = client.As<string>(); //复杂对象

        public override void OnException(ExceptionContext filterContext)
        {
            base.OnException(filterContext);
            client.EnqueueItemOnList("errorMsg", filterContext.Exception.ToString());

            //filterContext.HttpContext.Response.Redirect("/Error.html");
        }
    }
```

```
}
```

（2）修改 FilterConfig 类，注释掉原来的 HandleErrorAttribute，改为新建的自定义异常处理类 MyExceptionAttribute。

```
using My.RolePermission.WebApp.Filter;
using System.Web.Mvc;

namespace My.RolePermission.WebApp
{
    public class FilterConfig
    {
        public static void RegisterGlobalFilters(GlobalFilterCollection filters)
        {
            //filters.Add(new HandleErrorAttribute());
            filters.Add(new MyExceptionAttribute());
        }
    }
}
```

（3）Web 应用程序一启动，就要监听 Redis 队列中是否有异常日志，有的话就不断记录下来。修改 Global.asax，在 Application_Start()中添加调用 RecordLog 方法。RecordLog 方法的代码如下：

```
//采用分布式的方式记录日志
void RecordLog()
{
        ThreadPool.QueueUserWorkItem((a) =>
        {
            while (true)
            {
                if (MyExceptionAttribute.client.GetListCount("errorMsg") > 0)
                {
                    string ex = MyExceptionAttribute.client.
                    DequeueItemFromList("errorMsg");
                    ILog logger = LogManager.GetLogger("errorMsg");
                    logger.Error(ex);
                }
                else
                {
                    Thread.Sleep(3000);//如果队列中没有数据就休息，以免造成CPU 的占用
                }
            }
```

```
        });
}
```

12.5.3 授权

我们要防止客户绕过登录，直接操作系统后台界面的功能，首先就会考虑到使用 AOP。在 MVC 中，AOP 可以通过各种过滤器来实现，这里就用到了授权过滤器。

在 Filter 文件夹中新建类 MyAuthorizeAttribute，继承自类 AuthorizeAttribute，然后重写 OnAuthorization 方法。

```
using System.Web.Mvc;
using My.RolePermission.Model;

namespace My.RolePermission.WebApp.Filter
{
    /// <summary>
    /// 授权过滤器——在 Action 过滤器前执行
    /// </summary>
    public class MyAuthorizeAttribute: AuthorizeAttribute
    {
        public override void OnAuthorization(AuthorizationContext filterContext)
        {
            //注释掉父类方法，因为父类里的 OnAuthorization 方法会调用 ASP.NET 的授权验证机制
            //base.OnAuthorization(filterContext);

            if (filterContext.HttpContext.Request.Cookies["sessionId"] != null)
            {
                string sessionId = filterContext.HttpContext.Request.Cookies
                ["sessionId"].Value;//接收从 Cookie 中传递过来的 Memcache 的 key
                object obj = Common.MemcacheHelper.Get(sessionId);
                //根据 key 从 Memcache 中获取用户的信息
                if (obj != null)
                {
                    Account _Account = Common.
                    SerializerHelper.DeserializeToObject<Account>(obj.ToString());
                    //Common.MemcacheHelper.Set(sessionId, obj.ToString(),
                    //DateTime.Now.AddMinutes(20));//模拟滑动过期时间
                    if (_Account == null)
                    {
                        filterContext.HttpContext.Response.Redirect("/Error.html");
                        return;
                    }
                }
                else
```

```
                {
                    filterContext.HttpContext.Response.Redirect("/Login/Index");
                    return;
                }
            }
        }
    }
}
```

然后在需要授权的控制器类上面加上[MyAuthorize]标记即可。

12.5.4 增删改查

后台管理界面的增删改查操作很简单，请直接查看项目源代码。

12.6 运行项目

（1）创建数据库。打开 SQL Server 2012，右击“数据库”，选择“附加”，添加 RolePermission.mdf 文件。或者右击“数据库”，选择“还原数据库”，选择 Chapter12 目录中的“RolePermission”备份文件进行还原。

（2）运行 Memcached 服务。

（3）运行 Redis 服务。

（4）打开 RolePermissionSln.sln。

（5）运行 RolePermissionSln，结果如图 12-9 所示。

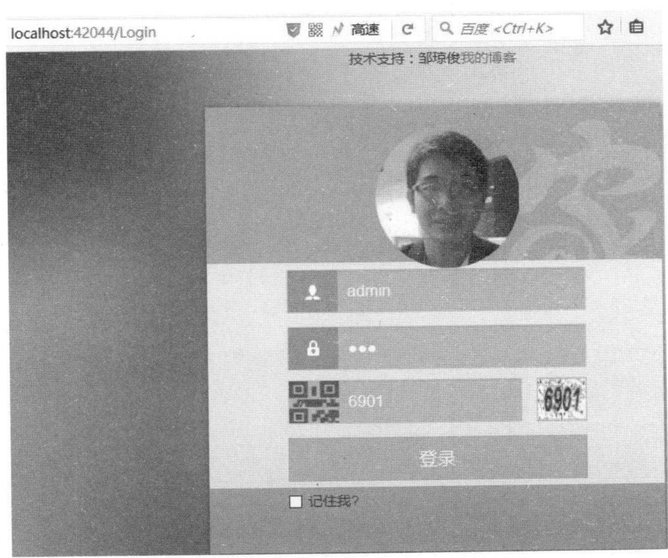

图 12-9

登录名为 admin，密码为 123。

登录成功后，单击“人员管理”，界面效果如图 12-10 所示。

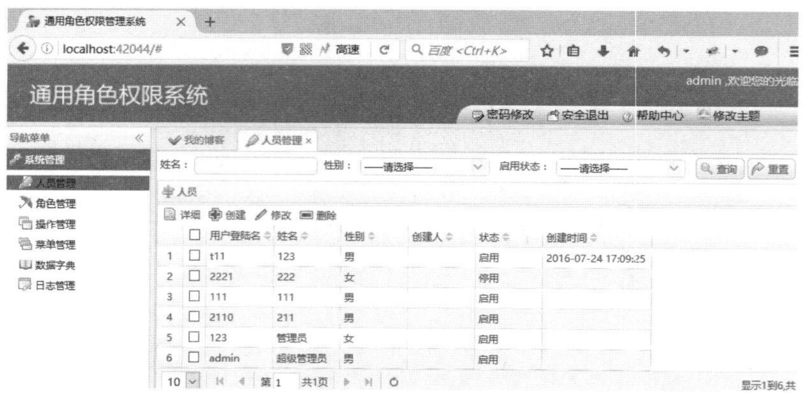

图 12-10